PERIODIC INSPECTION OF PRESSURIZED COMPONENTS

PERIODIC INSPECTION OF PRESSURIZED COMPONENTS

I Mech E CONFERENCE PUBLICATIONS 1979—4

Conference sponsored by the
Pressure Vessels Section of the
Applied Mechanics Group of
The Institution of Mechanical Engineers

London, 8-10 May 1979

Published by
Mechanical Engineering Publications Limited for
The Institution of Mechanical Engineers
LONDON

First published 1979

This volume is complete in itself

There is no supplementary discussion volume.

ISBN 0 85298 426 X

Printed and bound by The Burlington Press (Cambridge) Limited
Foxton, Royston, Hertfordshire

CONTENTS

DESIGN OF NUCLEAR PLANT PRESSURE COMPONENTS CONSIDERING EASIER IN-SERVICE INSPECTION

Y. ANDO
University of Tokyo, Japan

The MS of this paper was received at the Institution on 24 October 1978 and accepted for publication on 12 December 1978

SYNOPSIS   The design of nuclear pressure components with shorter weld seams, high accessibility and inspectability makes ISI easier and reduces personal radiation exposure. The development of high quality heavy section steel segments makes it possible to design RPV for easier ISI and several examples are described. Larger size containment vessel facilitates ISI and it has been realized by the development of new high tensile steel. The application of bent pipe and distance criteria for piping arrangement is expected to perform easier ISI.

INTRODUCTION

1.   Though great efforts for the development of ISI equipments have been made in various countries, there is no denying the existence of many restrictions on ISI from the view point of requirements specified in the code, such as ASME Code Section XI and JEAC 4205.

2.   In order to achieve the easier ISI, the accessibility to the pressure components, the reduction of welded points and the inspectability of welds must be taken into account in the early stage on plant planning and in the design stage of components. These feasible factors could make it possible to increase the structural integrity of components and the availability of nuclear power station.

3.   Through long time efforts, several steel works in Japan have succeeded in manufacturing high-quality heavy section steel material. Currently, the plates up to nearly 5 m in breadth and the forgings made from ingots up to 500 tons are produced for nuclear components. Those heavy section segments also have come to be used for reactor pressure vessels at domestic and overseas nuclear power stations. The use of large size segments enables us to design the vessels for easier ISI, some examples of which are shown in this paper.

4.   The use of the larger-size primary containment vessels, based on higher strength steels, has been an approach designed to increase the accessibility to PV components. The improvement of layout inside containment vessel could consequently afford to reduce both the personal radiation exposure and the time required for ISI.

5.   The inspection of piping is also an important part of ISI. As the use of bent pipe in place of elbow joint leads to the reduction of weld joints, the consideration of arrangement is found to improve the accessibility of piping very much.

REACTOR PRESSURE VESSEL DESIGN CONSIDERING EASIER IN-SERVICE INSPECTION

6.   Minimization of weld seam length reduces the inservice inspection time which corresponds to personal radiation exposure, and also improves the structural integrity of RPV. To the end, some Japanese steel works have developed the manufacture of large plates or forgings in heavy section steel for the purpose of improvement of PRV design.

7.   In April in 1977 at the IAEA IWG-RRPC Technical Committee Meeting, which was held in Kobe, three papers were presented about the design improvement of RPV. Since this conference, designs have been developed to the practical stage to take advantage of these developments. In this chapter several examples are to be shown.   (Refs. 1-3)

New design of RPV bottom head for 1100 MWe BWR

8.   The control rod drive mechanism in BWR has been fixed at the bottom of RPV, so that the inspection of the weld seams, located in the bottom head near this drive mechanism, sometimes ran into many troubles because of inaccessibility.

9.   Recently large SA-508 cl.3 forging segment for bottom section has been developed to minimize the number and length of weld seams to solve the above mentioned problem. In the new design, which has been adopted since TEPCO Fukushima II-2, the bottom section consists of two large forging segments and one weld seam. In order to avoid the complex weld structure, jointing the petals to the support skirt, from the conventional design, one piece of forging with skirt ring is used as shown in Fig. 1.

10.   In the conventional design, bottom dished segment was made of plates with welded joints, located within the CRD holes area which has made it almost impossible to inspect the weld seam with any non-destructive examination technique to meet ISI requirements.

11.   The comparison of main features between current and new design is shown in Table 1.

12.   While the normal forging technique is unsuitable to provide the present facilities, new techniques are established through extensive experiments to achieve the forging of large segment.

Table 1

Comparison of main features between current and new design of bottom head

	Current design	New design
Bottom disk	{ forged & welded 3 pieces	forged one piece
Petal	{ forged & welded 6 pieces	{ forged one piece
Mini skirt	welded to petal	
Weld seam length	100 %	34 %
Weld seams within CRD holes area	2 seams	none

The new forging process is a little different between steel works at the early stage of forging: one way is to forge the comparatively thin large blank, another to forge the comparatively thick one. After forging we need to form the wrought material into final shape using male die of half moon shape which rotates step by step as forming proceeds. Not only these techniques are fulfilled with the use of present forging machines, but also the mechanical properties and fracture toughness of formed material are homogeneous and can meet the specified requirements. Moreover, in the bottom section there exists only one circular welded joint, which can be inspected with ease and accuracy by the current ISI equipments and techniques.

Improvement of reactor pressure vessels for PWR

13. In the case of domestic PWR plants, the core region of reactor pressure vessel shell has been made of plate ASME SA-533 Gr. B cl.1. Based on the skillful and experienced techniques and capabilities of domestic material suppliers and vessel fabricators, the reactor vessels for PWR plants are designed in order to reduce the length of welded joints. For instance, in the case of 800 MWe class reactor vessels as illustrated in Fig. 2a, the length of weld seams in shell and head is reduced about 40 % being compared with current design. Also, many heavy section segments for overseas nuclear power plants are supplied from Japan very often.

14. Fig. 2b shows the present design of KWV's 4 loop 1300 MWe PWR RPV including vessel flange and nozzle belt. (Ref. 4) In the new design, one piece forged ring made from 400 t ingot is adopted in place of four pieces of bent forging electro slag welded to form a ring. The same philosophy is also applied to the design of bottom ring which was formerly made of plates, formed and welded. Thus the time of ISI was reduced together with easier performance.

15. In the new design of Cockerill 157" PWRPV shown in Fig. 2c, the reduction of weld seams is considered. The new design consists of three girth welds plus six set-on nozzles, while the conventional one consists of five girth welds with the corresponding longitudinal welds for plate construction plus six set-in nozzles to shell welds. It is reported that the geometry of vessel offers the possibility of ISI both from inside and outside, and nozzle can be examined without the removal of

internal, which is impractical in conventional PWR vessels. (Ref. 5)

Properties of heavy forging

16. The investigation on tensile properties in radial direction for the mono-block vessel flange made from 500 t ingot was performed by full wall thickness tensile test specimen with 535 mm in length. The results are shown in Fig.3 with fracture surface. The first break point was near the center portion of the wall thickness. The values tabulated in Fig.3 show excellent uniformity of mechanical properties through the wall thickness. The fracture surface shows the typical cup-and-cone fine ductile surface.

17. The nil ductility transition temperature distribution through thickness for 560 mm shell flange forging is compared with that for 300 mm SA-533 Gr. B. cl.1 plate in Fig.4. The NDT temperature of heavy forging shows the uniform distribution through wall thickness and its value is enough low even at the center of thickness.

18. Another comparison of NDT temperature of 200 mm SA-508 cl.3 bottom head spherical shell for forging indicates such low values as tabulated in Table 2.

IMPROVED CONTAINMENT VESSELS CONSIDERING EASIER IN-SERVICE INSPECTION

19. The feasibility studies on the standardization of LWR plants have been performed by the Ministry of International Trade and Industry. (Ref. 6) The basic aims are:
1) conducting adequate maintenance and inspection by the aid of automation and remote control,
2) reducing radiation exposure of employees,
3) improving reliability of plant and equipments,
4) reducing the time needed for regular inspections. The increase of capacity of containment vessels is expected to make easier the access to the components at ISI. The comparison between the standard and the improved vessels for 1100 MWe Mark I and II type BWR is shown in Fig.5.

20. As the diameters of containment vessels increase, new development of high strength steel is demanded which can be used without post weld heat treatment. The types of SA-516 Gr. 70 or 60 and SA-537 are usually used, for their strength levels

Table 2

Chemical composition and fracture toughness properties of 200 mm bottom dome forging (SA-508, cl. 3)

Chemical composition (Ladle, %)

C	Si	Mn	P	S	Ni	Cr	Mo	V	Al
0.19	0.31	1.41	0.004	0.004	0.95	0.17	0.49	0.007	0.017

Cu	Sn	Co	As	Sb
0.04	0.001	0.016	0.007	0.001

Fracture toughness

Location*	Depth	T_{NDT}(°C)	RT_{NDT}(°C)	vTr 35mils(°C)	$_vE_{max}$(kg-m)
Edge A	1/4 T	- 40	- 40	- 33	20.3
Edge B	"	- 40	- 40	- 37	22.5
Bottom	"	- 40	- 40	- 36	22.2
Edge A	Outer surface	- 50		- 48	22.8
	1/4 T	- 40		- 33	20.3
	1/2 T	- 35		- 28	19.1
	3/4 T	- 40		- 37	21.3
	Inner surface	- 40		- 53	19.7

* Longitude difference between edges A and B is 135°

Table 3

Chemical composition of SPV50 steel (%)

C	Mn	Si	P	S	Ni	Cr	Mo	V	$C_{eq.}$
0.16 MAX.	0.15 ≤ 0.55	0.90 ≤ 1.50	0.025 MAX.	0.020 MAX.	0.15 ≤ 2.60	0.30 MAX.	0.30 MAX.	0.10 MAX.	0.45 MAX.

$$C_{eq} = C + Mn/6 + Si/24 + Ni/40 + Cr/5 + Mo/4 + V/14$$

are 42 to 56 kg/mm² at tensile test. The new developed material JIS SPV50 has 62 kg/mm² tensile strength, 50 kg/mm² yield strength and chemical composition as shown in Table 3. (Ref. 7)

21. Series of ESSO tests with temperature gradient were performed using various specimens prepared with different welding procedure, material and heat treatment. From crack arrest fracture toughness Kca, limit stress was calculated for the postulated flaw, a semi-elliptic surface flaw, the size of which is 0.25 t x 1.5 t. The limit stress vs. temperature for various conditions is plotted in Fig. 6. SPV50 has better fracture toughness property than SA-516-70 without distinction of post weld heat treatment. Considering SPV50 has also better weldability, we can decide to adopt this kind of material to the new plants.

22. Besides SPV50, series of investigation can allow us to authorize the use of JIS SGV49 steel— equivalent to SA-516— up to 45 mm for containment vessel without post weld heat treatment. This steel is expected to be available for PWR, and actually enables us to enlarge the inner diameter of containment vessel for the standardized 800 MWe class PWR from about 38 m to 40 m.

23. The enlargement of containment vessel diameter contributes to provide wider area for ISI work, not only outside but also inside the steam generator compartment. The steam generator compartment which contains the reactor coolant loop consisting of the steam generator, the reactor coolant pump and the reactor coolant pipe, is very important at the time of periodic inspections during refueling shutdown. About 70 % of present ISI work being concentrated inside SG compartment, the effort for layout improvement has been directed toward the improved accessibility and the increased working efficiency inside SG compartment, including such items as compartment shape, access openings, stairs, gratings, pipings, trays, ducts, and supports. The figure 7 shows the samples of related improvement as follow:
1) providing two access openings instead of one, for a SG compartment.
2) providing stair, instead of ladder for each platform.
3) providing platforms optimized in Location, Area and Number, considering ISI from the early stage of the plant.
4) providing sufficient space around the reactor coolant pump through rearranging the small piping route and supports.

PIPING DESIGN CONSIDERING EASIER IN-SERVICE INSPECTION

24. As a large amount of piping is installed in the nuclear power plant, the inspection of piping occupies moderate portion of ISI. Especially, bending portion of piping increases the number of welding joints in the piping systems because the most of bendings are composed of fabricated elbows. The disadvantages of the use of elbow, from the view point of ISI, are
1) increase of weld seams,
2) poor inspectability of weld joint from elbow side because of abrupt thickness change in elbow,
3) large diameter elbow with longitudinal points.

25. To solve these disadvatages, bent pipe with small bent radius such as 2D or 3D has been developed using high frequency induction heating, where D is the diameter of pipe. Though the application of bent pipe is expected to replace 50 or 60 % of elbows in each plant, the ultrasonic examination at ISI can be easily conducted as well as the ordinary pipe joint examination, so that the examination time at ISI should be reduced considerably.

26. In the recent piping design of PWR plant, the following consideration has been involved for improvement of accessibility.
1) Establishment of distance criteria as shown in Fig. 8. Distances between two piping runs, between the piping and the wall and between weld point and brantch or support are to be comparatively longer.
(i.e. In the recent design, distance between two 3 inch O.D. pipes becomes 150 mm or 600 mm corresponding to the use of either manual or automatic UT inspection equipment, while in the conventional design it was decided to be 100 mm with consideration of space requirement for welding work.)
2) Installation of the grating and ladder and fixing up of removable thermal insulation.

CONCLUSIONS

27. The development of large scale heavy section steel made it possible to reduce the length of weld seam especially at the location which is difficult to inspect. Several examples of RPV design considering easier ISI were introduced and properties of large segments were found to be satisfactory when compared with the currently used material. The new RPV design can allow us to reduce inspection time and personal radiation exposure and to improve structural integrity of RPV.

28. Larger size containment vessels are adopted in Japanese new standard light water reactor design for the main purpose of easier ISI, making use of new high tensile steels with excellent mechanical properties and fracture toughness alone. The enlargement of containment vessel diameter suits for easier ISI not only outside, but also inside the steam generator compartment in PWR. The application of bent pipe in place of elbow joint is expected to save the examination time at ISI. The establishment of distance criteria for piping arrangement can also improve the accessibility.

ACKNOWLEDGEMENTS

29. The author would like to express his great thanks to the following persons and company who offered materials or have permission for this paper: Mr. K. Tomono (TEPCO), Mr. K. Ishikawa (IHI), Mr. T. Asai (MHI), Dr. S. Onodera and Mr. N. Tsukada (JSW), Mr. R. Sakai and Mr. H. Honjo (NSC), KWU AG and S. A. Cockerill.

REFERENCES

1. Ishikawa, K. and Tomita, J.,'Design of Nuclear Plant Components for Easier ISI', IAEA IWG-RRPC Kobe Meeting, April 1977.

2. Widart, J.,'Design of Reactor Pressure Vessels Considering Easier ISI', ditto.

3. Onodera, S., Nagata, N. and Tsukada, N.,'Larger Size, Integrated Type Steel Forgings as Intended for Easier ISI', ditto.

4. Onodera, S. et al.,'Advantages in Application of Integrated Flange Forgings for Reactor Vessel', 3rd MPA-Seminar, Stuttgart, September 1977.

5. Onodera, S., Widart, J. et al., 'Mono-Block Vessel Flange Forging for PWRPV 1,000 MWe ', 8th International Foremasters Meeting, Kyoto October 1977.

6. Japan Atomic Industrial Forum, 'Nuclear Power Development in Japan', Compiled from Atoms in Japan, October 1978.

7. Ando, Y. et al., 'Reports on the Applicability of SPV50 for Reactor Containment Vessel' (in Japanese), Generating Thermal Power Association, March 1977.

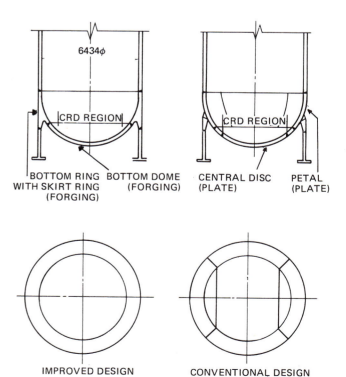

Fig. 1 RPV bottom design (1100 MWe BWR)

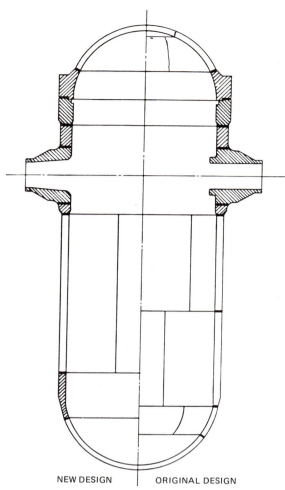

NEW DESIGN | ORIGINAL DESIGN

(note) Slash marks show solid forging.
Other portions are of plate forming.

a

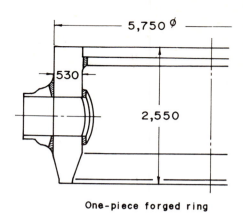

5,750 ⌀

530

2,550

One-piece forged ring

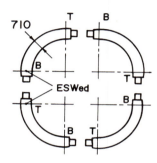

710

T B

B

ESWed

T

B

B T

Four pieces of bent forging,
electro-slag welded to form
a ring

b

Nozzle shell course

Weld

Nozzle

Original design
of nozzle

Nozzle shell course

Weld

Nozzle

New design
of nozzle

W.L.

Total weight : 305 ton | 350 ton

Original layout   New layout of
c of materials (WEC) materials (COCKERILL)

Fig. 2 Improvement of PWR pressure vessel design
    a 800 MWe PWR (MHI)
    b Combined vessel flange and nozzle belt forging of PWR PV, KWU/1300 MWe. Weight is
       165 ton with eight nozzle necks welded
    c WEC/157″ PWR PV: a comparison of material layout and nozzle design between original
       and new ones

5

Distance from the outer surface (mm)

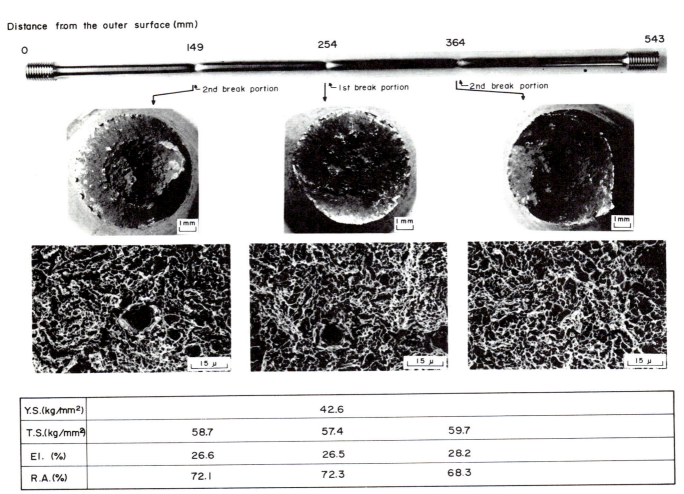

Y.S.(kg/mm²)	42.6		
T.S.(kg/mm²)	58.7	57.4	59.7
El. (%)	26.6	26.5	28.2
R.A.(%)	72.1	72.3	68.3

Fig. 3   Full thickness tensile properties of vessel flange made from 500 t ingot

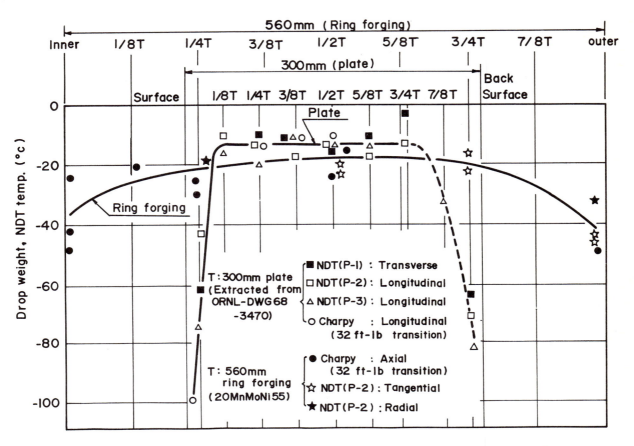

Fig. 4 NDT temperature distribution through wall thickness for 300 mm plate and 560 mm ring forging made from a 400 t ingot

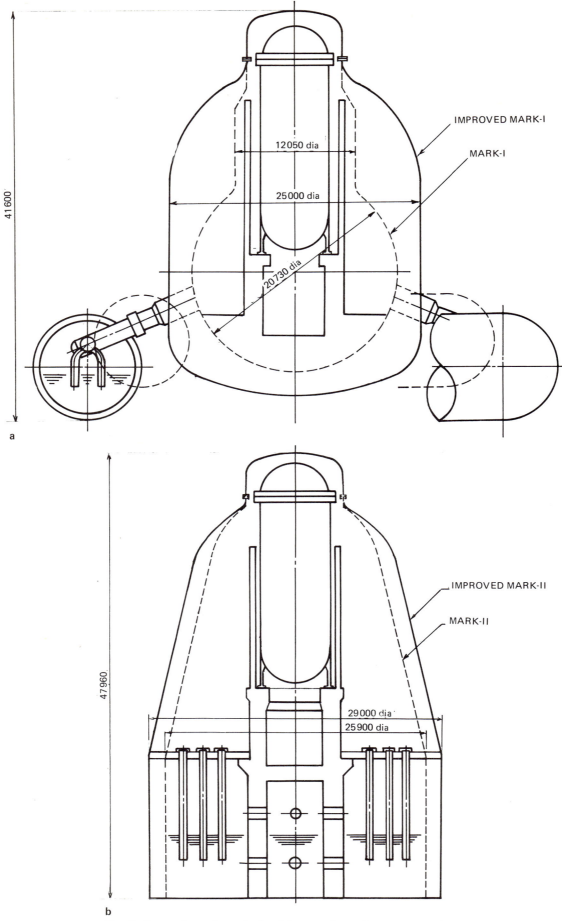

Fig. 5 Comparison of standard and improved containment vessels (1100 MWe BWR)
a Mark I type
b Mark II type

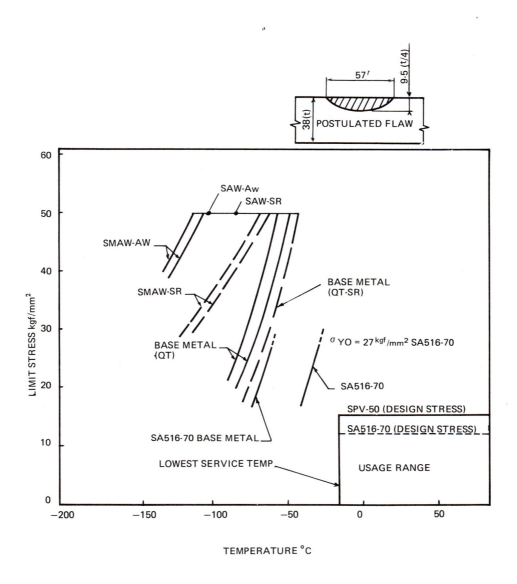

Fig. 6 Limit stress vs temperature (JIS G115 SPV50)

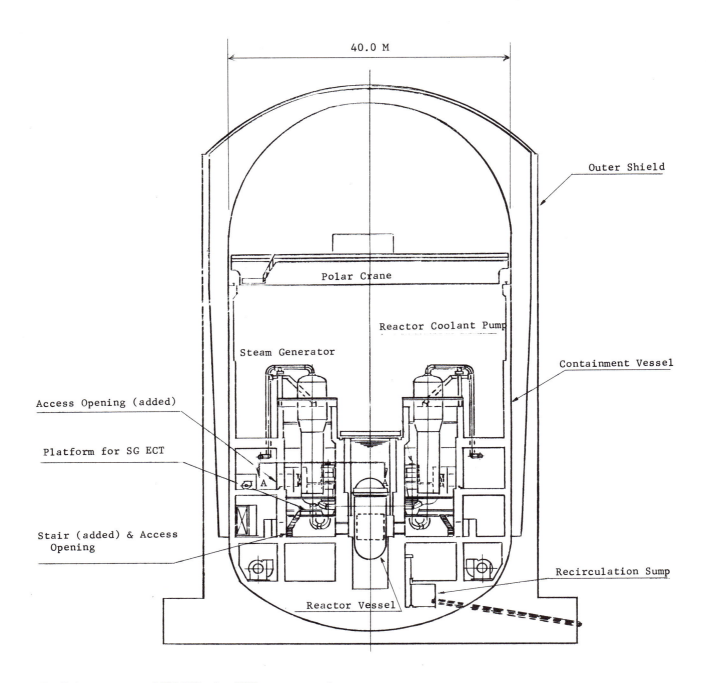

Fig. 7 Improvement of 800 MWe class PWR reactor containment

10

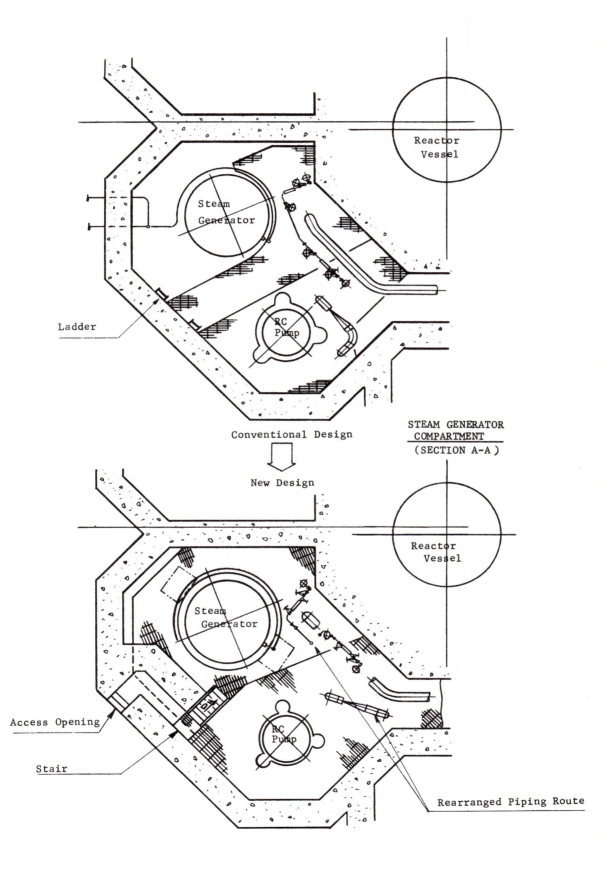

Reactor
Vessel

Steam
Generator

RC
Pump

Ladder

Conventional Design

STEAM GENERATOR
COMPARTMENT
(SECTION A-A)

New Design

Reactor
Vessel

Steam
Generator

RC
Pump

Access Opening

Stair

Rearranged Piping Route

Fig. 7   Improvement of 800 MWe class PWR reactor containment

11

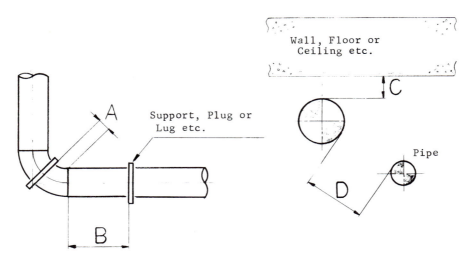

Unit:mm

DISTANCE	SIZE	ISI EQUIPMENT (UT)	
		MANUAL	AUTO
A	-	2T + 10	2T + 50
B	-	2T + 10	5.3T + 50
C	-	200	300
D	≤ 6 B	150	600
	≥ 6 B	200	

T: PIPE THICKNESS

Fig. 8 Improvement of piping arrangement considering easier ISI

REVISIONS TO THE ASME SECTION XI INSERVICE INSPECTION CODE TO ACCOMODATE ENFORCEMENT ON OLD AND NEW NUCLEAR POWER PLANTS

W. F. ANDERSON, PhD, Nuclear Regulatory Commission, Washington DC, USA
L. J. CHOCKIE, MSc, General Electric Company, San Jose, California, USA
and
W. O. PARKER, BSc, Duke Power Company, Charlotte, North Carolina, USA

The MS of this paper was received at the Institution on 9 October 1978 and accepted for publication on 20 October 1978

SYNOPSIS   The Codes effecting the program of inservice inspections in the United States are published by the American Society of Mechanical Engineers and adopted as a mandatory require- ment by Regulations of the U.S. Nuclear Regulatory Commission, as well as adoption by most of the jurisdictions in which the nuclear power plants are located. Initially, the Code was written to cover only the primary portion of the plant. During the ten years since the first edition was pub- lished, the coverage was broadened to include more of the pressure boundary of the plant, as well as the functional operational requirements of the pumps and valves in the system. Following this, refinements were made to reduce the amount of effort required where experience and studies dic- tated that a commensurate benefit was not being realized; and also at the same time examinational requirements were increased where it was determined that coverage was not adequate.

The modification to the rules contained in Section XI is shown in the Summer 1978 Addenda in recognition of the problems encountered when recent revisions of the Code were attempted to be implemented in the older plants as well as implemented in the plants currently undergoing con- struction. The modifications were necessary to accommodate the enforcement of the rules as mandatory requirements on both old and new plants.

INTRODUCTION

1.   Previous reports (1, 2, 3) presented at the first (1972), the second (1974), and the third (1976) conferences, Periodic Inspection of Pressure Vessels, described both the orig- inal safety philosophy of ASME Section XI, "Rules for Inservice Inspection of Nuclear Power Plant Components" (4) and the changes resulting from experience gained through its application.

2.   Mandatory application of the rules con- tained in Section XI is provided by the adoption of the Code by the Nuclear Regulatory Com- mission, as well as adoption by most of the jurisdictions (states) in which the nuclear power plants are situated. As the rules were revised from time to time to include coverage of more of the plant, to increase or decrease the amount of examination on particular components or systems, or to add new requirements added to determine which portion of a piping system should be selected for examination, problems began to materialize when the revisions were attempted to be implemented in the older plants. In addition, other problems manifested them- selves when the rules were imposed on new plants being constructed. As might be expected, not all the problems presented themselves at the same time, nor were the solutions always immediately apparent, since there were sev- eral instances when the solution to one problem would create another problem. For example, it is practical to utilize the calculations for cumulative usage factors to identify the piping joints to be selected for examinations on newer plants where such calculations have been made during the analysis of the piping. Such calcu- lations were never made on the older plants; accordingly, the criteria could be applied to the older plants in a practical manner without a re-analysis of the piping.

3.   To identify the problems, to resolve the issues, the ASME committee responsible for the Section XI of the Code and the staff of the Nuclear Regulatory Commission, as well as representative chief inspecting officers of the jurisdictions representing the various states, worked together in an attempt to provide reso- lution and solution to each of the problems identified. In some instances the solution was to modify an existing requirement in the Code, and in other instances modifications to the Regulations adopting the Code provided the solution.

4. The staff of the Nuclear Regulatory Commission identified the problems which were being encountered, compiled the list and, through their representative on the committee, suggested some solutions. Counter proposals were developed by members of the Committee in some cases; as for example, a counter-proposal to a suggestion that the NRC would "take exception" or "fail to recognize" certain provisions in the latest Code. The counter-proposal in this case was to state the "additional requirement" to be imposed to the provision in the Code rather than take exception. The purpose of the counter-proposal was in recognition that to take exception would be a precedent in the United States and may encounter many time-consuming difficulties, whereas stating the requirements as additions to requirements was the customary procedure and would entail the least amount of difficulty, providing solution in the most expeditious manner. Due to the complexities of the exemptions stated in the Code, certain exemptions were not permitted, thereby imposing additional requirements.

SAMPLING PLAN FOR CLASS ONE COMPONENTS

5. Revisions to the Code identified "components exempted from examination" and provided a philosophy, based upon statistical studies (5, 6, 7, 8) that a single loop (representing several identical loops) could be utilized to provide a representation of the behavior of the other loops. The problems that ensued were:

(1) The format which provided exemptions made it difficult for the inspectors to readily identify what was required to be inspected

(2) Criteria, exempting piping from inservice inspection, based upon location of postulated breaks was being applied too broadly and did not consider other consequences of failure such as the loss of coolant accident

(3) Sampling criteria should include a sufficiently broad and random sample to provide a high probability that "technical surprises" would be located before they progress too far.

6. Discussions at a special meeting, where the only agenda items were those to resolve the problem areas, resulted in the agreement that a strictly random sampling method is not the correct route to take, and that consideration should be given to high stressed welds.

The proposed limit of 2.4 S_m for primary plus secondary stress intensity provided a reasonable definition of the joints requiring examination. Accordingly, it was agreed that Section XI would be revised to require the following:

(1) All high stressed welds in each loop

(2) Terminal Ends in each loop

(3) All dissimilar metal welds, including high alloy steels to high nickel alloys, in each loop

(4) Additional welds, in a single loop, to make up a total of 25% of the total welds

7. In order to avoid continued revisions of the sampling programs in the nuclear plants, it was agreed that the Nuclear Regulatory Commission would distinguish between the "old" and the "new" plants and whether requirements above and beyond those contained in the Summer 1975 Addenda (the latest addenda the "old" plants were obligated to meet) would be imposed. Defining the difference between the "old" and "new" plants would have the same effect as the "grandfather" clause for the old plants; and it was further agreed that the tabulation of the "grandfathered" plants could be contained in future editions of Section XI to avoid a difference in requirements between the Nuclear Regulatory Commission and the jurisdictions.

VESSEL NOZZLES AND ADJACENT REGIONS

8. An exemption provided by the single loop sampling plan was certain nozzle to vessel welds, those nozzles at the ends of the loops excluded by the sample plan. It was readily agreed to revise the Code to again require that all of the vessel to nozzle welds would be required to be examined each inspection interval. Examination of all nozzle to vessel welds was the requirement prior to the single loop plan.

PIPE BRANCH CONNECTIONS

9. Revisions to the Construction Code (9) changed the size of pipe branch connection welds requiring examination by volumetric methods from four inches to two inches, nominal pipe size. It was agreed that the number should be revised to reflect the two inch size in the Summer 1978 Addenda to Section XI, with consideration of the grandfathered plants.

STEAM GENERATOR TUBING

10. Continuing problems with steam generator tubing make requirements for inservice

inspection of these components fluid and difficult to codify. A task group of the Code Committee is developing new requirements, and the Nuclear Regulatory Commission is handling these requirements on a case-by-case basis. Neither the Code Committee nor the NRC had any proposal to make which could resolve the issues. Consequently, it was agreed that no immediate revision be made to Section XI.

EXEMPTIONS AND REQUIREMENTS OF CLASS TWO SYSTEMS

11. The problem identified with the Class 1 systems was also noted for the Class 2 systems; that is, requirements were usually stipulated in what was not required to be examined rather than what was to be examined. This made it difficult for the Inspector to identify what was required to be examined. Additionally, it was agreed that the exclusion of static systems should be reviewed since the record showed that these systems have had cracking develop in service. Since the problem was quite complicated, a task group was appointed to prepare a rewrite to state in positive terms what was required to be examined.

12. The task group recommendation, which was subsequently approved by the committee, provided that static systems would be included in the examination program; the following criteria could be used to establish which systems were to be examined.

    (1) Systems which are static and whose flooded condition is 80% or less of the pressure at which they are required to operate.

    (2) Systems, other than Residual Heat Removal Systems and Emergency Core Cooling Systems, that operate at a pressure greater than 275 psig or at a temperature above 200 F.

    (3) Component connections (including nozzles in vessels and pumps) piping and associated valves and vessels (and their supports) that are greater than four inch nominal pipe size.

13. It was also agreed that the grandfathering of the plants from these requirements was to be proposed by the NRC staff for inclusion in the Regulations; and the task group was given the assignment to continue their study and comparisons, considering also that the amount of examinations for Class 2 systems should not overshadow the important Class 1 systems. The task group is to develop realistic examination requirements for Class 2 systems consistent with the safety requirements of the plant, recognizing also that high pressure hydrostatic tests of the Class 2 systems provide a further measure of integrity.

ULTRASONIC EXAMINATION

14. The evaluation of indications from the ultrasonic examination was evidenced as a point of concern by several members of the NRC staff. The recording requirements did not produce a record until the indication exceeded the level of 100% of the distance amplitude correction (DAC) curve; and at that point the indication had to be evaluated in terms of acceptance or rejection. Agreement was reached that the Code should be revised to require the recording of indications exceeding the 50% level, with acceptance or rejection based upon the 100% level.

SUMMARY

15. Revisions to the Code had created a situation which, when implemented on older plants, resulted in uncertainties as what was required to be examined, and uncertainties as to whether the examinations addressed the correct requirement. The problem areas were identified and a series of meetings devoted to resolving the problems. Revisions to the Code were approved for the Summer 1978 Addenda, and the revisions to the Nuclear Regulatory Commission Regulations were proposed which would implement the requirements in the Summer 1978 Addenda. Also, the proposed regulation would identify the series of plants to which the rules would apply.

16. In each instance, the problems of implementing revisions of the Code to older plants were identified and a concerted effort devoted to effecting a resolution. A resolution which would allow the correct edition and requirement of the Code to continue to be enforced on older plants, while at the same time allow revisions of the Code to be made applicable to the newer plants.

REFERENCES

1. BUSH, S.H. and MACCARY, R.R., "Development of Inservice Inspection Safety Philosophy for U.S.A. Nuclear Power Plants", I. Mech. E., Periodic Inspection of Pressure Vessels, 1972.

2. JOHNSON, W.P., BUSH, S.H. and MACCARY, R.R., "Augmented Scope of the 1974 ASME Section XI Code Inservice Inspection of Nuclear Power Plant Components", I. Mech. E., Periodic Inspection of Pressure Vessels, 1974.

3. CHOCKIE, L.J., BUSH, S.H. and MACCARY, R.R., "Extended Rules of the 1974 ASME Section XI Code, 'Inservice Inspection of Nuclear Power Plant Components' ", I. Mech. E., Periodic Inspection of Pressure Vessels, 1976.

4.   ASME Boiler and Pressure Vessel Code - Section XI "Rules for Inservice Inspection of Nuclear Power Plant Components" - 1977 Edition, inclusive of all published Addenda to Summer 1978.

5.   WASH-1285, Report by the Advisory Committee on Reactor Safeguards on "Integrity of Reactor Vessels for Light-Water Power Reactors," January 1974.

6.   WASH-1400, Reactor Safety Study - An Assessment of Accident Risks in U.S. Commercial Nuclear Power Plants, U.S. Atomic Energy Commission, October 1975.

7.   WASH-1318, Technical Report on "Analysis of Pressure Vessel Statistics from Fossil-Fueled Power Plant Service, and Assessment of Reactor Vessel Reliability in Nuclear Power Plant Service", May 1974, U.S. Atomic Energy Commission.

8.   ARNETT, L.M., "Optimization of Inservice Inspection of Pressure Vessels", presented at Conference on Nondestructive Testing in Nuclear Industry, ASM, Denver, CO, December 1975.

9.   ASME Boiler and Pressure Vessel Code - Section III - "Nuclear Power Plant Components", Published by American Society of Mechanical Engineers, New York, New York 10017

C23/79

ENHANCEMENT OF THE RELIABILITY OF REACTOR PRESSURE VESSELS BY IN-SERVICE INSPECTION

L. P. HARROP, BSc and A. B. LIDIARD, PhD, DSc, FInstP

The MS of this paper was received at the Institution on 14 November 1978 and accepted for publication on 18 December 1978

SYNOPSIS. The recently developed methods of analysis of the probability of failure of pressurized water reactor (P.W.R.) vessels are here used to estimate the benefits of (a single) in-service inspection. More particularly, the theoretical basis of certain recent suggestions for 'optimal' schedules of inspection (ASME XI Program A) is critically examined. In fact, both the magnitude of the benefit and its dependence upon the time of inspection are shown to be sensitive to detailed assumptions, particularly to the rates of crack growth in service. Use of the most realistic model presently available for P.W.R.'s indicates no well-defined optimum time of inspection but does show that this should be within the first ten years of service if the aim is to reduce the whole life chance of failure as far as possible. Further reductions can, of course, be attained by repetition of inspection later.

§1. INTRODUCTION

1. Section XI of the ASME Code provides a comprehensive set of rules for the in-service inspection of nuclear plants and is, in turn, the basis for various other national and international rules. Formerly these required examination of reactor pressure vessels and piping at equally spaced intervals throughout the life of the reactor. Recently, however, the rules have been amended so as also to allow an alternative sequence of inspections, one which has been arrived at by using a statistical or 'probabilistic' model of reactor vessel failure. The original sequence of equally-spaced inspections is referred to as "Program B" while the supposedly optimum sequence obtained from the statistical model is called "Program "A". These alternatives and the expected gains in the reliability of nuclear pressure vessels to be obtained by following Program A rather than Program B have been reviewed by Chockie (Refs. 1 and 2), while the statistical model and the calculations themselves are reported by Arnett (Ref. 3).

2. However, it should be noted that this statistical model is very restricted in several important ways. First, it does not allow for any statistical distribution of material toughness or crack growth rates. Second, it assumes that the reactor vessel is subject to only a single set of transient stresses. Third, it employs a very simple stress-intensity function within the framework of linear elastic fracture mechanics. Fourth, it does not allow for the fact that the vessel may be assumed to have passed a cold hydro test before entering service. The model must therefore be regarded as very rudimentary compared with those recently presented by ourselves and by others (Refs. 4-7) and it therefore seemed desirable to examine the basis for such suggested changes in inspection programmes by using these more developed probabilistic models. An additional reason for doing so is that although Arnett (Ref. 3) explicitly details his assumptions about the incidence of cracks in reactor vessels and about the efficiency with which the inspection techniques detect such cracks, he does not justify the functions which he uses to represent these quantities. Furthermore, we have shown elsewhere (Refs. 7, 8) that the predicted rates of vessel failure are sensitive to some of these functions.

3. The purpose of this paper is therefore to report a theoretical examination of the overall improvement in reliability which is predicted to follow from in-service inspection made at a variable time after the reactor vessel enters service. The basic model is that devised by Lidiard and Williams (Ref. 5) for the Marshall study of the integrity of P.W.R. pressure vessels (Ref. 10) but as developed later by Harrop (Refs. 9 and 11). In this form the calculations include the following features: (i) the separate loadings of the principal regions of the vessel (nozzles, beltline, upper and lower heads), (ii) statistical distributions of toughness, K_{Ic}, and of initial crack sizes, (iii) vessel transients according to Westinghouse design specification for normal, upset and test conditions, (iv) stress-intensity factors, K_I, calculated according to the methods prescribed in Section XI of the ASME code including the Irwin plastic-zone correction, (v) crack growth according to specified growth laws (representing fatigue or corrosion-fatigue), (vi) allowance for the fact that the statistical population of reactor vessels in service may be assumed to have passed a pre-service pressure test and (vii) vessel failure is deemed to occur either when K_I exceeds K_{Ic} or when the load on the net ligament exceeds the corresponding limit load, i.e. elastic-plastic failure criteria. Fuller

details of the model are given in Ref. 9 where some considerations of the benefits of in-service inspection upon vessel reliability have already been given. Here, however, we look at the matter in the same way as it is considered in Refs. 1-3, i.e. we look at the total chance that a vessel fails during its lifetime* and then evaluate the reduction in this chance which follows from making a complete inspection at (a variable) time, t.

§2. ENHANCEMENT OF VESSEL RELIABILITY

4. We have first of all particularized our assumptions so as to make a close comparison with Arnett's results and to examine their sensitivity to the physical assumptions. We can enter Arnett's assumptions about the incidence and size distribution of cracks and about the efficiency of detecting them during inspection into our calculations in exactly the form he uses them but this is not possible for that governing crack growth. Here Arnett uses a more elaborate equation than we have so far employed and one which cannot be inverted analytically, whereas such inversion is required by our present method of calculation. Nevertheless, to make the comparison as close as possible we have chosen the parameter, γ, in our crack growth equation so that this equation exactly reproduces the lifetime growth of a crack initially 1.6 in. deep under the other conditions specified by Arnett**. Fig. 1 reproduces Arnett's results for one particular example (his Fig. 4) and also shows our corresponding results, the difference being mainly due to the different representations of crack growth in the two calculations. The general form of the two curves is similar; in particular, the optimal time of inspection, i.e. the time of inspection which gives the greatest factor of improvement, is closely the same.

5. Nevertheless, the quantitative differences between the two curves in Fig. 1 suggest a sensitivity to crack growth rates. Figs. 2 and 3 therefore show the consequences of changing the coefficient γ in our crack growth law. In Fig. 2 γ is taken to be twice the value of Fig. 1, i.e. a very high value: the shape of the curve is not much changed but the magnitude of the factor of improvement is everywhere much less. On the other hand, if γ is taken to be smaller than in Fig. 1, we find that the maximum occurs at earlier times and that the factors of improvement may be very much larger. Thus, Fig. 3 shows the results when γ is taken to be only one-tenth as large as in Fig. 1, i.e. of a magnitude which may be typical of 'dry' fatigue crack growth in P.W.R. vessels. In this particular case the maximum has moved to such early times as to be no longer visible.

---

*It may be noted that this chance is the integral over the vessel lifetime of the "failure probability per unit time" or "failure rate", Q_f, which has been the subject of our previous examinations. As before, we take the vessel lifetime to be 40 years.

**In fact, under these conditions this size of crack just grows through the vessel wall (10 in. thick) in 40 years (Ref. 3, Table 1, case 2:1:2).

When γ is one-quarter of the value in Fig. 1, the maximum occurs at two years after entering service. Thus, Arnett's model predicts that both the factors of improvement and the optimal times of inspection are sensitive to the rates of crack growth in just the range of practical concern, i.e. in the range corresponding to fatigue crack growth under 'dry' and 'wet' conditions. Since knowledge of the rates of growth of cracks in practical situations is still rather imprecise it would thus seem fair to conclude that one should be cautious in accepting the precise quantitative predictions of Arnett's model.

6. In view of the inherent simplifications of Arnett's model it therefore seems sensible to examine the predictions of our model in its most developed form (Ref. 9). Fig. 4 shows the nature of these predictions both when it is assumed that the vessels in the population have all passed a pre-service pressure test (cold hydro test) and when it is not. All other assumptions are as in Fig. 8 of Ref. 9; in particular, the chance of missing cracks during the in-service inspection is taken to be the same function, B(x), suggested by Marshall et al (Ref. 10) for the chance of missing a crack of depth x during ultrasonic examination before service. It will be seen from Fig. 4 that the assumed change in statistical population resulting from the imposition of the cold hydro test exerts a strong effect on the factor of improvement but that in neither case is there a maximum in the curve - even though the crack growth rates, like those in Figs. 1 and 2, are those for 'wet' conditions. It is sometimes argued that the cold hydro test does not modify the statistical population of vessels in the way we have assumed (mainly by reason of the difficulties in defining precise failure criteria in the presence of so-called stable crack growth under load). Were this to be so then we would expect the line in Fig. 4 marked "without cold hydro test" to represent the limiting result for populations subject to a cold hydro test but in which stable crack growth could occur in the vessels. From these results, therefore, it will be seen that the overall benefit from an in-service inspection is greater if the inspection is conducted during the first ten years or less but that there is no well defined optimal time of inspection as suggested by the simple Arnett model. There may therefore be no good reason for following the particular, supposedly optimal schedule of inspections suggested by Chockie Refs. 1 and 2).

§3. SUMMARY AND CONCLUSION

7. In this paper we have examined the particular suggestions which have been made for optimising the benefits of in-service inspection of reactor pressure vessels (Refs. 1 and 2). These suggestions and the anticipated gains in vessel reliability have been derived from a very simple statistical model (Arnett, Ref. 3). Here we have first of all used the more highly developed model of Harrop, Lidiard and Williams (Refs. 5 and 9) to examine the sensitivity of Arnett's predictions to reasonable variations in the assumptions. This showed that both quantitative and qualitative features were sensitive to changes in several of these; in particular to (i) the rates

18

of crack growth and (ii) the efficiency of crack detection. Secondly, we made calculations of the gain in vessel reliability expected from our more fully developed model. The results were again both quantitatively and qualitatively different from Arnett's, although like his they indicated that the overall gain in reliability would be greater when the inspection was carried out earlier in the vessel's life rather than later (say, before ten years). Here, however, there were qualitative differences dependent upon application of a pre-service pressure test.

8. The broad features of these results are understandable from the theoretical results on failure *rates*, Q_f, previously published (Refs. 5-9). In general, these show that, in the absence of any in-service inspection, the failure rate, Q_f, is extremely small at early times, but that during some later interval of time it increases rapidly and then more slowly after that interval. This sort of variation of Q_f with time is, however, tied to the assumption that the vessels of the population have all successfully undergone a pre-service pressure test. If we do not make such an assumption, then Q_f is initially much larger and it increases steadily throughout the vessel life; in the later years the predicted values of Q_f are much the same in both cases. The effect of in-service inspection is to reduce the values of Q_f at all later times. Since the factor of improvement relates to the integral of Q_f over the whole life of the vessel, the forms of the two curves in Fig. 4 are thus easily understandable.

9. All the above discussion assumes a complete in-service inspection of the whole vessel. It is, of course, possible and indeed expected, that some regions of the vessel will contribute more than others to the failure rate (see, e.g., Fig. 9 of Ref. 9). In such situations it may only be necessary to conduct partial inspections to attain a given level of reliability. We showed an example of how this could be done in Ref. 9 (Fig. 10). However, the conclusions in such cases must be sensitive to the correctness of the model in each region and not simply overall. In particular, the incidence of cracks in the different regions of the vessel is one important uncertainty at present; in these and all previous calculations we have assumed that the cracks occur equally frequently and with the same size distribution in all the five principal regions of the P.W.R. vessel. Any marked departure from such uniformity might well invalidate our conclusion (Ref. 9) that it is more important to inspect the nozzle regions and the bottom head than it is to inspect the beltline and the top head. Better information on the frequency with which cracks occur in different regions of the vessel is thus one of our principal requirements at the present time.

REFERENCES

1. CHOCKIE L.M., BUSH S.H. and MACCARY R.R., Extended rules of the 1974 ASME Section XI Code "In-Service Inspection of Nuclear Power Plant Components". In Periodic Inspection of Pressurized Components (I. Mech. E., London, 1976) p.91.

2. CHOCKIE L.M., Codes, standards and practices and their influence on the reliability of nuclear plants. In Reliability Problems of Reactor Pressure Components, Vol. II (I.A.E.A., Vienna, 1978) p.119.

3. ARNETT L.M., Optimization of in-service inspection of pressure vessels, de Pont Savannah River Laboratory Report DP-1428 (1976).

4. NILSSON F., A model for fracture mechanical estimation of the failure probability of reactor pressure vessels. Proceedings of the 3rd International Conference on Pressure Vessel Technology, Pt. II Materials and Fabrication (American Society of Mechanical Engineers, New York, 1977) p.593.

5. LIDIARD A.B. and WILLIAMS M., A simplified analysis of pressure vessel reliability, J. Br. Nucl. Energy Soc. (1977) 16, 207.

6. LIDIARD A.B. and WILLIAMS M., A simplified analysis of the probability of failure of P.W.R. pressure vessels. In ICOSSAR '77, being Proceedings of the 2nd International Conference on Structural Safety and Reliability (Werner-Verlag, Düsseldorf, 1977) p.449.

7. LIDIARD A.B. and WILLIAMS M., The sensitivity of pressure vessel reliability to material and other factors. In Reliability Problems of Reactor Pressure Components, Vol. I (I.A.E.A., Vienna, 1978) p.233.

8. LIDIARD A.B. and WILLIAMS M. A theoretical analysis of the reliability of P.W.R. pressure vessels. In Tolerance of Flaws in Pressurized Components (I. Mech. E., London, 1978) p.1.

9. HARROP L.P. and LIDIARD A.B., The probability of failure of P.W.R. pressure vessels evaluated using elastic-plastic failure criteria. In Proceedings of the O.E.C.D. Specialist Meeting on Elastoplastic Fracture Mechanics (U.K.A.E.A., Risley, 1978) p.

10. MARSHALL W. et al. An assessment of the integrity of P.W.R. pressure vessels (U.K.A.E.A., Harwell, 1976).

11. HARROP L.P., The extension and application of a simplified model for the reliability of P.W.R. pressure vessels (to be issued).

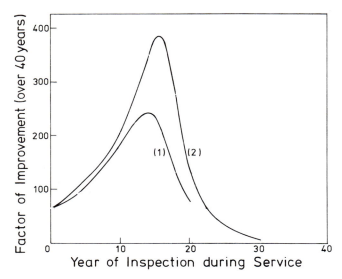

Fig. 1 Calculated factor by which the total chance that a vessel fails during its lifetime is reduced by making a complete in-service inspection at a time, t (factor of improvement). The curve labelled (1) is due to Arnett (Ref. 3, Fig. 4). The other curve (2) has been obtained by following the method of calculation of Ref. 9 and using the same assumptions as Arnett except that (i), as explained in the text, the crack growth law is different ($\gamma = 1.55 \times 10^{-2}$ in$^{-\nu}$ yr^{-1}; $\nu = 0.865$) and (ii), the toughness is taken to follow a Gaussian distribution (with the same mean value of 150 k.s.i. $\sqrt{\text{in}}$) with a coefficient of variation of 10%

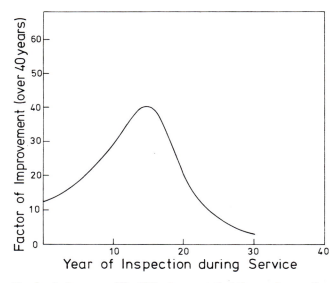

Fig. 2 As for curve (2) of Fig. 1 except that the crack growth coefficient is twice as large, i.e. $\gamma = 3.10 \times 10^{-2}$ in$^{-\nu}$ yr^{-1}

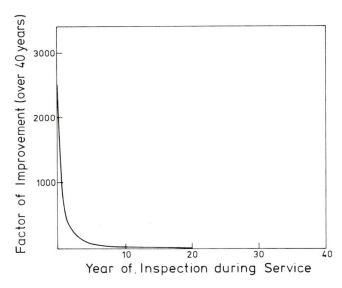

Fig. 3 As for curve (2) of Fig. 1 except that the crack growth coefficient is only one-tenth as large, i.e. $\gamma = 1.55 \times 10^{-3}$ in$^{-\nu}$ yr^{-1}. Broadly speaking this value corresponds to fatigue crack growth in a PWR under dry conditions whereas the values assumed in Figs 1 and 2 correspond to fatigue crack growth in contact with water

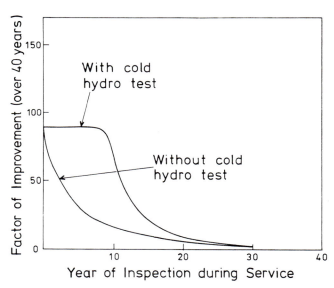

Fig. 4 Factor of improvement as a function of the time of inspection of a PWR vessel as calculated using the method of Harrop and Lidiard (Ref. 9). The detailed assumptions are as in Ref. 9, Fig. 8

C24/79

SPANISH EXPERIENCES ON ISI CODES AND STANDARDS

M. CERECEDA, Dr Ing, and A. ALONSO, Ing,
Tecnatom, Madrid, Spain

The MS of this paper was received at the Institution on 11 September 1978 and accepted for publication on 4 December 1978

SYNOPSIS   Implementation of inservice inspection programs in Spain is being performed applying Codes and Standards used in the country that provides the nuclear
steam supplier system. Code edition applied for each preservice or inservice inspection is carefully selected to satisfy the requirements of nuclear regulations,
updating, in some cases, this edition to later ones, in order to provide a more
realistic inservice inspection program within the established safety limits.

INTRODUCTION

1.   Spanish Nuclear Regulation at present does not cover specific requirements for some nuclear activities. In these cases, it is required to satisfy the applicable requirements of the country that has furnished the Nuclear Steam Supplier System (NSSS). One of these activities is the inservice inspection (ISI) of nuclear power plant components.

2.   Table 1 shows the Spanish nuclear program, consisting on thirteen plants supplied by United States (USA) and one plant by Germany. Tecnatom is at present, exception made of Garoña plant, responsible for their ISI programs development, which must satisfy either USA or Germany requirements.

3.   As it is known, ISI requirements in USA are established in USNRC 10CFR50 "Licensing of production and utilization facilities" (Ref. 1), refering to Section XI of ASME Boiler and Pressure Vessel Code (Ref. 2). The applicable edition for each plant has to be defined as a function of either, the date of issuance of the construction permit or the code editions applied to the component construction. It is also required, throughout the service life of the plant, to update, in defined periods of time, the ISI programs to comply with the requirements set forth in code editions in effect no more than 6 months prior to the start of each period. Refering to the German regulation, there are only general ISI requirements established in the Reactor Safety Commission Guidelines for Pressurized Water Reactors (RSK) (Ref. 3). Therefore, ISI programs in Spain have to be developed in accordance with Section XI or RSK requirements.

4.   This paper describes codes interpretations, exceptions taken and approaches used to develop such programs.

APPLICABLE CODE TO AN INSERVICE INSPECTION

5.   As a minimum, ISI programs for plants supplied by USA, shall comply the requirements established in the Section XI edition required by USNRC 10CFR50. For plants proceeding from Germany, ISI programs shall be developed in accordance with RSK guidelines. Due to the fact that these guidelines are nowadays too general, it could be very practical to use, in addition, a more specific ISI code to detail which components require inspection, type and frecuency of examinations, etc. The most helpful code at present that can be used in this approach, is Section XI of ASME Code. Therefore, it could be concluded that, in order to develop ISI programs for Spanish nuclear plants, Section XI of ASME Code is applied, in some cases as requirement (USA plants), and in others, as a helpful approach (German plants).

6.   The ISI program for José Cabrera plant, was developed in 1.971 for the reactor coolant pressure boundary components, in accordance with the 1.970 edition, performing its first inspection in 1.972. This inspection as well as the performed in 1.973, satisfied the Code requirements, exception made to the reactor pressure vessel, because the non availability of an appropriate and reliable equipment, the regenerative heat exchanger because the high radiation level and some portions of pipelines because inaccessibility. Therefore, in 1.973, the winter 72 requirements for Class 2 components were incorporated as

an effort to extend the volume of performable examinations. Lately, in accordance with USNRC, the ISI program has been updated to comply the requirements of winter 75 edition. The first inspection in accordance with this program, is going to be performed in 1.979, and it is expected that some exceptions to the program, will have to be taken, because the design of the plant did no foresee these examinations or because its performance could put the plant in an unsafe condition (welds interfered by support components, valves required to leak test without test connections, etc.). Refering to the reactor pressure vessel examination a remote UT equipment is being developed and it is expected that these examinations will be performed, at least partially, in 1.979-1.980).

7.   Preservice inspection programs for the second generation of Spanish plants (Almaraz, Lemóniz, Ascó and Cofrentes), whose construction permits were issued between 1.971 and 1.974, should meet the requirements set forth in Section XI editions in effect 6 months prior to the date of issuance of their construction permits. In accordance with the construction permit dates, these programs should be developed satisfying the requirements established in editions from Winter 72 (Almaraz unit I) to 1.974 edition (Cofrentes unit). Every Code edition included in this period of time, has some advantages in reference to the others. One of these is to perform UT examinations applying the criteria set forth in Appendix I (first time issued in summer 73). But to use this addendum could require to incorporate also the inservice testing of pumps and valves, which is a hard requirement for both, its preparation and performance, because the valves required to test are not clearly defined and the performance of tests could be critical path in the refueling shutdown. Other advantage, is to apply the new NDT evaluation and acceptance criteria which are being, from summer 73, gradually incorporated to the code. For these reasons, it was decided not to apply a single code edition, but to use the edition required by the USNRC regulations to determine the components required to inspection, examination techniques and evaluation and acceptance criteria. This approach may be justified having in mind that ISI programs have to be updated during the inservice life of of the plant to comply with the requirements of later addenda and allows some flexibility satisfying the ISI requirements. So that, a plant required by regulations to comply, as a minimum, with summer 73 requirements, establishes in its preservice program the intention to satisfy these requirements and, as much as possible, those established in subsequent addenda to summer 75. Other aspect to be taken into account is the inservice testing of valves. Programs for these tests were developed in accordance with requirements set forth in the 1.974 edition. In winter 77, these requirements have been widely modified in such a way that the number of valves to be tested could be reduced in a half. Although this addendum has not been already approved by the USNRC, their requirements could be used as an interpretation of the older ones, that means, winter 77 requirements are interpreted as an explanation of the 1.974 ones. This approach could be justified again because it has been taken to test the same valves in inservice than in preservice. Therefore it has been also used to develop the preservice inspection programs of these plants.

8.   The remaining plants have constructions permits issued after July 1.974. Therefore, their ISI programs have to be developed for each component in accordance with the requirements set forth in Section XI of editions of the ASME Code applied to the construction of the particular component. This means, that the preservice inspection program for a particular plant should be developed in accordance with several code editions, which is not practical enough. For this reason and having in mind that the ISI of any component may meet the requirements set forth in subsequent editions, it was decided to use, as minimum requirement, a single one selected as an envelop of those required by USNRC applying also the approaches described in the previous paragraph.

9.   Relating to the German plant, RSK guidelines have to be satisfied. As these guidelines establish general requirements, details of ISI programs (scope, examination methods, frecuency, etc.) are defined in Germany between the utility and the authorities. Therefore, Section XI has been taken as an approach to detail the ISI program of the German plant, selecting the edition to be applied, as it has been described above (Construction permit date after July 74).

COMPONENTS REQUIRED TO INSPECTION

10.   Once the Code edition has been determined, the next step in developing an ISI program, is to define plant systems and components required to inspection.

11.   Section XI, in general, establishes rules and requirements for inservice inspection of Classes 1, 2 and 3 pressure-retaining components (Classes 1, 2 and 3 respectively, of Section III) and inservice testing of pumps and valves in nuclear power plants.

12.   The plant Owner is responsible for determining the appropiate Class for each component of the plant. Classification criteria for Class 1 components are specified in 10CFR50 and, for other safety related components, in Regulatory Guide 1.26 (Ref. 4), although classification criteria specified in ANS-51.8 (Ref. 5) or ANS-22 (Ref. 6) may

be used. Regulatory Guide classifies in Quality Groups and ANS in Safety Classes, refering, both of them, to the construction Classes defined in ASME. ANS covers some type of components not covered by Regulatory Guide, but with these exceptions, both classifications require the same construction Class for each component. In one or other way, Classes for each component are defined by the Owner of the plant. This classification is reviewed, to determine which systems or parts thereof are strictly required to inspection, with the assumption that, only those pressure-retaining components containing radioactive material, water or steam, are required to inspection. This assumption has been taken, understanding that, Section XI refers to Class 1, 2 and 3 pressure-retaining components in accordance with Regulatory Guide 1.26 and that other systems, not covered by this guide, such as instrument and service air, diesel and auxiliary support systems, emergency and normal ventilation may be excluded, although they are constructed in accordance with any nuclear Class. This approach concurs with L.J. Chockie (Ref. 7) interpretation, who, in the referenced paper, establishes "it was not the intent of the committee to include in Section XI such systems not directly associated with the reactor coolant pressure boundary". Another reason to review the utility component classification, is, as Section XI recognises, that some components may be optionally constructed in accordance with upgraded requirements by reclassifying, for instance from Class 2 to Class 1. In these cases the examination requirements of the not upgraded Class (Class 2 in the example) apply.

Table 2 shows, for typical P and BWR plants, the systems that in accordance with the described interpretations are normally included on ISI programs.

13.   In the German plant, Section XI requirements are applied to detail the ISI program. Components in this plant are classified by the utility in five categories. Three of them nearly concur with those specified by USNRC and the other two are non nuclear and may be excluded. Therefore, each Class requirements of Section XI are applied to each "German" category, selecting the applicable code edition as in USA plants.

14.   Once that systems required to ISI and Section XI Classes have been clearly defined, exemption criteria (IWB-1120, and IWC-1220) are applied. These criteria, which allow to exempt from examination some components, are clearly enough specified and their application do not present any problem.

15.   In reference with the inservice testing of pumps, code requires to test Class 1, 2 and 3 pumps which are provided with an emergency power source. These requirements are clearly enough specified

and their application do not present any problem. In fact both turbine and motor driven pumps, have been included in the programs.

16.   In reference with the inservice testing of valves, the code requires to test Class 1, 2 and 3 valves which fall in the categories defined in Section XI. To determine which ones are required to test is not too easy, because of the lack of exact criteria defining testing categories and exemptions. Therefore, programs of inservice testing of valves have been developed to include, in each category, the following valves:

Category A

- Reactor coolant pressure boundary valves.
- Reactor containment isolation valves.
- Isolation valves on radioactive gas decay tanks.
- Isolation valves on radioactive products discharge lines to the environs.
- Accumulator isolation valves (PWR).

Category B

- Air or motor operated valves which do not belong to Category A.

Others categories (C, D and E).

Criteria to define these valves categories are clearly enough specified and their classification do not present any problem.

With this code interpretation the amount of valves requiring testing is, for a typical PWR, as follows:

CATEGORY		AMOUNT
A		165
AE		20
AC	(check)	65
AC	(relief)	35
B		190
C	(check)	135
C	(relief)	65
CB		5
E		200

In other words, as preservice testing of valves, the following tests have to be performed requiring the indicated man power.

TYPE OF TEST	AMOUNT	MAN POWER (man-hours)
Leak rate	285	1140
Exercising (A and B)	355	355
Exercising (C)	300	600

As it has been mentioned, code requirements (1.974 Edition) used to develop this approach have been widely modified in winter 77 addendum, requiring to test only those valves which have a specific function in shutting down the reactor or mitigating the consecuences of an accident. With these requirements the above approach could be modified as follows:

Category A

- Reactor containment isolation valves.
- Isolation valves on radiactive gas decay tanks.

Category B

- Motor or air operated valves which have to change their position to fulfill their function in shutting down the reactor or mitigating the consequences of an accident.

With this approach the amount of valves requiring testing should be decreased to a half. As at the same time these new requirements look much more practical it is recommended to use them, at least, as an interpretation of the older ones.

EVALUATION CRITERIA

17. Section XI evaluation criteria, for indications detected by NDT examination, have been widely modified with the issuance of the summer 73 addendum. Earliest criteria, before summer 73, refer to Appendix IX of Section III, that is, to the code used for construction of nuclear components, applying for both, RT and UT, the same acceptance limits. From summer 73 addendum, Section XI is incorporating their own evaluation criteria, starting with those applicable to the reactor beltline region and extending them, gradually with the subsequent editions to all of the areas required to examination. In general, it could be affirmed that the new criteria are more practical for defects detected on inservice and less restrictive than those used in construction. In the other hand, they are prepared to apply the fracture mechanics analysis contained in Appendix A of Section XI. For these reasons, to apply these new evaluation criteria should be strongly recommended. The question is if it is possible, that is, if it is acceptable to apply the last criteria to any inspection, which has to be performed in accordance with an earlier edition, (i.e. Almaraz I preservice inspection is being performed in accordance with winter 72. Can be any indication evaluated in accordance with summer 75 criteria?). In general any indication detected is evaluated with the criteria specified in the code edition applied in the development of the particular ISI program, and only, if the indication results non-acceptable, later evaluation criteria are applied. This approach has been taken understanding that, an non-acceptable indication could be accepted in the next ISI, due to changes on evaluation criteria originated by the ISI program updating.

18. Some actual cases, discussing the evaluation process followed in each one, are described in next paragraphs.

18.1 During an inservice inspection performed in José Cabrera plant using the winter 72 code edition, an indication was detected by UT in the head-to-flange weld of the reactor vessel clousure head. The response obtained (600% DAC) was indicative of a lack of side wall fusion of approximately 85 mm long and between 3 and 10 mm in depth. Both the defect characterization and sizing was a difficult task because the lack of as-built drawings and the surface condition of the head. Any way, the defect had to be considered unacceptable in accordance with the applicable ISI code, and in consequence, additional examinations to improve the defect sizing and characterization was performed before deciding if any repairing was required, and having in mind that this type of defects have to be originated during the welding process, a radiographic examination was performed and the obtained results compared with those obtained during fabrication. Additionally, a facture mechanics analysis (Ref. 8) was performed in a very consecutive way, assuming a defect of 100 mm long and 10 mm in width, obtaining, that the crack growth during one operation cicle did not suggest an unaceptable growth and that a 10 mm crack would not be critical assuming even quite pessimistic K_{Ic} values, and therefore, the defect was acceptable under ASME XI rules, or by any other LEFM approach. Therefore, the defect was considered acceptable for one operation cicle and additional and improved UT examinations recommended at the end of the cicle, to define more accurately the indication. In this way, meassurements of attenuation were performed on the reactor vessel head material using two 45-2 MH$_z$ transducers, obtaining a good two mode pitch and catch signal that showed only 1 db attenuation on 100 mm path, slightly better than the obtained on the reference block using 1 db per 145 mm. It was unexpected to find these results on a reflection through the vessel head cladding, while the calibration block was uncladded. With this correction, the indication would be reduced to 50% DAC. In the other hand, the defect character was reviewed detecting that the maximum reflection was obtained with an incidence angle with the direction of the weld preparation of 25º, which would suggest, that the defect could not be a lack of fusion. In addition the defect transversal dimension was more accurately measured, obtaining between 2 and 3 mm in width. Therefore, and after evaluating the indication in accordance with the summer 74 criteria, the defect was considered acceptable.

18.2 During an inservice inspection performed in José Cabrera plant, using the winter 72 edition, one indication was found by radiographic examination in one of the pressurizer safety line to safety valve welds. This line is a two inches stainless steel pipe (A312T316), nine mm. thick. The indication was characterized as a slag inclusion of six mm long and apparently with its ends cracked. In accordance with the applicable ISI Code this indication should be unacceptable. Therefore, additional examinations to improve the defect sizing and characterization was carried out, before deciding if its reparation was required.

These additional examination were performed by radiographic examination from different distances and angles and by ultrasonic examination using both, standard and focalized transducers of differents angles. The ultrasonic examination did not provide any additional information because the access and the external surface configuration, but the additional radiographies showed a slag inclusion of ten mm long and according to the different views obtained, it could be estimated a width of three mm and a depth between two and three mm and also it could be assumed that the cracked ends could be a very thin inclusion because they did appear with the same thickness in all films. With these data, an additional evaluation was conservatively performed, assuming, that the indication was enclosed in a 10 by 3,6 mm rectangle located in a plane forming 33º with the axis weld. This evaluation was performed in accordance with the criteria established in later code editions, resulting an aspect ratio $a/1 = 0,15$ and $a/t = 10,75\%$ being the maximum allowable a/t 11, 95%. Therefore, in accordance with this conservative evaluation the indication could be considered acceptable within an upper limit of 10%.

Additionally, the crack growth for one operation cicle was evaluated applying the crack growth criteria analysis (Ref. 9, 10 and 11) obtaining, with a very conservative assumptions, such a crack growth that provided an aspect ratio $a/1 = 0,145$ and $a/t = 11,5\%$ with a maximum allowable a/t, of 11,93% in this case.

In summary, the indication was considered acceptable recommending to repeat the examination of this weld after one operation cicle in order to better assess the calculated crack growth parameters.

18.3 During a PSI applying the winter 72 code edition, 25 indications were detected by liquid penetrant examination, which were, all of them, considered unacceptables in accordance with the applicable ISI code. These indications were also evaluated in accordance with the criteria specified in later code editions, resulting, that only 7 of them would be considered unacceptables. However, it was decided to repair all of them in order to avoid future records in subsequent inspections. After repairing, all of them were reinspected and considered acceptables.

CONCLUSIONS

19. Section XI is applied in the development of Spanish ISI programs, in ones cases as requirement (USA plants) and in others as an approach to detail ISI requirements (German plant).

20. Applicable code edition should be, within the regulations, carefully selected. Using a single edition to determine the inspection scope, examination methods, frecuency, etc. and later ones to determine examination techniques and evaluation criteria should be a very practical approach.

21. Scope of inservice testing of valves should be defined applying winter 77 criteria. Results obtained with this edition are much more practical than those obtained with earlier ones.

22. In our opinion, to evaluate the NDT indications in accordance with a single code edition is not realistic. It is recommended that non-acceptable indications in accordance with the applicable evaluation criteria be additionally inspected and evaluated with later criteria including evaluation by analytical procedures as required, before deciding its reparation.

REFERENCES

1. UNITED STATES NUCLEAR REGULATORY COMMISSION. "Rules and Regulations". Title 10-Chapter 1, Code of Federal Regulations-Energy Part 50- "Licensing of production and utilization facilities".

2. AMERICAN SOCIETY FOR MECHANICAL ENGINEERS. ASME Boiler and Pressure Vessel Code. Section XI "Rules for Inservice Inspection of Nuclear Power Plant Components".

3. REACTOR SAFETY INSTITUTE OF THE TÜV E.V. Reactor Safety Commission (RSK). "RSK Guidelines for Pressurized Water Reactors" (Translation) 1st Edition. April 1.974.

4. U.S. NUCLEAR REGULATORY COMMISSION. Regulatory Guide 1.26 "Quality Group Classifications and Standards for Water, Steam, and Radioactive Waste Containing Components of Nuclear Power Plants". REVISION 3 February 1.976.

5. AMERICAN NUCLEAR SOCIETY. ANS-51.8 (ANSI-N-18.2) "Nuclear Safety Criteria for the Design of Stationary Pressurized Water Reactor Plants" 1.973 and 1.975 Revision and Addendum.

6. AMERICAN NUCLEAR SOCIETY. ANS-22 (N-212) "Nuclear Safety Criteria for the Design of Stationary Boiling Water

Reactor Plants" Draft May 74.

7. CHOKIE L.J. "Use of Non-destructive Testing for Inservice Inspection of Reactor Pressure Components". IAEA Technical Committee, Kobe, Japan 1.974.

8. R. O'NEIL "Assessment of the ultrasonic inspection carried out at the reactor plant at the central nuclear José Cabrera, Zorita, Spain". UKAEA, SRD. February 1.974.

9. D.T. RASKE and C.F. CHENG "Fatigue crack Propagation in Types 304 and 308 Stainless Steel at Elevated Temperatures" NUTYBB 34(1) 1 - 126 (1.977) June 1.977.

10. LEE A. JAMES "Fatigue-Crack Propagation in a Cast Stainless Steel" NUTYBB 26(1) 1-118 (1.975) May 1.975.

11. T.W. CROOKER "The Role of Fracture Mechanics in Fatigue Design". ASME 76-DE-5. Conference and Show Chiccago I11. April 5-8, 1.976.

TABLE   1

SPANISH NUCLEAR PROGRAM

NAME OF PLANT	REACTOR TYPE	NSSS SUPPLIER	OUTPUT MW (e)	STARTUP DATE
JOSE CABRERA	PWR	USA	160	1.968
STA. M. GAROÑA	BWR	USA	460	1.971
VANDELLOS I	Gas-graph	France	500	1.972
ALMARAZ I	PWR	USA	900	1.979
ALMARAZ II	PWR	USA	900	1.980
LEMONIZ I	PWR	USA	900	1.980
LEMONIZ II	PWR	USA	900	1.981
ASCO I	PWR	USA	900	1.980
ASCO II	PWR	USA	900	1.982
COFRENTES	BWR	USA	940	1.981
VALDECABALLEROS I	BWR	USA	940	1.983
VALDECABALLEROS II	BWR	USA	940	1.985
TRILLO	PWR	Germany	997	1.983
VANDELLOS II	PWR	USA	980	1.984
SAYAGO	PWR	USA	1.000	1.985

TABLE   2

TYPICAL P AND BWR SYSTEMS REQUIRED TO ISI

PWR (USA)

- Reactor Coolant System.
- Residual Heat Removal System.
- Safety Injection System.
- Chemical and Volume Control System.
- Containment Spray System.
- Main Steam System.
- Feedwater System.
- Fuel Pool Cooling and Clean-up-System.
- Component Cooling System.
- Essential Service Water System.
- Gaseous Waste Processing System.

PWR (Germany)

- Reactor Coolant System.
- Life Steam System.
- Main and Emergency Feedwater Systems.
- Volume Control System.
- Chemical Feed System.
- Coolant Purification and Degassing System.
- Coolant Storage and Treatment System.
- Nuclear Pool Purification System.
- Residual Heat Removal System.
- Gaseous Waste Disposal System.
- Extra Borating System.
- Leakage Extraction System.
- Nuclear Equipment Drain System.
- Demineralized and Seal Water Supply System.
- Chilled Water System.
- Service Cooling Water System.
- Secured Closed Cooling Water System.
- Well Cooling Water System.

BWR (USA)

- Nuclear Boiler System.
- Recirculation System.
- Standby Liquid Control System.
- Hydraulic Control Rod Drive System.
- Residual Heat Removal System.
- Low Pressure Core Spray System.
- High Pressure Core Spray System.
- Reactor Core Insulation Cooling System.
- Reactor Water Clean-up System.
- Fuel Pool Cooling and Clean-up System.
- Essential Chilled Water System.
- Essential Service Water System.

PERIODIC INSPECTION OF NUCLEAR COMPONENTS IN SWITZERLAND

H. BASCHEK
Nuclear Power Plant Beznau, Switzerland

The MS of this paper was received at the Institution on 4 September 1978 and accepted for publication on 27 November 1978

SYNOPSIS   In 1976 the authorities and the plant owners decided after very intensive discussions, to establish a Swiss Code for Inservice Inspections, based on Section XI. The first part of this report deals with areas where the Swiss Code deviates from the requirements in other countries. The second part contains a very short report about experiences with inservice inspections in five Swiss nuclear power plants.

PART I

Regulatory Aspects

In Switzerland the era of commercial nuclear power started in 1965. To-day there are five plants in operation or under construction.

Table 1

Name	Electrical Output (MW)	Reaktor Type	NSSS Supplier	Status
Beznau I	350	PWR	W	in operation
Beznau II	350	PWR	W	in operation
Muehleberg	300	BWR	GE	in operation
Goesgen	900	PWR	KWU	start up
Leibstadt	900	BWR	GE	under construction

The organization of the Swiss regulatory bodies has been slightly modified since 1965, but in principle there is a federal agency responsible for nuclear safety aspects which report directly to the government and there is an official inspection agency for pressure retaining components, which acts on behalf of the nuclear safety agency as far as classified mechanical equipment is concerned.

All conventional unclassified pressurized components in a nuclear power plant are under the jurisdiction of the same agency by direct governmental mandate.

The authorities as well as the owners of nuclear power plants in Switzerland pay considerable attention to all inservice inspection aspects and the activities in that field were closely followed for many years. In the past the Swiss engineers often felt the need for a binding inservice inspection code. The question was whether to adapt ASME Section XI or to work-out new regulations, and in 1976 the authorities invited the plant owners to discuss the subject. It was decided to establish a Swiss Code hereafter referred to as the "Code", which would closely follow Section XI.

This option was favored for the following reasons:

- Some of the requirements in Section XI did not find full support in Switzerland, and on the other hand, some points were missed in Section XI which the Swiss liked to be included in the Code.

- In Switzerland regulations for the periodic inspection of pressure containing components exist for many years and were implemented by law in 1925. These regulations proved to be very useful and they have been successfully applied. No good reason exists to disregard them for nuclear power plants.

- It was considered very difficult to deal with Section XI, mainly because of the various references and addenda. The opinion dominated, that in the Code the contents should be reduced considerably and should be supplemented only if needed in future.

- Considerable editorial work was necessary anyway, since regulations and requirements which are to be implemented in Switzerland should be documented in an official form and language.

Lay-Out and Component Design

Considerable efforts are made to keep the radiation exposures as low as possible for all maintenance work, including inservice inspections. In order to meet this objective, requirements have to be established for the design and manufacturing of systems and components, included the surface conditions for areas to be examined. The requirements may differ from plant to plant, dependent on the inspection concept and/or on the selected test equipment. It may be advisable, for example, to specify special welding processes or heat treatments purely to improve the structure of the materials for ultrasonic testing.

Furthermore the Code states explicitly that the number of welds should be reduced to a minimum in particular for piping in areas with high radiation levels. This means for example that seamless pipes are preferred and welded elbows should be avoided.

Concerning the accessibility it is recommended that between components to be inspected and walls or other components a free space of 40 to 50 cm is available. In case of piping diameters larger than 1000 mm, the space should be 60 to 80 cm.

Obligatory Inspection and Inspectability

The Code distinguishes between areas which shall be inspected periodically and areas which may be inspected when the results of the periodic inspection require an extension of the test programme.

For the first category, removable insulation and permanently installed platforms are highly recommended. For the second category the areas have to be accessible and the design as well as the surface conditions shall allow the inspection by specified methods. Removable insulation is not required here and temporarily installed platforms may be used, if needed for accessibility.

One of the objectives for determining the areas of possible inspection is that early during the design stage the requirements for inservice inspections are properly considered.

Preservice Inspections

Prior to criticality a preservice or a base-line inspection must be performed. The preservice inspection includes all examinations which have to be performed periodically.

In addition to that for class I components the preservice inspection must be extended to:

- a volumetric examination of the base metal 100 % within the core-region of the reactor pressure vessel;

- all circumferential and longitudinal weld of the main-coolant piping respectively of the recirculation loops 100 % volumetric;

- all main coolant respectively recirculation pumps and all feedwater and main steam isolation valves according to the inspection program which asks for the examination of only one of these components per loop. The periodic inspection of all these valves and pumps is not mandatory.

Repair-welds can be subjected to a base-line inspection.

Examination in the workshop can be accepted as preservice inspection, if

a) all tests has been performed under the same conditions, using the same test method and techniques as specified for the inservice inspection;

b) recording was performed in the same way and the same method as specified for the inservice inspection;

c) in case of pressure vessels the hydro-tests had been performed prior to the examination.

Inspection-Programme and Test-Specifications

A general inservice inspection programme has to be submitted to the inspection agency for approval and to the nuclear safety agency for information. The programme should include the following:

- identification of the components,

- detailed description of the inspection area,

- information about the applied inspection techniques,

- detailed test-specifications with information about test-method, test equipment, calibration standards, sensitivity, system of recording, etc.,

- inspection instervals.

The general programme has to be approved at the latest before the preservice inspection is started, but some particular information are already needed to approve the components for manufacturing.

Each year and prior to the refueling shutdown an inspection program has to be submitted to the authorities by the plant owner.

Inspection Intervals and Periods for Class I and Class II Components

According to the Code three programs exist for class I and class II components:

- a programme for the reactor pressure vessel, which is modified compared to the inspection program A of Section XI, (Table 2 and Table 3),

- a programme for the Steam Generator Tubing which correspond to the Section XI-program for SG tubing,

- a programme for all other components, which correspond to the inspection program B of Section XI.

Alternate inspection programmes are not offered.

Inspection of Identical Loops

In 1975 the inspection philosophy of Section XI was modified. In case of a system having a number of similar loops with identical functions, inservice examination are concentrated upon one of the loops only. The present edition of Section XI gives preference to an increased inspection scope of one loop compared to random examinations of all the loops. This philosophy is not fully acknowledged in Switzerland. Though for multiple-loop systems slightly different inspection requirements apply for components and piping and though even alternatives are offered for the piping, the general and preferred trend is to reduce the overall number of examinations but to select the welds with high stress levels, independent of the loop.

Inspection Methods, Inspection Area

The inspection area of a weld includes the weld itself and on both sides of the weld a zone of the base material "t" mm wide. "t" is 13 mm or in minimum half of the wall-thickness.

For volumetric inspection the ultrasonic test method is prefered.

A weld is considered to satisfy the ultrasonic testing inspectability requirements if the following conditions are fulfilled:

1. for defects perpendicular to the weld-axis testing must be possible from both circumferential directions.

2. For defects longitudinal to the weld-axis the following alternatives exist:

2.1 from both sides of the weld with 45° US probe, full skip, (Fig. 1),

2.2 from both sides of the weld with 60° US probe, half skip and with additional surface examination, (Fig. 2),

2.3 from one side of the weld with 45° US probe full skip and with additional surface axamination, (Fig. 3).

The techniques and methods used for an US examination should enable the detection of defects which may occur during the plant life or which already exist and have to be watched. For a reactor pressure vessel, for example, the most likely orientation and location of possible defects should be defined first and prior to the selection of an ISI system. A volumetric inspection system should then be selected which is able to detect these possible defects with an adequate sensitivity.

Up to now multiple-function probe systems are used in Switzerland to meet the requirements. To-day such systems may include:

- 45° transmitter and receiver for transfer control,

- 45° pulse-echo probes, angle beam,

- 45° probes in tandem,

- 70° pulse-echo probes, angle beam,

- 70° inclined longitudinal wave probes,

- 90° pulse-echo probes, straight beam.

Sketches are shown in Fig. 4.

The Code does not specify any particular ultrasonic testing technique or system.

For surface examination a suitable and approved method should be used.

Visual inspections are considered as very important, but they are not mandatory for all surface areas.

Surface Inspections

Surface inspections on welds is not required if volumetric examination by ultrasonic methods is satisfactorily performed, however, dissimilar-metal welds must be surface-inspected independent of any volumetric test methods. A surface inspection of bolts is not required if it can be demonstrated, that the ultrasonic testing equipment detects a defect in the thread root having a depth of 2 mm.

Detection of corrosive attack

Particular attention is paid to the subject corrosion, and not only to stress corrosion cracking of stainless steel. For Class II and in principle also for Class III components the Code requests random wall thickness measurements in areas where wastage type corrosion is expected.

Hydro Tests

Hydro tests for inservice inspections is one of the frequently discussed subjects. Though the term "hydro-test" is only indirectly defined in the Code, in Switzerland the general understanding is, that the pressure for a "hydro" should be higher than the design pressure. The factor is considered to be between 1.1 and 1.5.

The Code does not request hydro-test for Class I components. For Class II and Class III components the hydro-test is only permitted as an alternate ISI method in case other inspection procedures cannot be applied.

In Switzerland the discussion over the usefulness of hydro-tests continues.

Presently the main arguments are:

Table 4

Pros:

- Stress intensities at the crack-tips are decreased by stressing the material locally above yield.

- If there is a critical defect, it should fail during hydro and not during plant operation.

- Integral test possible.

Contras:

- At critical locations the stresses caused by temperature-transients may be higher than the stresses caused during a hydro-test. The hydro-test as a preventive protection method against pressure vessel failure during operation is questionable.

- The material is locally stressed above yield during the first hydro-test. Repetition is unnecessary.

- During a hydro-test the possibility of a crack-propagation without vessel-failure cannot be excluded.

- Plant outage is extended.

Acoustic Emission Techniques

The Code does not mention Acoustic Emission Techniques. In Switzerland the opinion dominates, that this method may supplement, but cannot substitute presently used ISI methods. The technique is not suitable to detect cracks during operation but may be useful for the detection of loose or vibrating parts.

Steam Generator Tube Inspections

The Code requires that 3 % of the tubes per steam generator are inspected for cracks and wastage within the first two years after initial plant start-up. During the second interval, 2 to 4 year after plant commissioning, 1 1/2 % of the tubes per steam generator have to be tested. The following intervals will start every three years and only 1 1/2 % of all tubes have to be inspected. Furthermore it is permitted then to carry-out the inspection on one steam generator only.

On purpose the Code provides flexibility, if defects or irregularities are detected during inspection. Furthermore eddy current testing is not mandatory should tube leakage require an unscheduled plant shutdown.

Class 2 Components

According to the Code Class 2 components are divided into two categories:

- category 2.1 covers systems which are considered to have a low potential for risks, in case of failure. (Sufficient redundancy, low temperatures, low pressures, no significant vibration, etc.);

- category 2.2 includes systems and subsystems, failure of which may effect the nuclear safety.

Typical for category 2.2 systems are

- Secondary side of steam-generators with the connected pipes up to the isolation valves;

- Suction headers in emergency core cooling and residual heat removal systems including the pipe-penetrations through the containment and the pipe-connections to the suppression pool.

The inspection requirements for category 2.2 components are very similar to those for class 1 components.

For category 2.1 components corrosion is considered to be the most potential risk. Therefore emphasis is put on visual inspection from the inside.

Class 3 Components

These are divided into the following three categories:

3.1 All boilers and pressure vessels for which periodic inspections are required by the old regulations;

3.2 Auxiliary systems, which are important for the operation of safety-systems, including the components of these auxiliary systems as far as

they do not already belong to category 3.1;

3.3 Vessels, tanks and systems, which are classified only because of their radioactive content.

Based on the old regulations, inspections from the outside have to be performed during plant operation in specified intervals. This applies to category 3.1 and 3.2. The old regulations also specify visual inspections from the inside, but only category 3.1 components have to comply with this requirements. If an inspection from the inside is not possible or in case of high radiation levels highly undesirable, random volumetric examinations from the outer surface or hydro-tests have to be considered as alternative inspection methods.

The intervals for inspections from the outside and from the inside can be doubled in special cases. For category 3.3 inservice inspections are not a requirement.

Other Aspects

The Code does not mention that base-metal ligaments between two neighbouring control rod drive penetrations must be tested volumetrically. Furthermore it is not required, that a reactor pressure vessel has to be inspected from the inside as well as from the outside.

The Code does not deal with functional tests of pumps and valves, they remain part of the technical specifications only. But the future version of the Code may cover the testing of safety valves.

The following areas are intended to be included in the Code but are still under discussion:

- Evaluation of test results, (permissible defects);

- Repairs;

- Hydro and leakage tests;

- Replacement of components.

PART II

Beznau I and II

Beznau I was ordered in 1965, when inservice inspections had not yet become a subject of discussions in Switzerland. Beznau II is identical to Beznau I. Between the two plants there was a construction interval of about two years.

In 1969 an action-plan was established with the following objectives:

- to study the feasibility of a meaningful ISI program;

- to work out an examination plan for the main components within the reactor coolant boundary;

- to procure equipment for a volumetric preservice inspection for the unirradiated Beznau II reactor pressure vessel.

At Beznau particular emphasis was put on the feasibility studies. The intention was to get an answer to the following questions:

- Which areas can be inspected with mechanized equpment representing the state of the art at that time and with the "as build" geometry ?

- Which areas should be examined considering peak stresses, fracture mechanics behaviour and "as-built" quality ?

- Which are the critical and which are the detectable defect sizes ?

One of the encouraging results was, that many of the areas, which should be inspected were also found suitable for an ultrasonic examination. Hence in 1970 equipment for ultrasonic inservice inspection of the reactor pressure vessel was ordered. In particular the mechanical part of the equipment look very similar to that used for other PWR pressure vessels nowadays.

In 1971 the pre-service examination on the Beznau II reactor vessel was performed successfully. The first part of the base-line inspection on the irradiated Beznau I pressure vessel was done in 1974 and the second part in 1976. Three points are significant enough to be reported:

1. The comprehensive examination did not reveal any unacceptable indications.

2. The applied ultrasonic system is capable to detect reflectors corresponding to a substitute flaw size of 4 mm $\emptyset$ (circular reflector) with guaranteed reproducibility. The accuracy of the positioning system is better than $\pm$ 6 mm (disassembly and reerection of the equipment included). These statements are based on the practical experience at the Beznau I plant. An acceptable UT-indication in the shell which was not removed during the fabrication process but reported in the QA-documentation, was found again during the 1974 baseline examination. A re-scanning of the same area was performed in 1976 and the discontinuity was detected again, without a change in signal amplitude.

3. Safe control of the manipulating equipment is not fully assured if the operator is insufficiently trained or if the check-list is incomplete or if the water quality in the reactor cavity is poor.

The main coolant piping at Beznau is of stainless steel and consists of hot extruded straight pipe sections and cast bends. All welds have been examined using the liquid-penetrant procedure on the outer surface. Since liquid-penetrant surface inspection from the inside is not possible due to radiation problems, efforts were made to find ultrasonic transducers suitable to penetrate very course grained austenitic structures (weld- and cast-metal) with a thickness up to 80 mm.

In the past years the results were rather disappointing. Tests were performed with various transducers, as for example standard or focused types, such of low frequency and of inclined longitudinal wave mode. Satisfactory results (calibration reflector 5 mm $\emptyset$ have been achieved only on wall thickness up to approx. 40 mm.

Because of the significant importance of the subject a new effort was made during the refueling shutdown summer 1978. The preliminary results were very encouraging.

Much experience has been collected in the field of steam-generator tube inspections. The test equipment was improved step by step during the past years. The present status can be considered as an optimum in respect to costs, inspection time and radiation exposures.

Muehleberg

In Muehleberg some of the outer pressure vessel surface is accessible, and from the inside only the upper sections of the vessel can be reached. Specially designed ultrasonic test equipment was ordered for testing from the inside. The manipulator is able to reach to following:

- the upper three circumferential vessel welds,

- the welds between the vessel and the four steam outlet nozzles.

Furthermore the following areas are volumetrically tested from the outside of the vessel:

- the welds between the vessel and the four feedwater nozzles;

- the welds between the vessel and the two recirculation suction nozzles;

- the inner radii of these nozzles.

A combination of single probe and tandem technique is used for these ultrasonic tests.

Various safe-end welds are inspected using liquid-penetrant on the outer surface.

The inside preservice inspection was performed in 1971 and the first inservice inspections in 1974. A manipulator for the nozzle inspections from the outside was used for the first time in 1975.

Goesgen

The preservice inspections were performed in 1977 and beginning of 1978.

The requirements of the Code could be met in most cases.

The equipment for the preservice inspection of the reactor pressure vessel is quite similar to that used in Beznau, but much more sophisticated. The sensitivity of the UT equipment is specified to be better than a 3 mm $\emptyset$ reflector for the single probes and a 7 mm $\emptyset$ reflector for the tandem system.

During the initial primary system hydro-test the acoustic-emission technique was used. Emphasis was put on an in-line evaluation system. In correlation with the ultrasonic test results no significant indications were found.

Leibstadt

It is expected that this plant can meet nearly all requirements of the new Code.

The plant is equipped with a GE-BWR-6 Nuclear Steam Supply System. The basic concept of the pressure vessel has not changed considerably during the last 10 years. Inservice inspections of a BWR vessel remains a difficult task mainly due to the great number of nozzles. Some of them are placed at locations which are not very favourable for the currently available ISI equipment. But before the fabrication started, it was possible to improve the situation considerably by making minor design modifications. A particularly difficult situation exists at the bottom head, which consists of four segments welded together and which is penetrated by 149 nozzles for the control rod drives.

An ISI of BWR pressure vessels is feasible from the outside only. Leibstadt will use ultrasonic ISI equipment which consists of field installed track racks carrying a fixture on wheels on to which the ultrasonic transducers are mounted by means of expandable and rotating arms. Thus almost any location on the cylindrical shell can be reached.

Due to the difficult accessibility of some areas of the RPV surface special provisions were necessary. The tracks had to be intersected because various nozzles interfered with the tracks.

A transfer car on a circumferential rail transports the UT-scanning equipment between the intersected vertical rails. The access opening in the concrete shielding allows a direct contact with the UT-equipment on the carriage (Fig. 5).

For the inspection of the nozzle-to-pipe and pipe-to-pipe welds a special scanning device had to be developed. It will be mounted through the annulus between pipe and the openings in the biological shield.

For the bottom inspection curved rails are installed underneath the vessel bottom head, between the CRD housing rows. These rails also carry a vehicle with UT scanners mounted on it inspecting the ligaments between the control rod penetrations. The scanner can also be moved in the direction perpendicular to the rail in order to extend the examination area.

A further rail is mounted on the inner surface of the support skirt. It's purpose is to transport a special vehicle which allows the examination of the weld between the bottom head and the forged intersection ring (Y-shaped cross-section). Fig. 6 shows the arrangement of the various rails.

Acknowledgement:

This report only reflects considerable efforts carried out by others, who are experts and work for nuclear power plant owners or for the Swiss authorities. I like to address my thanks to all of them.

Table 2

Swiss Code

Inspektion Program for Pressure Vessel

Inspection Interval	Inspection Period Calendar years of Plant Service	Minimum Examination Completed %	Maximum Examination Credited %
1	3	100	100
2	10	100	100
3	13	16	34
	17	50	67
	20	100	100
4	23	16	34
	27	50	67
	30	100	100
5	33	16	34
	37	50	100
	40	100	—

Table 3

Section XI

Inspection Program B

Inspection Interval	Inspection Period Calendar years of Plant Service	Minimum Examination Completed %	Maximum Examination Credited %
1	3	100	100
2	7	33	62
	10	100	100
3	13	16	34
	17	40	50
	20	66	75
	23	100	100
4	27	8	16
	30	25	34
	33	50	67
	37	75	100
	40	100	—

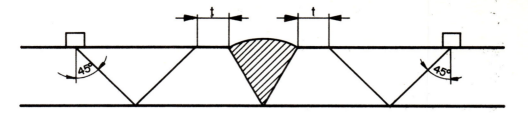

Fig. 1

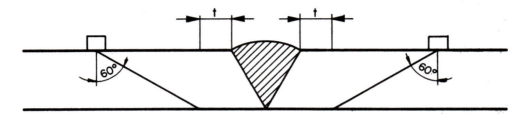

Fig. 2

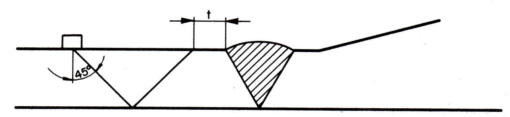

Fig. 3

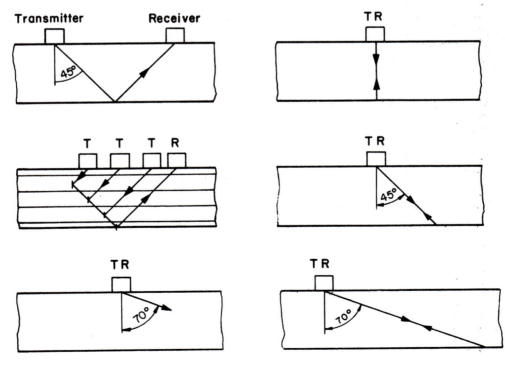

Fig. 4

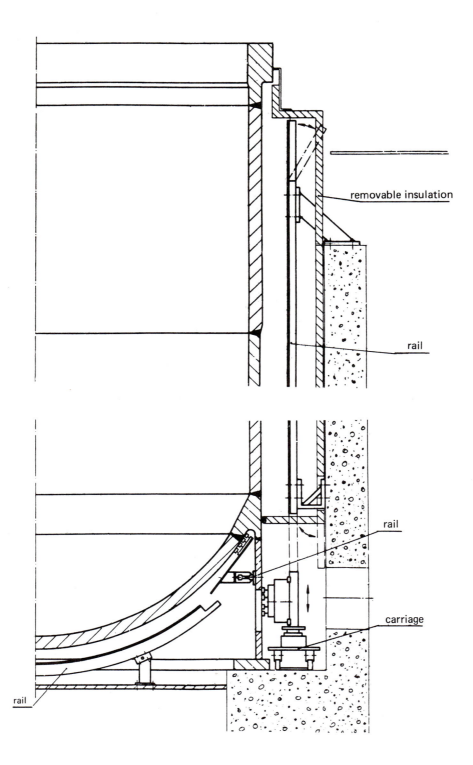

Fig. 5

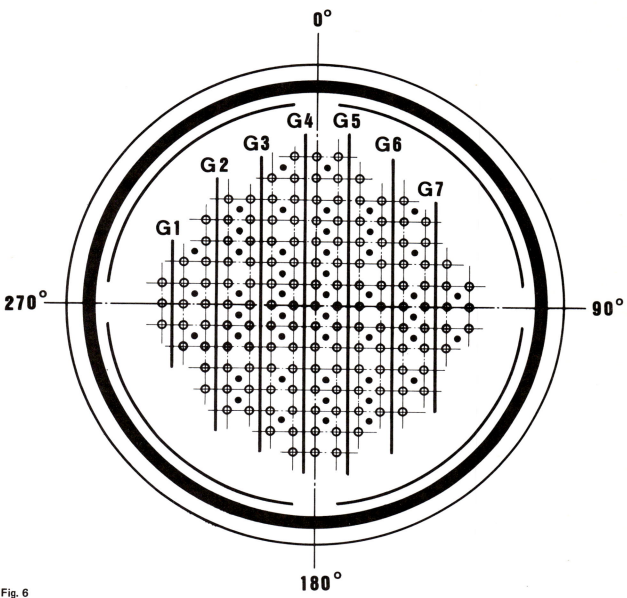

Fig. 6

C26/79

PROBLEMS ENCOUNTERED IN APPLICATION OF NDT FOR PWR INSPECTION

J. SAMMAN, and G. FOREST,
EdF/GdL, 93206 Saint-Denis, France

The MS of this paper was received at the Institution on 30 November 1978 and accepted for publication on 18 January 1979

SYNOPSIS After a brief review of French Regulation on In Service Inspection of the Primary Circuit of PWR and the description of EdF programme, the authors discuss the choice of NDT methods generally used for defect evaluation. Specific examples of this choice are given by the use of focussed Ultrasonic beams for clad thick welds inspection, radiographic examination of bimetallic weldments and acoustic detection of leakages.

INTRODUCTION

1. Electricité de France has a long experience in carrying In Service Inspection on fossil fuel boilers, steam generators and turbines. Availability and safety of installations are the main target of this inspection, which at the same time has to fulfil the requirements of 1926 regulation on steam pressurized vessels. With Nuclear Gas Cooled Reactors, inspection programmes were extended to cover CO_2 gas piping components in addition to the conventional steam components. Since reinforced concrete is used to build the Reactor Pressure Vessel (RPV) we use some specific method applied in dam construction to check its safety : i.e. pendulus measurement, sonic extensometer...(1) The development of PWR in France was accompanied by a new regulation of the nuclear pressurized vessels, in which the radioactive fluid containment leads the legistation to more stringent requirements. Regulation of 26 Feb.1974 defined the philosophy and requirements in conception, construction, operation and inspection of nuclear reactor. Fundamental differences between this regulation and the section XI of ASME Code is the large freedom given to the user of the reactor to define what has to be inspected and how (programme and methods), but on the other hand, he has to prove to Authorities that he has performed a complete and conservative inspection, and has maintened full safety of the reactor. This philosophy is similar to criminal rights practice, where in France, the defendant has to prove his innocence in opposition to Anglosaxon way, where the prosecution has to prove that defendant is guilty in not respecting the law.

2. REGULATORY REQUIREMENTS

2.1. Primary Circuit Limits for PWR

In Service Inspection has to be applied to the cooling system which contains the fluids directly in contact with radioactive fuel elements. This include the RPV, Steam Generators (SG), Pressuriser, Primary Pumps and the main piping of cooling loops. In other piping the extent is limited to the second automatic valves which can safely and quickly isolate the reactor coolant system. Small piping of less than 25mm ID is excluded from the system boundary, except steam generator tubing, reactor incore instrumentation tubing and pressuriser's heating rods penetration. In addition, the shell to tube plate welds in the secondary part of SG and the fly-wheel of the primary pumps should be inspected with the same conditions as the Primary Circuit for safety reasons.

2.2 Periodicity and type of inspection

Two types of inspection are foreseen by the regulation :

2.2.1. Complete Inspection (CI)
This inspection should cover 100% of the inspection programme and has to be performed in conjunction with the primary circuit hydrotest (1.20 the RPV calculated pressure). The first one occurs at the end of the construction : The Preservice Complete Inspection. It is considered as a finger-print for inservice inspections.The second inspection occurs at the first refuelling outage, with condition that it should happen not later than 30 months after the first inspection. Intervals between later inspections should not be longer than 10 years.

2.2.2. Biannual Periodic Inspection (BPI)
Partial inspection programme has to be performed as often as necessary to maintain the safety of the reactor between two CI. Intervals between partial inspections should not exceed 2 years. EdF intend to examine some typical components in each annual refuelling outage in order to cover 100% of these components in the 10 years interval. Other components will be examined only when they have to be disassembled for maintenance purposes, with a periodicity of 3 to 4 years.

2.2.3. Consequences :
Table I indicates the ISI extent for two elements of the RPV during the design life of the reactor. It appears in both example that soon after starting the reactor they reach a high value of inspection percentage. This strategy was

chosen to cover the numerous risks of
" non-critical failures " generally encountered in the " wear in phase " of
the reactor design life, so called by
Jordan(2).It is obvious that the probabilistic failure rate of safe-end weldments are higher than those of cylindrical shell weldments. Thus, during the
" useful life phase " a biannual inspection by rotation, loop by loop, of welds
is forecast in order to achieve a higher percentage inspection and shorter
intervals.

2.3. Sensitivity level of inspection methods

There is no acceptance/repair criterion
for defects in the French regulation.
Requirements are based on the evolution
of existing defects between two successive inspections or detection of new
defects in previously sound regions. To
insure this capability it is necessary
to define what defects can develop under
normal operating conditions and under
accident conditions and then apply an
NDT method with sufficent sensitivity to
detect and evaluate the defect with a
good reliability factor. Unfortunately
this theoretical argument can not be
applied in advance to NDT procedures for
all areas submitted to ISI. Otherwise,
the regulation requires that NDT method
should be able to detect and evaluate
all defects which could be rejected by
construction specifications. This means
that the sentivity level applied in ISI
should be set as low as that used in
shop testing. We find in table II a comparison for ultrasonic testing of welds
between EDF requirements (C.P.F.C.) and
ASME code Section III requirements. This
table shows the very low recording level
required in shop testing, in France.
This sensitivity should guarantee the
good quality of weldments. In ISI this
level leads to note evaluate and characterize an excessive number of indications. Generally, all these indications
are far from the critical defects dimensions, but small variation in procedure
application (human factors) can bring
the operator to overestimate their
geometric dimensions. This could be analysed as dangerous evolution of the defects.

2.4. Inspection Availability of the reactor

If the inspection programme can not be
applied on some components due to reduced
accessibility, geometric design, insulation... etc, the user is mutually responsible with the builder in this situation. They have to modify the design to
allow the possibility of complete inspection of all components included in
the primary circuit. This argument can be
easily applied by the regulatory authority
because there are few nuclear reactor
users in France, and all of them have a
public or national character.(Commissariat à l'Energie Atomique, National Navy,
EDF).   But difficulty still exists in
the meaning of Complete Inspection.The
actual position can be understood as a

complete access to either the outside
or the inside surface of all pressurized vessel and piping.

3. INSPECTION METHODS

Since 1973 an ISI programme was prepared for
PWR. Some modifications were introduced after
publication of the 1974 decree. This programme defines the extent and the methods that
have to be applied during the Complete Inspection. At the end of 1978 the preservice
inspection of 8 reactors of 900 MWe was achieved on the base of this programme. One Complete ISI was   performed on the 10 years old
franco-belgian PWR (300 MWe). The NonDestructive Testing methods imposed by this programme
can be resumed in the following :
- Remote ultrasonic test of all pressure retain weldments of the RPV.
- Volumetric examination (Ultrasonic or Radiography)of weldments in other pressure vessel.
- Radiographic test of all bimetallic weldments.
- Surface examination of supports and attache-
  ments on all vessels and piping by Dye Penetrant test, and visually under water pressure for other weldments in austenitic piping.
- Eddy Current test of steam generator tubes.
- Acoustic Emission Leak Detection test on all
  penetrations in the vessels (Control rods,
  instrumentation, heating rods).
- Remote TV examination of inside surface of
  all vessels.
Figure 1 shows a schematic diagram of a one
loop PWR Coolant System. The extent of ISI is
indicated with different shaded areas corresponding to different methods.
Application of these methods raises some problems due to different factors : environment,
access, surface finish, metallurgical or physical parameters, interpretation of results,
comparison with fabrication acceptance requirements...etc. For each specific problem a
study or a research programme was initiated in
order to find the best suitable solution. We
shall present in this paper some of these
solutions, but we are   aware   that all solutions of technical problems are a compromise
between several requirements.

4. ADAPTATION OF NDT METHODS TO PROBLEMS

4.1. Reproducibility of manual ultrasonic test

Due to the requirements described in §
2.3. the sensitivity of instruments had
to be set to the same level for all metal
thickness.Distance Amplitude Curves (DAC)
of 2mm diameter side drilled holes were
used. Dimensions of the defects are measured by the -6 dB fall method. To be sure
that evaluation of the acceptable defects
left at the end of the construction shall
be the same during the useful life of the
component, and in order to
determine growth or detect new defects, it
is necessary to adjust the sensitivity on
the same calibration block for each ISI.
In this order, each power production center (a group of 2 or 4 reactors) was
endowed with a set of calibration blocks
representative of the geometry of the
inspected areas (Welds, Ligaments, Studs,
Nuts...). Some of them are very heavy as

shown in figure 2, and these are
kept at our Central Laboratory. The second condition for a good reproducibility is in using transducers with the same
beam spread. The acoustic field of each
transducer used is mapped in water by
measuring acoustic pressure along its
beam axis and in two planes situated at
1/4 and 3/4 of the equivalent metal path
in wich it will propagate when it is
used for a specific examination. Figure
3, shows a routine transducer card.
Selection of transducers having the same
commercial labels is done before an inspection in order to reject those presenting a beam spread different from the
one used at the preservice inspection.

4.2. Ultrasonic Wave Transformation :

Generally this phenomenon is considered
as a trap for ultrasonic examination.
It gives false indications, in some
cases inseparable from defect indications. We have used this phenomenon to
search for cracks growing in the primary
pump fly wheel keyway corner. As shown
in figure 4 a shear wave is emitted from
transducer A and scans tangentially the
bottom surface of the keyway when the
transducer is moved. If there is a crack
initiated on this particular surface the
shear wave changes to a compression
mode. This last wave is detected by a
receiver located in position B. The sensitivity is checked on fine EDM notches
of 2mm deep 35mm long machined in a 600Kg
weight   calibration block. Checking is
made in the laboratory, and a transfer
method with light weight block is used
to recalibrate electronic instrument on
site. No other methods can detect these
dangerous cracks without disassembling
the fly-wheel from its shaft. (3)

4.3. Inspection of Cladded Welds :

It is recognized that junction surface
between ferritic materials and austenitic welded cladding produces an important modification of the acoustic beam
generated by the transducer which pene-
trates the metal through the cladding.
Evaluation of the defects can be seriously affected by their position related
to cladding orientation or to the cladding waves. This was particulary demonstrated for 25mm 2MHz flat transducer,
used on a 500x500x210mm ASTM SA 533 grad
B, cladded block, in which 2mm diametre
holes were drilled at different depths,
on each of its four faces. Echos from
reflectors situated at the same depth
varied by 12 dB when the holes were drilled parallel or perpendicular to the lines of the strip deposited cladding.(4).
This situation is that of our Reactor
Pressure Vessel where access to the
welds is practically impossible from the
outside surface, but the inside surface
is clad with strip deposited Cr Ni
alloyed stainless steel. The requirements
to construct identical power reactors
restrains the tendency to modify those
designs, mainly for the reinforced
concrete support and the biological

shielding. A research programme undertaken for a certain number of years at
Saclay's Center by Advanced Technology
Section of the Commissariat à l'Energie
Atomique (CEA) shows that in using low
frequency and large diameter focusing
transducer the effect of cladding can be
reduced to an acceptable value of sensitivity variation (2dB) (5). An homoge-
nuous cylindrical acoustic beam of known
diameter and length called " sting " can
be produced in the metal with small modification of its shape after penetrating
the cladding. Then evaluation of defects
is done with good reliability and reproducibility. EdF adopted this method for
ultrasonic inspection of RPV welds of
the major 900MW level reactors. ISI
machines were constructed by CEA to perform this examination from the inside
surface of the vessel in immersion technique. (6). A preliminary test programme
was achieved on the block described above. An ultrasonic C scan of the side
drilled holes was done with longitudinal
and transverse waves 1 and 2MHz focused
transducers through the clad and
unclad surfaces. Results showed an homo-
genuous representation of each reflector and the difference in sensitivity
was of the same order of value
for the both surfaces (4). The
disadvantage of this method is that it
is necessary to build several transducers adjusted to different curvature
and depth in the metal in order
to obtain the proper sting in the volume
to be inspected. For instance, ultrasonic examination of the RPV bottom
welds (2 circular welds between bottom
head, transition area, and cylindrical
shell) is done with 11 focused transducers (see figure 5). Their diameters
vary from 60 to 120mm. Also the mechanical device needs to be sufficiently
heavy to realise scanning movement with
all their connected parts.

4.4. Defects Sizing with Different Ultrasonic
Paths :

Examination of thick welds was done in
manufacturer's shop by High Energy Radiography and by manual ultrasonic test.
Comparison of these results with preservice focused ultrasonic examination was
done to satisfy requirements exposed in
§ 2.3. This comparison showed a weak
correlation in nozzles welds because the
ultrasonic path and angle between the
beam and the weld's chamfer were quite
differents. Essentially, defects are
detected in shop test from outside surface at short distance, as show in figure 6a. In ISI focused transducers are
placed in the nozzle bore, and the beam
path is about 500mm long. In the weld
region the ultrasonic "sting" has a
diameter of 20mm. Then depending on the
method used, detection and sizing defects
are not similar. At request of Regulatory Authority a programme was initiated
to compare the two methods on six
RPV nozzle welds at manufacturer's shop.
Evaluation of indications was done

41

according to construction specifications for the manual test,and with the 6dB level successive scan method,described by Saglio et al.(5),for the focussed beam probe. Metallurgical investigations were done on areas of some interesting indications, by multiple thin excavations accompanied by dye penetrant and magnetic particles tests. Figure 6b give an example where we can see the C scan of the indication at different sensitivity levels and the actual size of the defect. We can note that after 2 scanning levels no new indications are revealed, the contour is only magnified by a dimension equivalent to the half focal diameter of the ultrasonic beam. On the other hand, this programme shows that with different ultrasonic path all important indications (near or over the acceptance level) are detected by the two methods. For small acceptable indications detected by each method, only a few of them are common to both .

4.5. Radiographic Examination of Safe-Ends :

Radiography was generally discarded from ISI examination because of its sensitivity to ambient radioactivity which can produce scattered radiations on the films unsignificant for defect detection and, in some cases, generates important unsharpness of image,and in this way masks fine fatigue or corrosion cracks. This analysis is justified if fine working conditions are not adapted to this specific environment, and if the objective is not to obtain the best quality image possible, but only to satisfy specification requirements. Several years of experience in Gas Cooled Reactors leads us to use this technique when it can give more significant informations than others and in taking the following precautions:
- Front and back filtering with 2mm lead (more for back if necessary).
- Double fine grain films in sandwich with lead screens.
- Priority to Iridium sources.
- Ambient dose rate lower than 20% of the calculated incident radiation on the film.
With these conditions we perform examination of austenitic weldments where ultrasonic test is very difficult for several reasons related to grain size, dendrite orientation, sound speed variation...(8). In heterogenuous welds of Safe-Ends, these problems become harder. We can find 5 material configuration :
- Forging or casting low alloys ferritic steel nozzles.
- 24-12 CrNi alloy first layer buttering (AISI 309 equivalent).
- 18-10 CrNi alloy second layer buttering (AISI 308 equivalent).
- 22 CND17/12 weld.
- 22 CND17/12 forged pipe.(AISI 316 L equivalent).
Radiographic examination applied to these welds at on-site assembling phase, shows a good sensitivity of the method to detect fine cracks. Precise excavations before repairing confirm these indications.

In ISI, examination of RPV safe-ends is done by putting the isotope source in the center of the nozzle. Films are placed on the outside surface for a panoramic exposure. All this operation is done with vessel filled with water. A special tire inner tube is inflated around the source to remove water from the weld region. This mechanism is mounted on the inspection machine arm and remotely controled. For Steam Generator safe-ends a special manipulator placed from the outside, through the man-hole, drives the source to the axis of the piping. Operation is done in dry condition. For small diameter S.E. the double wall technique is used with five exposures. (OD 150mm)

4.6. Acoustic Emission Leak Detection :

Weldments of penetration on the vessels have only a tightning function. The tubes penetration are made of Inconnel material and they are fixed on the inner surface of the vessel by a small fillet, welded on an Inconnel buttering. No conventionnal NDT method can be used after the end of assembling or for ISI to check their cleanliness. Leakage detection by visual examination needs the removing of heat-insulation. This is a very difficult operation due to the numerous penetrations in the vessels spherical head. The narrow access and high level dose rate encountered in these areas makes the operation more difficult. Acoustic emission transducers placed in easy access areas can detect leaking assembly when the vessel is pressurized. Numerous applications of acoustic emission detection of cracks disturbed by fluid leakage are known,but this technique was not used for these purposes. However specific applications were developed for nuclear and non nuclear components (valves, gas pipe-lines, Candu Fuel Channels) (8). Laboratory tests done at Centre Technique de l'Industrie Mécanique under EDF contract (9) on artificial leaks demonstrate that acoustic energy measured by RMS values of the signal can be considered as approximatively proportionnal to leak flow flux :
$$E = K . (P X F) - t$$
were P is the differential pressure, F the leak flow rate, K the constant and t the threshold level due to electronic noise and transducer sensitivity. This relationship is to be applied only on small leakages and relatively high water pressure, where the flow along the leak channel can be considered as turbulent mode. Figure 7 shows experimental curves of this relationship with different section artificial leaks. In these experiments the transducer is placed near the leakage on the same metallic block. In vessel tests ,the transducers must be placed far from the components to be inspected for the reasons exposed above; then the signal is attenuated by losses of energy. These losses are due to the signal propagation along the metal path, to the crossing of metallurgical or geometrical discontinuitiesor to the energy

diffusion around obstacles (other penetrations). Then the residual signal reaching the transducer should be sufficent to characterise the lower limit for acceptable leakage. Attenuation measurements, on three different power reactors, in actual situation of vessels after the end of assembling, were done by simulated acoustic waves propagation. Acoustic emitters or artificial leak blocks were sealed to the component surface and receivers were located in different places in order to quantify losses. The final transducers sites were chosen with an acceptable loss of acoustic energy (lower than 10dB) in the 100-300KHz frequency range. Figure 8 gives attenuation measurement and values encountered in the bottom head of the pressuriser. Leakages are supposed to be in the junction between penetration and heating rod, or between penetration and bottom head.

5. CONCLUSION

It is necessary to improve reliability of NDT methods required for in service inspection of nuclear reactors. Each type of reactor has its specific problems, and the solution of these problems leads one to develop conventional methods or to use new ones.
Solutions must take in consideration other points than the technical aspects like :
-training personnel in applying these methods
-limitation of radiation exposure
-minimizing outage duration.

REFERENCES
1. N. BEAUJOINT Auscultation des caisson en béton précontraints de récateurs nucléaires à EDF Doc. EDF/DTG réf. NB/NS19.466.

2. G.M. JORDAN The influence of frequency and reliability of in-service inspection on reactor pressure Vessel disruptive failure probability 3rd Conf.on Perio.Insp.of Press. Comp. London Sept.76.

3. J.P. BUTIN, G. COQUILLAY Volant d'inertie monobloc de pompes primaires. Mise au point d'une méthode de contrôle ultrasonore.
Doc. EDF ref. D5151/76753.

4. J. HUARD et al. Inspection en service des cuves de réacteurs. Contrôle ultrasonore du bloc de référence EDF.
Doc. CEA ref. STANT 108 bis.

5. R. SAGLIO et al. Determination of defect characteristics using focused probes.
Mat. Eval. Jan 1978 p. 62-66.

6. M.T. DESTRIBATS et al. Inspection en service des réacteurs à eau préssurisée. Colloque International AIEA Vienna October 1977.

7. L. ADLER et al. Ultrasonic Characterization of Austenitic Welds 1st Int. Symp. on Ultr. Mat. Char. Gaithersburg June 1978.

8. G.J. DAU A review of on line leak detection methods for reactor systems.
3rd Conf.Perio.Insp.Press.Comp. London Sept. 1976

9. P. DUMOUSSEAU Etude de la transmission du signal dans les tubes d'enceinte des mécanismes de grappe courte, la partie inférieure de la cuve et le fond du préssuriseur.
Rapport CETIM N° 3.449.7/0443 et 0491.

TABLE I

ISI extent during design life time for RPV components

Element	Frequency	Total after 2 years	Total after 12 years	Total after 40 years
Circular Welds	Each complete Inspection	200 %	300 %	500 %
Safe-Ends	Each CI + 1 S.E each biannual inspection	200 %	400 %	900 %

TABLE II

Specifications for shop ultrasonic testing of thick Welds

	EDF-CPFC Classe 1 paragraphe 5 et annexe 3	ASME Section III Class 1 NB 5330 and Appendix T 530
Calibration	Side drilled Holes Ø 2mm for $t \geqslant 100mm$	Side drilled Holes Ø 6,3mm (1/4") for $100(4")<t\leqslant150mm$ (6") Ø 9,5mm (3/8") for $200(8")<t\leqslant250mm(10")$
Reference Curve	Holes each 20 mm	Holes at t/4 - 3t/4 - 5t/4 ,...
Recording Level	0,25 DAC	0,20 DAC
Acceptance Standards	No cracks, or incomplete penetration or lack of fusion Volumic No indication >DAC No indication >0,5 DAC if length > 40mm	No cracks or incomplete penetration Volumic No indication >DAC if length >19mm (3/4") for $t>57mm$ $(2^{1/4}")$

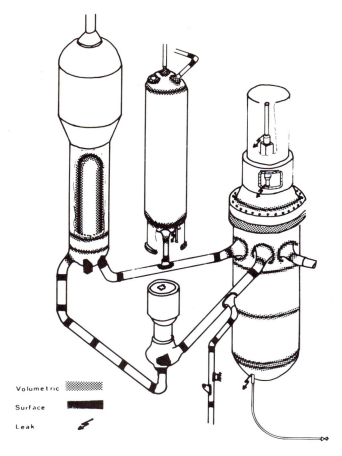

Volumetric ▨

Surface ▬

Leak

Fig. 1 PWR coolant systems with examination methods and extent

Fig. 2 Calibration blocks for RPV ligaments and primary pump flywheel

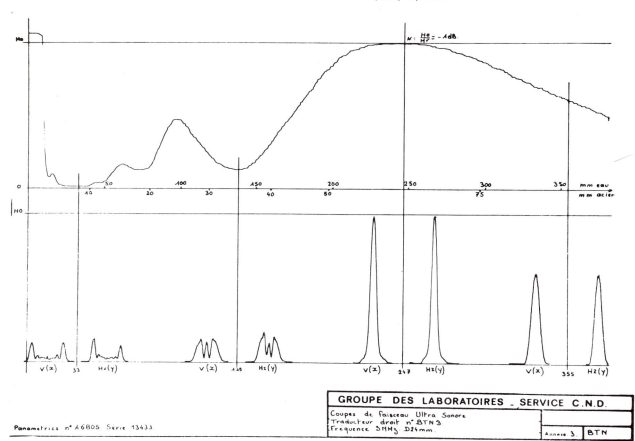

Panametrics n° A6805 Série 13433

GROUPE DES LABORATOIRES _ SERVICE C.N.D.

Coupes de faisceau Ultra Sonore
Traducteur droit n° BTN 3
Fréquence 5MHz D24mm.

Annexe 3. BTN

Fig. 3 Routine card for acoustic field measurement of transducers used for ISI

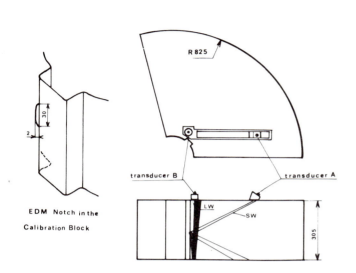

EDM Notch in the
Calibration Block

transducer B

transducer A

R 825

Fig. 4 Detection of cracks in keyway corner by ultrasonic wave transformation

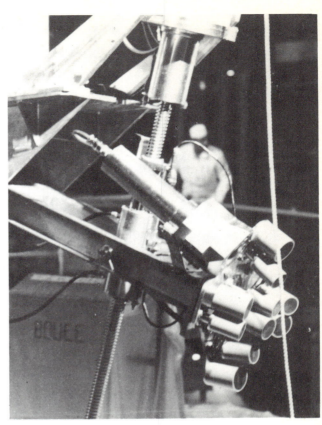

Fig. 5 Ultrasonic plate for RPV bottom welds examination

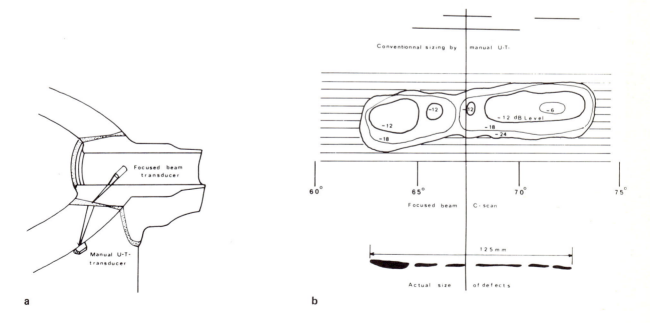

a

Focused beam transducer

Manual U-T-transducer

b

Conventionnal sizing by manual U-T-

−12 dB Level

60°   65°   70°   75°

Focused beam C-scan

125 mm

Actual size of defects

Fig. 6 Comparison for UT defect sizing between manual scanning and focused beam scanning

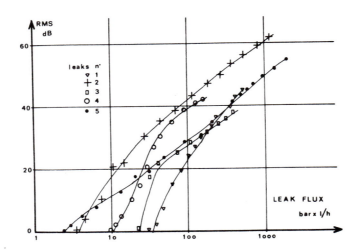

Fig. 7 Relationship between acoustic energy signal (RMS) and leakage flux (pressure × flow rate) for artificial leaks

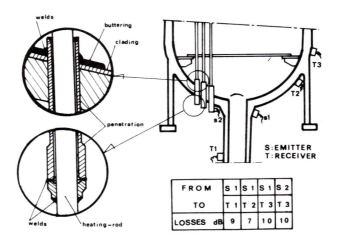

FROM	S 1	S 1	S 1	S 2
TO	T 1	T 2	T 3	T 3
LOSSES dB	9	7	10	10

Fig. 8 Simulated acoustic propagation in pressuriser heating rods penetration

C27/79

INSERVICE INSPECTIONS OF REACTOR PRESSURE COMPONENTS AND PIPINGS IN FINLAND

P. KAUPPINEN, MSc(Eng), and J. FORSTÉN, DTech
Technical Research Centre of Finland, Metals Leboratory, SF-02150 Espoo 15, Finland

The MS of this paper was received at the Institution on 22 September 1978 and accepted for publication on 22 November 1978

SYNOPSIS   The Finnish inservice inspection requirements and the inservice inspection programmes are described in this paper. The planning and performance of the inservice inspections as well as the inspection techniques are discussed briefly. The inservice inspections of the reactor pressure vessels of the Loviisa reactors are performed by VTT using mechanized equipment constructed in the USSR. The reactor pressure vessel is normally inspected from the outside and thus the fuel and reactor internals do not have to be removed. The various possibilities to inspect the reactor pressure vessels of Loviisa reactors are presented. The experience of the preservice inspections and of the first inservice inspections is reported.

INTRODUCTION

1.   A summary of the Finnish nuclear programme and the associated pre- and inservice inspections are given in Table 1.

2.   The Technical Research Centre of Finland (VTT) has performed the preservice and inservice inspections of the main components (primary circuit, steam generators and other components belonging to the safety classes I and II) of the Loviisa pressurized water reactor (VVER-440) and of all components and pipings excluding the reactor pressure vessel of the Olkiluoto boiling water reactor (Asea-Atom 660 MWe). The preservice inspection of the Loviisa-1 reactor was performed in 1976, and the inservice inspections in the winters 1978 and 1979. The preservice inspection of the Olkiluoto-1 was performed in 1977 and of Loviisa-2 in 1978. In the following a presentation is given of the way in which the inservice inspections of the reactor pressure components and pipings are carried out.

INSERVICE INSPECTION REQUIREMENTS

3.   The authorities in Finland have established general inservice inspection requirements in a collection of guidelines (so called YVL-guides). The requirements follow closely the practice outlined in ASME XI. The scope, volume and examination routines of the inservice inspection programme are scrutinized by the authorities to be in accordance with the YVL-guides and the Finnish pressure vessel code. The inspections may not start before the authorities' acceptance has been obtained. Possible changes and amendments must also be accepted before implementation (Ref. 1).

4.   The inservice inspection programme is presented in three stages as Programmes I, II and III. Programme I is included in the preliminary safety analysis report, which is a part of the construction permit application. Programme I includes thus the policy and general outlines of the inservice inspection programme. At this stage, e.g., the accessibility and applicability of the inspection methods are checked.

5.   Programme II must be presented and accepted before the operational permit is issued. This programme covers all items that will be inspected and is regarded as a summary of the inservice inspections. The principles describing the selection of the particular items for each annual revision must be given as well as a general description of the inspection procedures. In addition, programme II includes drawings, descriptions of the systems, flow sheets, etc. as well as the principles to be used in defect sizing and evaluation.

6.   Programme III shall contain the technical details of the inservice inspections (plans, procedures, detailed drawings, etc.) and it must be presented not later than one month before the inservice inspections are initiated.

7.   The three-stage programme results in a flexible system which can be adjusted to the technical development relatively easily as well as to changes in standards, requirements, etc. The utilities have the overall responsibility for the inservice inspections. They also do the planning and preparation of programmes I and II. The inspection company prepares programme III (Ref. 1).

GENERAL OUTLINES OF THE INSERVICE INSPECTIONS

8.   The preparation of the ultrasonic, radiographic, liquid penetrant, magnetic particle procedures, and other instructions used in the inservice inspections start as soon as the inservice inspection contract is finalized. In order to find the optimal inspection procedures even tests with, e.g., different ultrasonic probes have been carried out before the inspection procedures have been finalized. The detailed methods used for defect location and sizing are also settled. All programme III documents

are checked by VTT's QA group to be in accordance with the requirements and thereafter presented to the utility. The documents must be accepted by the utility before the inspections can start.

9. The manual inspections are usually carried out by several teams formed of two well qualified inspectors. All inspection records are checked daily by the inspection leader, and deviation reports are immediately handed over to the utility if the indications exceed the reporting level. The organization and performance of inservice inspections are described in detail in Ref. 2.

10. During the inservice inspections VTT's QA group audits the activities at the offices and at the site and issues a report describing the conformity with the plans, procedures and VTT's quality assurance manual etc. The quality assurance activities have had a positive self-disciplinary effect on the inspections.

11. The final inservice inspection report is issued not later than 60 days after the completion of the inspections. This final report includes a descriptive summary of the inspection techniques used as well as a summary of the inspection results. All weld identification drawings, technical certificates and calibration documents, certifications of the personnel's qualifications, materials certification (chemical analysis of, e.g., liquid penetrants) as well as the deviation reports are included as appendices in the final inservice inspection report. The calibration records, inspection records, wall thickness measurements, measurements of ultrasonic attenuation, etc. are separately sent to the utility. The utility receives only copies as all original records are stored by VTT. (Ref. 2).

Manual inspection

12. The main volumetric inspection method used in the inservice inspection is the ultrasonic examination which generally is performed using the pulse-echo technique (Ref. 1). The ultrasonic inspection including the calibration is performed in accordance with the ASME Code practice. In order to satisfy the reproducibility demand, the settings and condition of the inspection equipments are continuously supervised and recorded. The ultrasonic devices are checked every third month according to the ASME XI requirements. All new ultrasonic transducers are checked before they are taken into use and later re-checked at least every sixth month.

13. To the welds of the austenitic primary circuit in the Loviisa reactor no standard ultrasonic inspection technique could be applied due to very strong ultrasonic attenuation in the welds. The ultrasonic inspection could, however be performed in a satisfactory way using a refracted longitudinal angle beam technique. The angle beam transducers operating in a longitudinal wave mode were of the double crystal type with both crystals focussed towards the same point. (Ref. 1, 3, 4, 5).

14. The surface examination is made by magnetic particle method whenever possible. For austenitic materials the liquid penetrant examination is used (Ref. 1).

Mechanized remote inspections

15. The inservice inspection of the Loviisa reactor pressure vessel is performed by VTT using mainly a mechanized equipment. This equipment is designed and manufactured in the Soviet Union (Atomenergoexport) and owned by the utility (Imatran Voima Oy). The mechanized equipment can be used for
i) inspection of the reactor pressure vessel from the outside,
ii) inspection of the reactor pressure vessel from the inside and
iii) inspections of the nozzles of the reactor pressure vessel.
The main features of the different systems can be seen in Fig. 1.

16. External inspections

The cylindrical shells and the circumferential welds of the reactor pressure vessel are inspected volumetrically with an ultrasonic equipment and visually with a television equipment. The ultrasonic inspection can in certain areas be performed automatically according to a preprogrammed sequence. The ultrasonic inspection can be carried out in a semiautomatic mode and of course on the operator's instructions. The ultrasonic inspection from the outside is performed by the contact method. Water with corrosion inhibitors is used as a couplant. The transducer module carrying various ultrasonic probes is mounted on the top of a telescope mast. A rotating carriage placed below the pressure vessel moves the telescope mast around it. The transducer module consists of angle probes (39^0, 50^0, 65^0), a transmitter-receiver probe (0^0), and two tandem probes (39^0) with a transmitter and four receiver probes. The probes are connected to a 13-channel ultrasonic flaw detection system. For recording and documentation multichannel strip chart recorders are used. All indications exceeding the preset reporting level are recorded automatically. The location and depth of the defects can be determined from the strip charts.

17. The calibration of the ultrasonic equipment is made against flat-bottom holes of different diameters depending on the wall thickness. The calibration is carried out immediately prior to the start of the inspection.

18. The visual inspection is carried out with the television camera, which is mounted on the telescope. A video recorder is used for documentation of the visual inspection results. If necessary, paint and rust on the surface can be removed with a stripping device, which also can be mounted on the telescope mast. The surface cleaning has in some cases also proved to be beneficial for the ultrasonic inspection as better contact has been received.

19. The ultrasonic inspection of the bottom head of the reactor pressure vessel is manual and performed by ordinary equipment. The rotating carriage and shielding below the reactor pressure vessel are used to provide access and radiation shield for the inspector.

20. Internal inspections

The visual inspection of the inner surface

(cladding) of the reactor pressure vessel is performed with a remote television system. The underwater camera is inserted into the vessel with help of a console and a central mast. The inspection results are recorded on video tape. A periscopic inspection system is used for more closer inspections of some parts.

21. Inspection of nozzles and upper circumferential welds

For the nozzle to safe end and safe end to pipe welds radiographic inspection with Co^{60} and Ir^{192} is performed as is schematically shown in Fig. 2 and described in Ref. 6.

22. The base metal close to the safe end welds is inspected using a track guided ultrasonic device as seen in Fig. 3. The transducer module contains angle probes (39^0, 50^0) and a transmitter-receiver probe (0^0). No tandem technique is applied in this region. The transducer module moves around the nozzle and scans in the sideway direction. The same flaw detection and recording system as for the outside inspections is used here, too.

23. The flange to vessel shell weld as well as the nozzle shell welds are ultrasonically inspected utilizing a rail guided device seen in Fig. 4. The type of transducers is the same as used in the outside inspections. This device can also be connected to the universal flaw detection and recording system.

24. Evaluation of the mechanized inspection

The mechanized inspection system used in the Loviisa reactor was tailor-made for this particular purpose. Evaluating the merits of the system based on the preservice inspections and the inservice inspections (external ultrasonic and visual inspection, ultrasonic inspection of nozzles) the following highlights have been recognized:
- The ultrasonic inspection can be started before the reactor closure head is removed.
- The ultrasonic inspection can be performed in parallel with the refueling.
- The stainless steel cladding of the reactor pressure vessel has a minimal interference with the ultrasonic inspection.
- Outside and inside inspections of the reactor pressure vessel and the nozzles can be performed simultaneously.

EXPERIENCE

25. The preservice inspections of the Finnish reactors have included 700...1500 inspection items and required 800...1000 man days (Ref. 2). The first inservice inspection of the Loviisa reactor was performed in accordance with the ASME XI Program B and included about 100 items which needed about 200 mandays.

26. During the preservice inspections some changes in the inspection programmes have been necessary due to unfavourable inspection conditions. Sometimes additional work like grinding or removal of paint had to be performed in order to make the specified inspections possible (Ref. 2).

27. A list of items requiring special arrangements like scaffolds, disassembling of restraints, removal of insulations etc. is made during the preservice inspections. This list is of great value in planning the inservice inspection.

28. The utility's overall planning of all activities during a refuelling period seems very important since for example the opening of a pump, valve or steam generator only for inspection purposes might not be an optimal solution. In this respect a closer collaboration between utility, inspection company and authorities as well as between the utilities' operational staff and those responsible for the inservice inspections must be achieved.

29. Although the Finnish utilities have the responsibility for the radiation protection of the inspection personnel, VTT kept a daily individual exposure record in order to be able to react as soon as possible if the radiation dose approaches the maximum allowable dose. No transfer of personnel to nonradiation work was needed during the first inservice inspection of the Loviisa reactor. The total dose obtained by VTT's personnel was 2,5 Rem. The largest dose received by any inspector was 220 mRem and the largest daily dose was 70 mRem obtained from the purification system (ion exchange resins) of the primary circuit (Ref. 2).

30. All inspectors had personal Tl-dosimeters and the radiation level was monitored by dose rate and dose meters. The utility had arranged radiation protection education. In addition, many inspectors had passed a voluntary radiation protection examination arranged by the Institute of Radiation Protection.

31. It is our opinion that the preservice and inservice inspections should be carried out by the same competent personnel. Especially working in radiation areas is shortened if the inspectors are acquainted with the surroundings and working areas. Inspectors with experience from periodic inspections in conventional power plants must be specially trained for the nuclear power stations as the inspection practice differs considerably. During the preservice inspection a large number of inspectors should be acquainted with the nuclear power plant and the inspections so that possible changes in personnel due to changes of employment or due to radiation exposure can be replaced with experienced inspectors (Ref. 2).

32. It is our opinion that there are areas in which technical development or improvements might lead to more reliable and faster inspections. These areas are:
- Mechanization. In order to speed up the inspections, to reduce the radiation doses, and to improve reproducibility continuous efforts to mechanize especially the ultrasonic inspections of the piping and some other items now inspected manually are needed. When mechanizing the inspection, attention should also be paid to the subsequent data processing.
- Reliability of the NDT-inspections. The different NDT-techniques and the applied procedures do not guarantee that all defects which are considered serious are observed.

The reliability of the presently used NDT-techniques should be determined and correlated to destructive defect characterization. This presumably will lead to improvements in the procedures and inspection techniques.

REFERENCES

1. FORSTEN J., JUVA A. and SÄRKINIEMI P. The inservice (preservice) inspection of the Finnish nuclear power plants. The IAEA Technical Committee Meeting. Kobe, Japan, 1977-04-25...27.

2. KAUPPINEN P., SÄRKINIEMI P. and FORSTEN J. Organization and performance of the inservice inspection of nuclear power plants in Finland. The Eurotest Expert Meeting on "Inservice inspection of nuclear power plants". Brussels, Belgium, 1978-04-05.

3. JUVA A. Ultrasonic inspection of austenitic welds in the primary circuit of the Loviisa I nuclear power plant. Published in the Proceedings of the Specialists' Meeting on the Ultrasonic Inspection of Reactor Components. UKAEA Engineering and Materials Lab. Risley, Warrington, England, 1977.

4. JUVA A. and HAAVISTO M. Ultrasonic inspection of austenitic stainless steel welds. (In Russian). Finnish-Soviet Symposium on Atomic Energy. Moscow 1977-10-12.

5. JUVA A. and HAAVISTO M. On the effects of microstructure on the attenuation of ultrasonic waves in austenitic stainless steel. British Journal of NDT 19 (6) p. 293-297.

6. DEKOROV A.S., MAJOROV A.N. and FIRSTOV V.G. Some features of the isotope radiography of welded joints of a nuclear reactor under the condition of the radiation background. Eighth International NDT-conference, Warsaw, June 1973.

Table 1   The Finnish nuclear programme and the main contractors for the inservice inspections

Name	Type	MWe	Main Contractor	Date of regular power	Inservice inspection	
					Reactor pressure vessel	Piping
Loviisa I	PWR	440	Atomenergoexport	1977	VTT	VTT, Polartest
Loviisa II	PWR	440	"	1979	VTT	VTT
Olkiluoto I	BWR	660	Asea – Atom	1978	TRC	VTT
Olkiluoto II	BWR	660	"	1980	TRC	

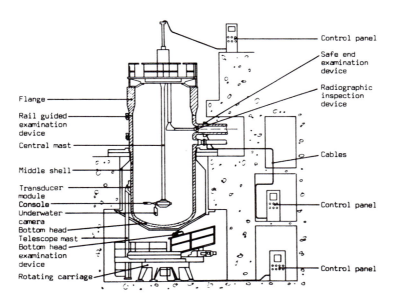

Fig. 1   The inspection possibilities of the Loviisa reactor pressure vessel

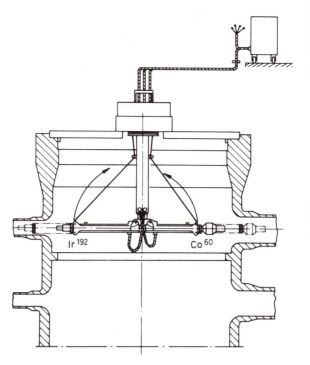

Fig. 2   The system used for radiographic examination of nozzle welds

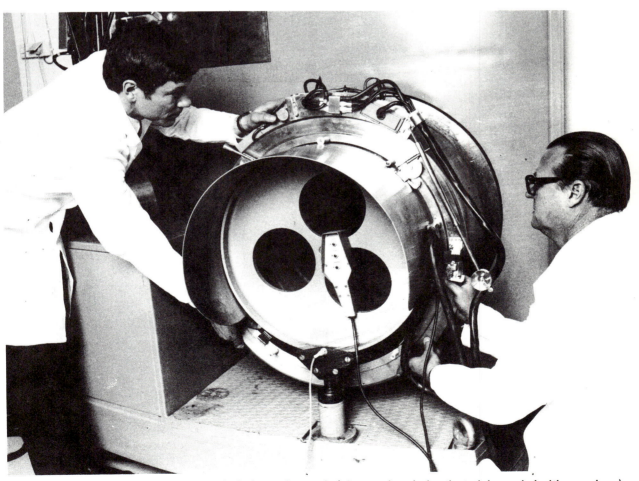

Fig. 3   Attaching the track guided ultrasonic device to the nozzle (picture taken during the training period with a mock-up)

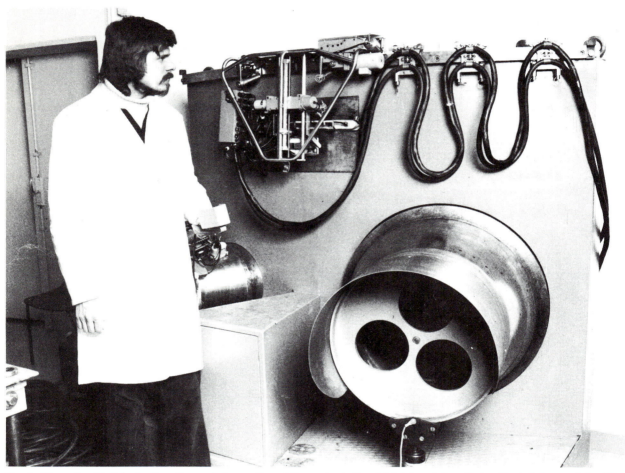

Fig. 4 The rail guided device for ultrasonic inspection of flange to shell weld (picture taken during the training period with a mock-up)

C28/79

THE IN-SERVICE INSPECTION ASPECTS OF STEAM GENERATOR REPAIR AND REPLACEMENT

L. R. KATZ, BS, MS
Westinghouse Pressurized Water Reactor Systems Division
and
P. P. DeROSA, BS
Westinghouse Nuclear Steam Generation Division

SYNOPSIS   Plans currently are being formulated in the U.S. to repair nuclear steam generators in the field. These repairs range from simple tube plugging to retubing or steam generator lower assembly replacement. The rules governing these repairs and the inspection requirements prior to and following the repairs are covered by several documents including U.S. Nuclear Regulatory Guide 1.83, ASME Section III and ASME Section XI. Although addressed in part by these documents, these repair procedures are not covered in detail by any of them. In the implementation of these procedures, several questions concerning the interpretation of these rules have been raised and subsequently have been resolved by the ASME and the NRC. Future additions and revisions to these code documents are required to further clarify the rules which apply to these procedures.

INTRODUCTION

1.   Operating experience with nuclear power reactors during the past ten years has, in certain cases, exhibited steam generator tube degradation. In 1976, A. J. Birkle (Ref. 1) highlighted the rules incorporated into the ASME Code to address this problem. In the past two years, these code rules have been further expanded but are not yet completely adequate to cover this subject.

2.   The degradation which has been experienced includes tube wastage (thinning), cracking and denting. As a result of these phenomena, various corrective procedures have been considered. Selective tube plugging has been chosen as the major course of action. This procedure is based upon a chosen limit of wall reduction determined by eddy-current testing. It can also be based upon an analytical prediction of tube denting progression confirmed by inner diameter measurements. Although tube plugging and reduction of heat transfer design margins have not required a change of rating of any plant in the U.S., contingency planning to preclude or limit the extent of such derating is underway.

3.   As a result of these contingency plans, the nuclear industry has embarked upon several alternative programs. These programs range in magnitude from a scheduled plan of tube plugging within the normal heat transfer design margins of the steam generators, to the more extensive alternatives of tube repair or tube bundle replacement. One approach is to replace the entire lower portion of the steam generator including the channel head, tube sheet, tube bundle and lower shell. A second approach under consideration is to retube steam generators in-place. Retubing involves replacing the tube bundle and associated hardware only. Another approach is the repair of degraded tubes by a

sleeving method. In this method, a honed sleeve is inserted and expanded into the straight portion of the steam generator tube to provide a leak-tight joint in a degraded area of the tube.

4. More than 65 steam generators have been repaired by the tube plugging method in the U.S. at this time. An additional two steam generators have been repaired by tube sleeving. Although no steam generators have been retubed as yet, this method is currently being considered by some utilities. In addition, several utilities are planning steam generator replacement.

5. In anticipation of the need for field repairs to steam generators, the U.S. Nuclear Regulatory Commission (NRC) in 1975 published Regulatory Guide 1.83, entitled "Inservice Inspection of Pressurized Water Reactor Steam Generator Tubes". Regulatory Guide 1.83 became the first standard for establishing a preservice and inservice inspection program and repair rules for steam generator tubes. Although nonmandatory, this guide was then cited by the NRC to all plant owners as a minimum requirement of their operating license.

6. In January, 1976, the first rules for inspection of steam generator tubes were published in the "Winter 1975 Addendum of the 1974 Edition of Section XI". Additional rules for this inspection were added in the "Winter 1976 Addendum" and the "1977 Edition of the Code".

7. Although the rules for inspection of steam generator tubes in Section XI and Regulatory Guide 1.83 are not in complete agreement at this time, work is continuing in the various working groups of the Code Committee to resolve differences. Several conflicting areas were discovered in applying the existing rules to the inspection and repair of steam generator tubes. As a result of these conflicts, several code

inquiries have been brought to the attention of the various Section XI Code Committees for resolution.

8. The following discussion outlines the repair procedures on U-tube type steam generators, emphasizing the inspection and repair requirements which must be completed to satisfy the various regulatory documents as well as those special requirements which may be imposed by the utilities and contractors. Each requirement, including special inspection techniques and required code interpretations, are highlighted for those planning similar actions.

TUBE PLUGGING

9. The most widely adopted technique presently in use to remove tubes from service is tube plugging. This technique includes both manual welding and explosive welding. In order to determine the number and location of tubes that need to be plugged, a periodic non-destructive inspection utilizing an eddy-current technique is performed. In order to minimize down time and limit radiation exposure, several automated eddy-current probe positioning fixtures have been designed. Figure 1 illustrates a typical eddy-current probe positioning fixture inside a steam generator channel head.

10. The rules for the inservice inspection of steam generator tubing are covered in two documents, namely, Section XI of the ASME Boiler and Pressure Vessel Code and Nuclear Regulatory Commission Regulatory Guide 1.83. Regulatory Guide 1.83 was published in 1975 to guide utilities in the selection and implementation of a steam generator tube inspection program. In 1976, Section XI rules for inspection of steam generator tubes were published to incorporate those rules which the Code Committee deemed applicable in the Regulatory Guide into a mandatory inspection code. There are still several areas of difference between the two documents. Two major differences are:
(1) The regulatory guide requires a preservice inspection of all steam generator tubes by the eddy-current technique prior to the initial service; the code waives the preservice inspection in lieu of the shop inspection by UT or eddy-current techniques of the tubing prior to forming the U-bends.
(2) The regulatory guide requires that for all inservice inspections subsequent to the first inservice inspection, the minimum number of tubes required to be inspected is 2 to 4 times the number required by the code.
The various code committees are currently working with representatives of the NRC to resolve the differences between these documents. In the interim, operating licenses are issued by the NRC based upon general compliance with the regulatory guide.

11. Although Regulatory Guide 1.83 defines the tube inspection criteria, including method, frequency and extent of inspection, it does not provide guidance with regard to acceptance standards. After publication of Regulatory Guide 1.83 in 1975, plugging criteria for degraded steam generator tubes were established

between the NRC and the plant owners on a case-by-case basis, using analyses and tests to establish the maximum tube degradation that could be tolerated. In order to standardize the methodology for determining plugging criteria, the NRC published Regulatory Guide 1.121 entitled "Bases for Plugging Degraded PWR Steam Generator Tubes", dated August, 1976. This regulatory guide is now in wide use by plant owners in preparation of their technical specifications to the NRC on the operability and inservice inspection of steam generators. Regulatory Guide 1.121 establishes the operational limits for steam generators, whose tubes have been subjected to degradation. These limits are based on the following:
(1) The maximum degradation which can be accommodated while the tubes sustain the imposed loadings under normal operating conditions and postulated accident conditions.
(2) An operational allowance for degradation between inspections.
(3) The crack size permitted to meet the leakage limit allowed per steam generator by the technical specifications of the particular plant operating license.

12. After the tube plugging criteria of Regulatory Guide 1.121 have been applied, the tube plugging plan for a particular plant application is submitted to the NRC for approval. Although tube plugging techniques are not limited to welding methods, Section XI in Article 4000 does contain rules for both manual and explosive tube welding techniques, which have been accepted by the NRC. If methods other than manual or explosive welding techniques are chosen for tube plugging, the alternate methods are subject to NRC approval on a case-by-case basis.

13. Section XI requires that the following non-destructive examinations be applied to tubes plugged by the manual welding method:
(1) A visual examination in accordance with subparagraph IWA-2210.
(2) A surface examination in accordance with subparagraph IWB-2220.

14. If the explosive tube plugging technique is used, the following non-destructive examinations are required:
(1) A liquid penetrant examination of the cladding and tube-to-tube sheet welds of the test assembly for procedure qualification in accordance with subparagraph NB5350 of Section III.
(2) A hydrostatic test, a metallographic examination and a visual examination of each plug to tube weld in the test assembly for procedure qualification. Upon application in the field, a visual examination in accordance with Article 9 of Section V of the ASME Code to evaluate detonation and the location of the installed plug and a leak test to check bonding of the plugs are also completed.

15. As indicated in the above description, the rules and requirements for eddy-current examination and tube plugging of degraded steam generator tubing are not available in a single source document but included in several NRC and ASME Code documents. Figure 2 summarizes the procedure and source for utilizing these rules.

REPLACEMENT

16. Replacement of a steam generator can take several forms. The most extensive program is to replace an entire steam generator unit. Other options include replacing only the tube bundle, the tube bundle including the tube sheet, or the entire lower section of the steam generator including the channel head. The replacement programs under most recent consideration are replacement of the entire lower section and replacement of the tube bundle, known as retubing.

17. Replacement procedures have certain distinct advantages as compared to other repair methods as follows:
(1) The unit can be manufactured in parallel with operation of the nuclear plant.
(2) Demonstrated shop practices can be used in manufacture of the replacement unit.
(3) Since major subassemblies are being replaced, design improvements can be readily incorporated.

18. The major disadvantage of replacement programs is related to access provisions for removal of the old unit and erection of the new unit. Providing access within the plant containment may result in the need to remove concrete shielding walls and portions of the containment structure. Because of these requirements, certain plants may experience down time for major replacement programs. In order to limit down time, the replacement program is usually specified such that the largest component must pass through the existing plant containment equipment hatch. This requirement may dictate a replacement program in which the lower section of the steam generator is replaced and the larger diameter upper section, including the steam drying equipment, is used again.

19. Figure 3 shows a typical lower assembly replacement unit. The maximum diameter of the replacement unit is established by the transition cone parting line. The sequence of operation required to implement a lower assembly replacement program follows:
(1) Cut all of the external piping connected to the steam generator including:
    (a) The reactor coolant piping to the inlet and outlet nozzles.
    (b) The main steam piping to the upper head.
    (c) The main feedwater piping at the feedwater nozzles.
(2) Separate the upper assembly from the lower assembly by cutting the transition cone.
(3) Move the upper assembly into storage on the operating deck.
(4) Lift out and remove the lower assembly from the containment.
(5) Prepare the weld ends of the reactor coolant, main steam and main feedwater piping.
(6) Move the new lower assembly into the containment and lift into place on the existing supports.
(7) Move the upper assembly from storage into position on the new lower assembly and reweld the transition cone.
(8) Reweld the main coolant piping, main steam piping and main feedwater piping to the steam generator.

(9) Complete all required ASME Code tests and inspections.

20. The rules for steam generator replacement are covered by several codes and standards including ASME Section III, ASME Section XI and the federal law covering the quality assurance requirements for nuclear applications entitled "Quality Assurance Criteria for Nuclear Power Plants and Fuel Reprocessing Plants", (10CFR50 - Appendix B). The general applicability of each of these documents follows:
(1) ASME Section III - Subsections NA, NB and NC cover the general rules and specific requirements for the design, construction and testing of Class 1 and Class 2 nuclear components.
(2) ASME Section XI - Subparagraphs IWA-7000, IWB-7000 and IWC-7000 contains the general rules and specific requirements for the replacement of Class 1 and Class 2 nuclear components and parts. Subparagraphs IWA-4000, IWB-4000 and IWC-4000 contain the general rules and specific requirements for making repairs to Class 1 and Class 2 nuclear components and parts by welding.
(3) 10CFR50 - Appendix B - This document outlines the requirements for quality control to be applied to the design, manufacture and installation of all Class 1 and Class 2 nuclear components.

21. Some of the most important rules contained in these documents which pertain to the replacement of steam generators are:
(1) Replaced components may be designed, constructed and tested to the rules of the ASME Code edition and addendum which applied to the original component. The replaced component may also be designed and constructed to any later ASME Code edition and addendum to the ASME Code.
(2) The replaced component or part must contain the applicable ASME Code stamp.
(3) The installation of the replaced component or part may be completed by any qualified organization. The organization is not required to have an ASME Code stamp.
(4) The organization installing replacement components and parts in the field must have a quality assurance program which meets the requirements of 10CFR50 -Appendix B.
(5) All welding completed during the installation of replaced components must meet either the rules of the applicable ASME Section III construction code or the rules of the applicable ASME Section XI inservice inspection code.
(6) All field welds which are made to implement the replacement are subject to the same qualifications and non-destructive examination as are required by the construction code for welds performed in the shop.
(7) A field hydrostatic test must be performed on the replaced steam generator in accordance with the rules of ASME Section XI.
(8) A preservice examination of all welds made to implement the replacement must be completed in accordance with ASME Section XI.
(9) A subsequent inservice inspection must be completed on all welds made to implement the replacement, if such welds are included in the category of welds requiring inservice inspection in ASME Section XI.
(10) The design of the replacement unit must be compatible, to the extent practicable, with the latest version of the ASME Section XI from the standpoint of accessibility and inspectability.

(11) The design of the replacement unit must be compatible with Regulatory Guides 1.83 and 1.121 from the standpoint of accessibility for eddy-current testing and possible future plugging of steam generator tubes.
(12) Tubing that has been replaced will be preservice inspected by the eddy-current method in accordance with Regulatory Guide 1.83.

22. From the above description, it is evident that the rules and requirements for steam generator replacement programs are not clearly and are not concisely spelled out in a single document.

RETUBE

23. Another steam generator repair method being evaluated in the U.S. is referred to as retube. Retube is a method designed to remove the tube bundle of a steam generator in-service and replace it with a new bundle. The bundle consists of a wrapper, support plates and tubes. This process utilizes remote tooling to minimize plant down time and radiation exposure to personnel. The various operations to accomplish retube are:
(1) Decontaminate the channel head.
(2) Cut the steam drum in the cylindrical portion of the upper head (Figure 4).
(3) Remove the steam drum with upper internals and place in a storage stand within the containment.
(4) Provide two access manways in the lower shell just above the tube sheet (Figure 4).
(5) Use the manways as access to cut the tubes just above the tube sheet and raise the tube bundle into a shielded cask for removal from the containment.
(6) Cut the tube-to-tube sheet welds and remove the tube stubs from the tube sheet.
(7) Refurbish and inspect the tube sheet in preparation for installing the new tube bundle.
(8) Install the wrapper-tube support plate assembly into the lower assembly (Figure 5).
(9) Replace tubes (Figure 5).
(10) Perform tube-to-tube sheet welds.
(11) Replace the steam drum onto the lower assembly and perform the closure weld.
(12) Complete all required ASME Code tests and inspections.

24. Retube procedures have certain distinct advantages when compared to other repair methods. They are:
(1) No major concrete removal or containment modifications are required to move retube components in or out of the containment. This feature has the additional advantage of minimizing plant down time.
(2) Major Class 1 component pressure boundaries such as the primary coolant piping, are not breached.

25. The rules for steam generator retube are currently not completely addressed by existing code and regulatory documents. The existing rules which are applicable, however, include ASME Section III, ASME Section XI and 10CFR50 - Appendix B. The rules covered in these documents which pertain to the retube of steam generators are the same as those listed for replacements, except for Item 2. In a retubing program, the replaced components contain no pressure retaining welds and require no code stamp.

26. Another important consideration in the retubing of a steam generator is to assess the condition of the existing tube sheet to accept processes necessary for retubing. Although examination of the tube sheet is not within the scope of Section XI, a tube sheet evaluation and inservice inspection program is included in the retube effort.

27. The evaluation program consists of two phases. The first, a preliminary inspection phase, is to take place at an outage prior to the actual retubing effort. This inspection utilizes various eddy-current probes as well as selective tube pulling and visual inspections. The second phase consists of inspections and modifications performed during the retubing outage.

28. Specific details of each phase are:

Phase one - prior outage

(1) A preliminary eddy-current inspection of the tube sheet.
(2) A selective pulling of the tubes to obtain statistical data from visual and eddy-current inspections of bare tube holes.

Phase two - retubing outage

(1) After removal of the tubes, the tube sheet undergoes a thorough cleaning and decontamination. This cleaning includes removal of the sludge down to the tube sheet surface.
(2) A liquid penetrant and visual inspection is made of the surface of the secondary side of the tube sheet.
(3) A refurbishment of the tube sheet is performed. This procedure consists of honing and reconditioning of the tube holes in preparation for tube expansion and tube-to-tube sheet welding.

CODE INTERPRETATIONS

29. The planning and design efforts on steam generator repair and replacement have progressed more rapidly than the writing and publication of code rules covering this effort. The lack of definition in the current codes has resulted in the need for several important code interpretations.

30. The first covers the use of thermal gouging of metal during the repair or replacement procedure. Thermal gouging by the air-arc method was deemed to be the most efficient method for cutting through the pressure boundaries of the steam generator to effect repairs in the field. The advantages of thermal gouging over other processes include the speed of metal removal and the need for only small torches better adapted to the limited access afforded by the steam generator enclosures. The 1977 Edition of Section XI in Paragraph IWB-4322(a) specifically prohibited the use of thermal gouging to effect repairs. On the other hand this same edition and addenda of Section XI permits the

use of the methods and procedures of the original construction code (Section III) to make field repairs. Section III in Paragraph NB 4453.1 permits the use of thermal gouging. It contains the additional requirement that the area prepared for repair by thermal gouging must be inspected by either the magnetic particle or dye penetrant methods. As a result of this apparent conflict between Section III and Section XI, a code inquiry was submitted to the Section XI Subgroup on Water Cooled Systems. The following is the official inquiry and ASME reply:

Question

31. Is it permissible under the rules of Section XI, Division 1, to use the thermal removal method for back gouging in preparation for welding during repairs?

Reply

32. Section XI, Division 1, IWA-4000 and IWB-4000 have recently been revised to allow use of the thermal removal method for back gouging welds in the preparation for welding during repairs. These revisions will appear in the Summer 1979 Addenda to Section XI, Division 1, of the Code. The use of later editions of the code require the concurrence of both the Nuclear Regulatory Commission and the jurisdictional authority for the state in which the repair is being performed.

33. The Summer 1978 Addenda to Section XI does permit the use of the thermal removal method for making.repairs with the following additional requirements:
(1) Preheating is required.
(2) A minimum of 1/16 inch of material must be removed by a mechanical method from the thermally processed area.

34. These requirements are over and above those specified in Section III because of the possible adverse conditions in the field as compared to the shop.

35. The second code interpretation for steam generator repair and replacement pertains to the non-destructive examination methods which apply to repair welds completed within the jurisdiction of Section XI. Section XI permits repairs to be completed to either the rules of IWA-4000 or the original Construction Code. If the Construction Code rules are used, the examination requirement for repair welds is radiography. In addition, Section XI requires that a preservice examination be completed on all repair welds utilizing a method which will be subsequently used during inservice inspection. In most cases, this perservice examination is performed using ultrasonic methods. Since ultrasonic techniques are ultimately required as the inspection method for these repair welds, a question was raised by a repair contractor as to the necessity for radiography. Field radiography of thick, pressure barrier welds has disadvantages. They are the need for a large high powered gamma source and the need for temporary shielding. As a result, a code inquiry was submitted to the Section XI Subgroup on Water Cooled Systems. The following is the official inquiry and ASME reply:

Question

36. Is it permissible for repair welds within the jurisdiction of Section XI, completed under the rules of Section III of the Code, to be non-destructively examined by the ultrasonic method as a substitute for the examination required by Section III?

Reply

37. It is not within the authority of Section XI to revise those examination requirements specified in another code section.

38. Although the substitutions of ultrasonic techniques for radiographic techniques are prohibited by the above code inquiry on legalistic terms, the exclusive use of ultrasonic techniques appears to be technically sound. Another approach which may be ultimately taken on this question is to request the regulatory agency responsible for the repair to grant an exemption from the need for radiographic examination based on the technical rather than the legal aspects of the issue.

39. A third code interpretation on steam generator repairs pertains to the requirements for subsequent inservice inspection of certain repair welds made on the secondary side of the generator. In several of the techniques being planned, metal cutting operations and subsequent weld repairs are required on joints which are originally cut only to allow access for actual repair or replacement of components. Section XI, as currently written, requires a preservice and subsequent inservice inspection of all repair welds. No distinction is made on welds which are only required as a result of providing access for such repairs. As a result of this lack of detail in the current code wording, an inquiry was directed to Section XI. The following is the official code inquiry and ASME reply:

Question

40. When a cutting or welding operation is performed in a Class 2 component or system within the scope of Section XI, Division 1, is it necessary to perform subsequent preservice and inservice examinations after the welding repair is completed?

Reply

41. If the area affected by welding or cutting is within the scope of Table IWC-2500-1, the preservice and inservice examinations are required.

42. If the same inquiry were made on a similar weld on the primary side of the generator, the answer would have been quite different. All welds in a Class 1 component require preservice

examinations even though they are not scheduled for routine inservice examinations. An access type weld on the primary side, therefore, would require a preservice examination but it does not require subsequent inservice examination.

43. The current words in Section XI are not as yet complete or concise concerning the exact rules for specific component repairs and replacements. Frequent requests for code interpretations in any particular area of the code is usually considered good reason by the code committees for code changes to clarify these issues.

CONCLUSIONS

44. This paper presents the following conclusions:
(1) None of the currently proposed repair and replacement procedures for steam generators are completely covered by either the ASME Codes or the pertinent NRC documents.
(2) Several code interpretations have been formulated by ASME in response to request for clarifications of the code in the area of steam generator repair and replacement.
(3) The rules of Section XI and Regulatory Guide 1.83 are in conflict on the subject of steam generator tube plugging.
(4) Additions and revisions to Section XI are required to clarify the rules pertaining to steam generator repair and replacement.

REFERENCE

1. Birkle, A. J., "PWR Steam Generator Inservice Inspection", Third Conference on <u>Periodic Inspection of Pressurized Components</u>, London, September 20-22, 1976. The Institution of Mechanical Engineers, London and New York, 1977, pp. 61-65.

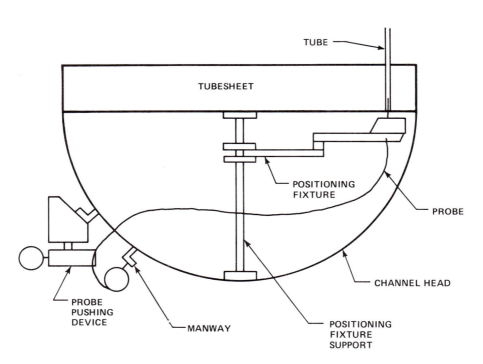

Fig. 1 Typical eddy-current positioning fixture

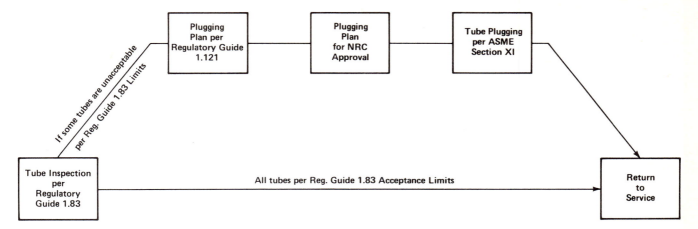

Fig. 2   Steam generator tube inspection and plugging procedures

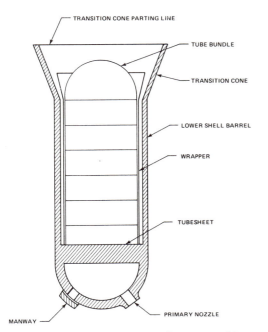

Fig. 3   Typical steam generator lower assembly

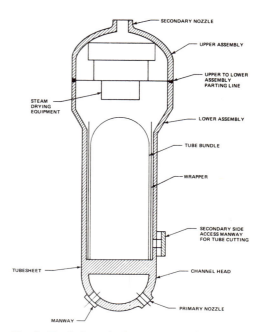

Fig. 4   Typical retubed steam generator

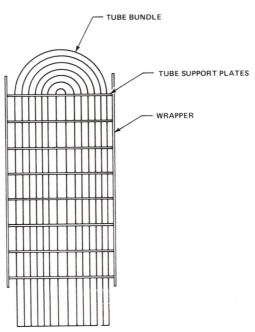

Fig. 5   Steam generator retube assembly

C29/79
FRENCH DEVELOPMENTS AND EXPERIENCE IN THE FIELD OF INSERVICE INSPECTION

R. SAGLIO, M. T. DESTRIBATS, M. PIGEON, M. ROULE, and A. M. TOUFFAIT
CEA CEN Saclay, Gif/Yvette, France

The MS of this paper was received at the Institution on 13 September 1978 and accepted for publication on 1 December 1978

SYNOPSIS : The French PWR nuclear plant program was at the origin of a large amount of R and D work in the field of inservice inspection.
The actions which were undertaken may be split up into different levels :
- the regulatory level, the R and D level, the design level, the flaw evaluation level.

The first results of pre and inservice inspections are presented. The experience gained by French Atomic Energy Commission with new techniques like focussed ultrasonics transducers and multi frequencies Eddy current apparatus are discussed.

I. INTRODUCTION

In-service inspections, or more precisely periodic in-service inspections of reactor pressure components are based mainly on non destructive testing techniques. However, the specific requirements of such inspections as well as the peculiar need for preventing catastrophic failures associated with strong radioactive releases have led to new developments.
In France as in the other countries, the efforts were going on in various directions.
A large R and D program was set-up to solve the problems associated with the specific type of water reactors developed in France.
The paper will give an idea of the various aspects of these efforts.

II. DEVELOPMENT OF METHODS AND DEVICES SPECIFIC FOR I.S.I.

Before beginning the R & D work in the field of methods specific to inservice inspection, a survey was made of all the methods used in the various countries.
Specific problems have been discovered and only partly solved. Among them can be cited :

- the necessity to provide access for testing personnel,
- the necessity to use remote testing devices,
- the difficulties associated with the presence of the austenitic stainless steel cladding on the inner surface of the PWR and BWR pressure vessels with ultrasonic testing (generally considered as the unique solution for remote volumetric examination).

However, three notions, presently considered as fundamental in France, were not yet familiar:

- the necessity to follow the possible evolution of the flaws,
- the long term reproducibility of the required tests able to give objective inspection results,
- the necessity to get a good idea of the defect size.

The R and D work was mainly related with these three notions and the various steps are explained in the following chapters.

II.1 Inspection of the pressure vessel

II.1.1 Focussed probes

The various steps of the research work were as follows :
- recognition of the diffraction effects as the main sources of troubles due to the austenitic

stainless steel cladding.
This point was presented in reference 1 and 2,
- systematic study of the acoustic field after it has crossed the cladding layer. The results are as follows :

. lower the frequency of transducers,
. increase their damping factor (thus increase their frequency spectrum),
. increase their diameter to reduce the diffraction grating effect,
. focalize the beam to get improved lateral resolution even at low frequency.

In fact this last point was the origin of most of the further developments : e.g. - systematic study of the properties of focussed beams using longitudinal and transverse oblique modes, as well as search for simple means to prevent aberrations due to the obliquity of the beam or to the curvature of the pieces to be tested.

A fundamental choice was also made : the necessity to use immersion testing to ensure a reproducibility independent of the surface conditions.

The net result of this research work was the development of small computer programs, aimed at determining the characteristics of the focussed probes to be used in any circumstances.

The advantages resulting from the use of such probes may be summarized as follows :

- existence of a focal zone called effective ultrasonic beam (E.U.B.) having the following properties :
  . constant diameter (at 6 dB drop) along the length of this effective beam,
  . constant sensitivity (at 6 dB drop) along the same length.

- possibility of use a constant gain without any adjustment to take into account the absorption due to the austenitic stainless steel cladding and also the depth of the ultrasonic path : a given depth may be "split up" into several layers having equal depth, and overlapping each other, each layer being surveyed by a transducer with a given effective beam (E.U.B.), similar for each layer. Only the sensibility may be different for each layer, but can be adjusted in advance, without readjustment during the testing process.

- long term reproducibility easily achieved
This is considered in France as a very important feature : the immersion technique cannot take into account all the deviations which can occur in the testing line, particularly

those due to the variations in the transducers characteristics.

The experience has shown the extreme difficulty to build up two identical probes, specially angle probes.

On the contrary, it is easy to manufacture at each time (taking into account the foreseeable and the unforeseeable evolution of the electronics), focussed transducers having determined effective ultrasonic beam characteristics. Moreover these characteristics can be predetermined, and measured. They also may be adjusted when not corresponding exactly to the specifications.

This means that at each time during the life of the reactor, all modifications will be attributed only to the flaws and not to the testing line. This is a fundamental result which gives all its meaning to the periodic inspection procedure.

- the sizing of the detected flaws is now possible, with a predetermined accuracy depending only on the E.U.B. diameter. This also may be considered as a fundamental property which precludes all the arbitrary analyses generally provided for by the specifications.
  This very important characteristic is fully explained in reference 3 and 4 .

- The existence of a corrective structure (lens mirror...) provides for the corrections of aberrations due to curved interfaces ; when the ultrasonic beam impiges on a spherical, or cylindrical surface for instance, this surface acts as a lens which may be divergent or convergent. Generally this phenomenon is ignored but may be very important for instance to test nozzle welds, safe-ends welds and studs (pressure vessel cover).

- Finally an analytical study has shown that the probability of detecting misoriented defects is larger using focussed probes. This also may be considered as a fundamental result for the safety : the critical flaw sizes given by the fracture mechanics analysis are large compared to the minimum detectable flaw size, and large and plane defects may be totally missed by other methods as long as they are not properly oriented relatively to the beam axis (see reference 5 and 6).

However, it may be pointed out that the use of focussed transducers cannot solve all the problems. For instance crack-like defects under high enough compression may escape even with the improved signal-to-noise ratio achieved with focussed probes. The same occurs also for inclusions, when their characteristics are almost the same as those of the surrounding matrix. But these are physical limitations which cannot be overcome with ultrasonic.

II.1.2 Application to pressure vessel inspection

Figure 1 shows the remote controlled device developed for this inspection. It consists of a sliding mast bearing a rotating platform on which several tools may be mounted. The device is supported by three legs on the vessel flange, guided by the dummy studs used to remove the cover.

In addition a sliding arm is also provided, bearing :
- a set of transducers to test the nozzle welds
- a set of transducers to test the safe-end welds
- a gamma source (^{192}Ir) also for the safe-end welds.

A special tool is provided for the testing of the flange ligaments as well as for the flange to cylindrical shell weld.

The device is centered in the middle of the pressure vessel by means of three legs powered by hydraulic jacks.

The design of the machine and the type of indexing device used allows a positioning accuracy better than 4 mm which is consistent with the diameter of the beam chosen.

Visual examination of the entire internal surface of the pressure vessel is provided using a TV camera equipped with remotely controlled focussing and zooming.

At this time three tools simular to the one presented in picture 2 are existing in France. The control panel utilising a process computer is presented in picture 3. These tools were utilised for the preservice inspection of :
- FESSENHEIM 1 and 2,
- BUGEY 2, 3, 4 and 5,
- TRICASTIN 1,
- GRAVELINES 1,
- DAMPIERRE 1,
and for the inservice inspection of CHOOZ.

II.1.3 Ultrasonic apparatus

A special ultrasonic apparatus is used. It is provided for energizing 10 transducers sequentially no matter what the order is and for the display of signals coming either from one or several transducers. The recording gate is thus easily adjusted on each channel.

Signals coming from the gates are recorded both on an events recorder (on paper) and on a magnetic tape recorder.

C-Scan displays can be achieved by reading the tapes and recording on a fast strip chart recorder.

A five channel device was also developed to display the B-Scan information. The five channels correspond to the five directions of testing.

This type of analysis is particularly useful for the determination of the exact in-depth position of the detected flaws and gives very valuable information for the interpretation of the signal. The images on the 5 scopes are recorded on magnetic tape together with information about the position of the transducers for further visualization.

A computer analyzing device named STADUS-PRODUS developped by French Atomic Energy commission is also used. This new device will be described in detail at the Melbourne Conference.

II.1.4 Testing of the studs

Focussed ultrasonic immersion testing is also used for the studs. This testing is made on a special device using the internal hole provided for stress measurement. A special probe with a high degree of aberration correction was developed to detect possible cracks at the root of the threaded part of the studs.

A standard notch 0.3 mm is depth can be easily detected with an excellent signal to noise ratio.

Eddy currents testing is also used for this purpose (see later).

II.1.5 Safe-ends testing

This testing is performed using both ultrasonics and gammagraphy.

II.1.5.1 Ultrasonics

It is well known that ultrasonic testing of austenitic stainless steel weldings offers great difficulties.

This is generally attributed to grain size and large absorption due to the structure. Another assumption has been made : due to large differences in the velocity of sound from one grain to another strong beam deviations occur at the grain boundaries leading to phase mismatch in the beam and to interference attenuation. This effect can be reduced by decreasing the number of grains insonified and thus focussing the transducer. This effectively reduces considerably the background noise, together with the use of longitudinal oblique waves at low frequency instead of transverse waves.

Nevertheless the sensitivity achieved is less than with ferritic steel and it was decided to perform also gammagraphic testing.

II.1.5.2 Gammagraphy

^{192}Ir was chosen as source for its better contrast capabilities. To prevent strong diffusion of the beam by water, the source placed on the nozzle axis is surrounded by a rubber inner tube inflated by air, figure 4. The film with lead screens is wrapped at the OD of the safe-ends. Relatively fine grained film is used leading to exposure time of the order of 1 or 2 hours. Standard AFNOR IQI is used to check the image quality.

II.2 Inspection of the steam generator

II.2.1 Eddy currents developments for tube testing

As the steam generator is the largest heat transfer area between the primary and secondary circuits, its integrity is of prime importance for the safety of the installation, either during normal operation or during the postulated accidents (LOCA or steam line rupture). The number of problems arising with these steam generators all around the world seems to indicate a lack of effectiveness of the testing methods presently used.
The typical damages which can be found in such tubes are for example :
- thinning by corrosion (OD or ID),
- crud deposits (eventually magnetic cruds)
- fatigue cracking,
- shocks in the vicinity of the tubes support plates or of the antivibratory bars,
- denting, this phenomenon being the result of combined actions : crud deposits, corrosion and vibration, the net result being a reduction of the internal cross-section of the tube.

Eddy currents testing is the most frequently applied method to test S.G. tubes. But a mono-frequency apparatus has not the capability to extract all the information content of the signal.

As a matter of fact the tube sheet, the tube support plates and the antivibratory bars give their contribution to the Eddy current signal as well as other spurious parameters like for instance cruds deposits.

The solution lies in the use of a multifrequency device (figure 5).

This device is fully described in reference 7 It was developed by French Atomic Energy Commission and is now commercialized by INTERCONTROLE who is a subsidary Society of French A.E.C.

In the device developed, 3 frequencies are used ; it gives the possibility :

- to detect flaws on the external surface of the tube in the tube sheet, even in the area where the tube has been expanded, and even in the simultaneous presence of internal diameter variations as can be found with some of the manufacturing processes used (pilgrim pass),

- to remove spurious noise corresponding to magnetic inclusions in the tube metal or to magnetic cruds on the OD,

- to measure the wall thickness using the absolute channel, and thus to detect overall corrosion or local corrosion (corrosion pitting if present is detected on the differential channel).

Moreover the discrimination between OD and ID flaws is quite easy and gives the possibility to determine the flaw depth.

All the information stored on magnetic tape can be replayed back with new mixing if needed in a way to eliminate spurious signals unknown at the time of inspection. This apparatus which improved considerably the capability of Eddy currents was used for the preservice inspection of :
- FESSENHEIM 1 and 2,
- BUGEY 2, 3, 4 and 5,
- TRICASTIN 1 and 2,
- GRAVELINES 1 and 2,
- DAMPIERRE 1 and 2,

and also for inservice inspection of :
- DOEL 1 and 2,
- TIHANGE 1,
- GINNA (Rochester Gas and Electric Company).

II.2.2 Automatic tube testing devices

It is well known that in the vicinity of the steam generator lower head the radiation dose may be important. Thus it is necessary to reduce the exposure of the testing personnel.
As the entire length of the bended tube even of those with the smallest radius of curvature is tested using special coils, a device was designed to insert and remove the internal coils. Data acquisition is made both during the direct and the reverse travel of the probe (figure 6).
To prevent manipulation the move from one tube to another is entirely automatic thanks to a device called "finger walker". The time required to transfer the coil from one tube to another one is less than five seconds (figure 7) This finger-walker is controled by a process computer allowing a fast and safe inspection.

II.2.3 Inspection of the nozzle-to-pipe transition welds

These welds are made on site and are difficult to test (dissimilar metal welds). Nevertheless both gammagraphy and ultrasonics are used. A special device is under development for the automatic ultrasonic testing from the outside surface.
It is essentially made of a water tank equipped with rollers and guided by a rail (figure 8). A motor is provided for the circular movement. The scanning of the inside focussed probe is achieved by an hydraulic jack. A silicon rubber seal gasket prevents water leakage - A-B or C-Scan representations can also be used in this application for better flaw analysis.

II.3 Inspection of the pressurizer

Among all the welds to be inspected on the pressurizer one is particularly difficult to test. This weld is situated at the bottom of the vessel (figure 9). The distance from the accessible surface is at least 700 mm, which prevents the use of standard contact probes.

Due to access limitations it was impossible to scan the transducer along the cylindrical support skirt. A solution was found which makes use of an oscillating focussed beam. Thus an oscillating mirror is used to send back to the weld a focussed beam coming from an immersed transducer and scan the total depth of the weld. This is another example of what may be done by "shaping" the ultrasonic beam and taking into account all the parameters involved in the problem like unaccessibility, curvature of the piece, etc... figure 10.

II.4 Special devices

Several other improvements are under study or will be implemented in the future plants. As an example studs and nuts may often be Eddy currents tested instead of visual testing normally performed. Small portable devices were developed specially for that purpose (figure 11).

III. EVALUATION OF DEFECTS

III.1 Ultrasonics

Presently, the evaluation of the detected flaws is one of the most difficult problems in ultrasonic testing. The only parameters available to characterize them are : the time of flight, the amplitude of the signal and, to a less extent the angle of the beam (which is not always well determined). To increase the number of parameters several tests are generally performed under various incidence angles. Even in this case it is really difficult to give objective appreciation on the shape, size and also position of the defects. It has been shown in chapter II how the use of the focussed probes may greatly improve the reliability of the sizing process. It may be pointed out that for a piece with a given wall thickness the accuracy achieved is independent of the flaw depth (even with several transducers because they have the same effective ultrasonic beam characteristics) and remain constant with time (reproducibility of transducer fabrication).

A rule was theoretically found and experimentally confirmed to perform this accurate measurement. It is very close to the 6 dB drop normally used. It utilized a step by step process in relation with the effective ultrasonic beam. This process is fully explained in references [3], [4], [8].

Confirmation of this process were obtained on real defects in relation with national and international experiments (E.P.R.I. and European programs).

III.2 Eddy currents

Chapter II.2.1 has shown the development of signal analysis during Eddy currents testing. The new method used can be considered as being a step forward to objective evaluation of some certain types of defects. However several parameters are still difficult to determine.

A large program is now underway to improve detectability and characterisation of defect in the denting. A catalog of defects is under production in a way to develop a computer analysis system.

III.3 Gammagraphy

It has been shown why gammagraphy is also used for the safe-ends testing. However if before the start-up of the plant no particular difficulties occur, the situation will be totally different after a certain operation time. The natural radioactivity developed into the pressure boundary will be a strong drawback. The first step to solve the problem was to choose ^{192}Ir as a source because of its wide energy spectrum different from that of ^{60}Co which is present in the material. It is hoped that it will give better contrast and thus better film quality. The next step will be to use films desensitized to the ^{60}Co radiation and not to that of ^{192}Ir. Similarly the source strength was chosen as a compromise between keeping exposure time and geometric unsharpness as low as possible. The 1st results seem to indicate that faster films could be used to limit the exposure time.

IV. EXPERIENCE OF ISI ON OPERATION REACTORS

Presently on the pressure vessels 10 preservice inspections have been performed :
- 1 on a small PWR prototype reactor, the CAP
- 9 on 900 MW Westinghouse type PWR's, FESSENHEIM 1 and 2, BUGEY 2,3,4 and 5, TRICASTIN 1, GRAVE-LINES 1 & DAMPIERRE 1. However 3 inspections on operating plants took place :
- 1 on the CAP reactor after 1 year of operation : this inspection gave the opportunity to check the long term reproducibility claimed for the focussed probe methods. This very important property was strongly confirmed.

- 1 on the SENA reactor which was built in cooperation by Belgium and France about 10 years ago. At the time of construction no provision was made for in-service inspection. A special machine was then designed and built to suit the particular requirements (and also those of the 900 MW type). This machine which incorporates the whole of the necessary equipment to test circular welds, nozzles and

safe-ends welds as well as visual inspection
by TV camera and γ source for complementary
safe-ends radiographic testing, is computer
controlled.
In this particular case where the supporting
of the pressure vessel is achieved by separate
supporting structures (instead of nozzles),
the testing of these supports has shown that
all around the pressure vessel diameter the
back echo after crossing through the austene-
tic stainless steel cladding (which was of the
old type) was constant at better than ± 1.5 dB
This would not be possible with standard
probes.
In this case it was not possible to compare
the results with those of a preservice inspection since it was not performed at the time
of the start-up.

- one on FESSENHEIM 1 during the first refuelling on the steam generators.

    14 preservice inspections and 6 inservice
    inspections have been performed. This repre-
    senting four thousands kilometers (4000 km)
    of tubes !

CONCLUSIONS AND RECOMMENDATIONS

It has been shown how the requirements of
periodic in-service inspection of reactor
components have lead to new developments in the
field of non destructive testing. New methods
and devices were designed and fabricated which
represent a step forward in the objective knowledge of the flaws.

The question arises to know if these methods
and equipment must also be used during shop
testing of nuclear components. It is the opinion
of the authors that the future development of
nuclear industry and the related safety requirements are a strong incentive to use them, as
long as they did not reveal any drawback not
seen by now.

Thus it is necessary to get more and more experience with these new tools. One way to acquire
this experience is to use these methods and
equipment extensively in the shop testing in
parallel with the standard means, and to increase the number of destructive testing leading
to size flaw measurements.

Such expensive destructive testings are the
unique way to make valuable correlation with
actual defects. Several programs are now going
on in France and in other countries and it is
hoped that international cooperation may reduce
the time required to implement these powerful
new methods.

LIST OF REFERENCES

[1] SAGLIO (R) - PROT (AC)
"Improvements in ultrasonic testing methods
of welds, specially in the presence of
austenitic stainless steel cladding"
- Second International Conference on
pressure vessel technology
SAN ANTONIO - 30th september 1973

[2] "Routine checkout"
- Conference on periodic inspection of
pressurized components
LONDON 4-6 june 1974
8th conference publication 1974 - C116/74
Institut

[3] SAGLIO (R) - TOUFFAIT (AM) - PROT (AC)
"Détermination de caractéristiques des
défauts de soudure à l'aide de traducteurs
focalisés"
COPENHAGUE 4-9 juillet 1977

[4] SAGLIO (R) - TOUFFAIT (AM) - PROT (AC)
"Evaluation of defects by means of
focussed probes"
Materials Evaluation
January 1978

[5] SAGLIO (R)
"How to improve the probability to detect
large badly oriented defect ?"
NDT international
September 1976

[6] SAGLIO (R) - DE VADDER (D)
"Determination of orientation and size
of badly oriented defects by means of
focussed probes"
first international symposium on ultrasonic materials characterisation
GAITHERSBURG 7-9 june 1978

[7] SAGLIO (R) - PIGEON (M) - KITSON (B)
"Multiple frequencies Eddy current
apparatus"
Conference on NDT in the Nuclear Industry
SALT LAKE CITY - February 1978

[8] SAGLIO (R) - ROULE (M) - DESTRIBATS (MT)
TOUFFAIT (AM)
"Utilisation des transducteurs focalisés
pour la représentation et le dimensionne-
ment des défauts"
Colloque Ultrasonic Characterization of
Materials
GAITHERSBURG 7-9 juin 1978

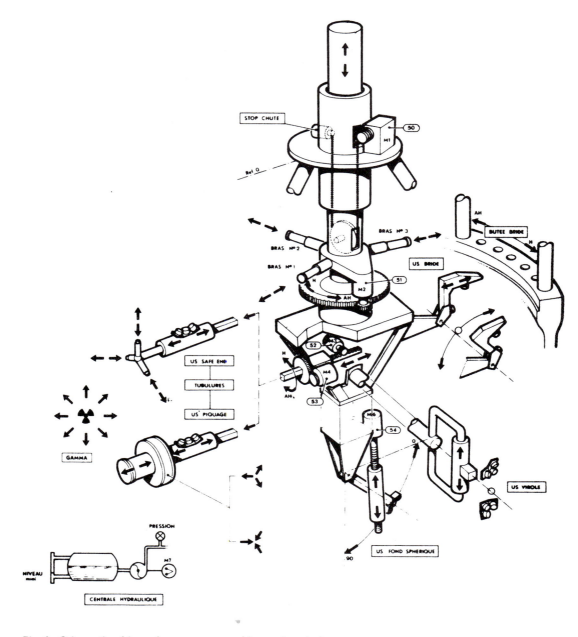

Fig. 1 Schematic of inservice pressure vessel inspection device

Fig. 2 Inservice inspection tool

Fig. 3 Control panel of MIS

Fig. 4   Gammagraphic device for safe-ends testing

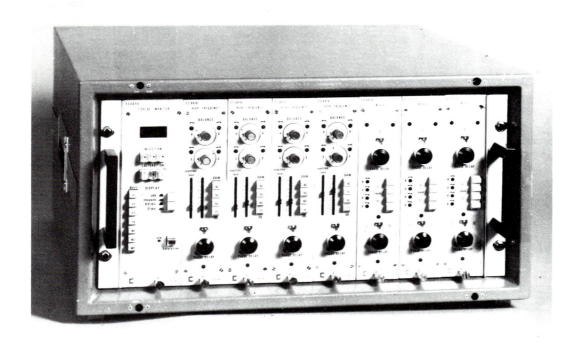

Fig. 5   Multifrequencies IC3FA Eddy currents apparatus

Fig. 6 Eddy current push pull device

Fig. 7 Finger walker device

Fig. 8   Nozzle to pipe transition weld testing device

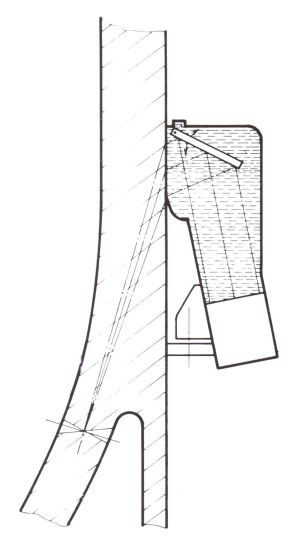

Fig. 9   Sketch of the pressurizer lower head weld

Fig. 10   Pressurizer lower head weld testing device

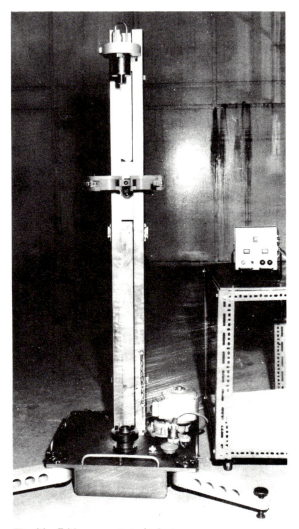

Fig. 11   Eddy current studs devices

C30/79

EXPERIENCE FROM THE IN-SERVICE INSPECTIONS OF THE PWR STEAM GENERATORS

D. DOBBENI and C. van MELSEN
Laborelec, Belgium

The MS of this paper was received at the Institution on 5 December 1978 and accepted for publication on 18 January 1979

SYNOPSIS

In the pressurized water reactors (P.W.R.), the steam generators must simultaneously allow a good heat transfer between primary and secondary coolants and prevent the radioactive primary water from polluting the secondary circuit ; as the wall thickness of the tubes is less than 1.5 mm their integrity is of prime importance for the safety of the power plant. Eddy Current testing is the most frequently applied method for tube inspection.
The need for in-service inspection of the PWR steam generators has been recognized and the Belgian licensing authority has made periodic inspection mandatory.
An automated Eddy Current system has been developped by LABORELEC and is currently being used in the field for the in-service examination of the Belgian PWR.
The experience to date has demonstrated the practicality of utilizing the multifrequency technique under field conditions and the ability of the data processing and analysis system to obtain the necessary information required by Section XI of the ASME Boiler and Pressure Vessel Code.

The detailed preparation of the examination is apt to cut down actual inspection time and, in many cases, enables substantial cost to be saved.

The fundamentals of multifrequency detection are briefly reviewed and illustrated with a simple example which is extraction of the E.C. signal of the tube integrity under tube sheet and tube support plates.

A short account of the evolution on methodology and equipments being used in Belgium is given along with organizational details such as planning, personnel, training, etc... Future trends are discussed in the conclusion. A description of the most recent Eddy Current instrumentation is found in appendix.

1. INTRODUCTION.

In a PWR reactor, the primary and secondary coolant systems are separated at the level of steam generators. The least failure leads to reactor shutdown and contributes to reactor downtime.
The importance of water chemistry in the secondary circuit was soon studied.

Two methods were initially implemented "sodium-phosphate treatment" and "all volatile treatment (AVT)".
Since late 1974, all the power plants have adopted AVT, and no widespread tube failures impair this choice. However, the corrosion phenomena in S.G. are so complex that even a perfect AVT does not prevent tube failures.

Remote and automatic testing becomes more valuable as radiation exposure increases due to :
1° higher radiation fields after longer cumulative plant operating time
2° more thorough and frequent inspections.

For these reasons, multifrequency Eddy Current testing must be regarded as the best method of inspection : one single movement of the probe in the tube provides all the necessary data for further signal treatment. The speed of the inspection is increased and thereby reduces the SG unavailability.
By developing and/or improving remote control methods, the operators are exposed to less irradiation. The organization of the EC inspection is of prime importance since the quality of materials preparation and operators training improve the quality of tube failure detection to some extent and reduces the inspection time. These two parameters affect the general operating cost of the SG during the power plant lifetime.

2. FUNDAMENTAL OF EDDY CURRENT MULTIFREQUENCY DETECTION

2.1 Generality

Electromagnetic induction is the basic principle of the E.C. test. Varying electromagnetic fields produces Eddy Current flow in metal test object. This current flow sets up a secondary electromagnetic field which induces electrical signals within the test coil. The test object electrical characteristics are represented by these electrical signals. Both electrical and magnetic characteristics of the test object are of importance. Further, E.C. density in the test object is maximum at the surface nearest to the coil and rapidly decreases as the depht in the test object increases. This phenomenon is known as the skin effect. The secondary magnetic field can be measured by the test coil by means of an impedance bridge.

The Eddy Current signal then appears as a change in the image impedance of the coil system. This induces authors to study impedance variations instead of current variations. The image coil impedance is a function of :
- equipment characteristics :
  . frequency of the primary electromagnetic field
  . probe design
  . fill-factor of the probe (or distance between the coils and the test object)
- test object characteristics :
  . electrical and magnetic properties
  . dimensions
  . temperature
  . cracks, ...
E.C. measuring units are sensing small impedance changes in the coils. The E.C. analysis unit then deals with identifying the essential parameters (imaging the test object integrity) from the complex experimental patterns.

2.2 E.C. measuring units

Owing to the design of U-tube bundle of the steam generator the measurements must be performed with an internal probe. The instrumentation needs six basic functions : excitation, modulation, signal filtering, signal analysis and signal display. Further, several test coil configurations are possible : one single coil (absolute), two differential coils, with or without shielding,... To improve the response of E.C. signals, the differential coil is preferred. It discriminates against slowly varying conditions and short local variations (as cracks). The coils are designed so that a bridge null signal is obtained when each individual coil meet the same test object conditions.
Such a configuration however is inadequate to measure the thickness of deposits affecting the tube sheet on the secondary side as well as possibly slowly varying defects. A special bridge setup allows this. Even if this E.C. signal does not provide true one-coil (absolute mode) detection, it gives very similar results.

Each frequency gives particular patterns with their own phase and amplitude characteristics. This feature is discussed for its application in the next paragraph. The wide sensitivity of the electromagnetic non destructive test leads variables of little or no interest to produce signals which could mask others coming from injuries in the tube integrity. The parameter separating capability is improved by the increase of the test frequencies number. The first approach of a multifrequency system was performed by successively testing the same tubes with different single frequencies.

Later, two probes were run at the same time using each an individual E.C. measuring device. All these methods did not provide the separating capability because there was no simultaneity of the E.C. signals at each frequency. The multifrequency E.C. equipment is presently the only feasible technical solution. The excitation of the probe by several test frequencies injected simultaneously or by multiplexing ensures this separating capability. It should be pointed out that the simultaneous injection allows a faster probe motion, which considerably shortens the inspection time. This technique was of prime importance for the in-service inspection of the steam generator, even if more complex equipments and procedures are presently required.

2.3 E.C.T. analysis system

The analysis must obtain the maximum information about the state of the steam generators from the experimental results gathered at each inspection and further follow their evolution.
In this respect various experimental parameters must be considered :
- diameter and wall thickness of the tubes
- electrical conductivity and magnetic permeability of the material
- frequencies used for the test
- fill factor
- characteristics of the probes (type of winding, length and diameter of the coils,...)

A nonexhaustive list of the numerous signals resulting from a SG inspection are listed hereafter. Some signals are labeled normal (N) and are due to a normal SG design. Other are benign ones (ND), i.e. no cause of serious concern. The dangerous defects (D) have lead to tube failures in some plants. Finally, the last type of defects (?) are likely to cause leaks but have so far not actually caused such a failure.

a) Manufacturing defects of tubes
- ND : various types of metallurgical noises (due to the pilgrim step rolling mill)
- ND : resumption during rolling (too short rolling bank)
- ND : magnetic oxides on the internal surface of tubes (resulting from annealing in badly controlled surrounding)
- ? : magnetic inclusions (precipitates or tool splinters in tubes)
- ? : rolling defects (clefts, flaws,...)

b) Manufacturing signals or defects of the SG
- N : magnetic signals originated from the tube sheet, the tube support plates and the antivibration bars
- N : bending limits
- N : rolled expanded zones
- N : clad on the primary side of the tube-sheet
- ? : mounting shocks
- ND : too short rolled expanded zones
- ? : discontinuities in rolled expanded zones (no overlapping in the successive operations)
- ? : wrinkles in the tube just above the tube sheet

- D : shocks due to foreign objects in the secondary system
- D : cracks at the limit of the rolled expanded zone
- D : denting at the limit of the tube sheet and the tube support plates
- D : tube crackings due to local pollution
- D : longitudinal crackings in the straight parts
- D : vibrations of tubes
- D : generalized thinning (use of phosphates)
- D : attrition on the antivibration bars
- D : fretting corrosion near the anti-vibration bars
- D : cracking and rupture of tube support plates
- D : cracks in the annular gap
- D : corrosion of the tube sheet and of the tube support plates
- D : external or internal pitting (with or without bottle shape)
- D : internal cracks at the limits of bending
- D : other internal cracks

Again this list is not exhaustive and reflects the defects observed so far.

The signals recorded during the examination of a SG tube are complex, due to the combination of several factors ; e.g. for the SG currently in operation in Belgium, the end of the rolled expanded zone of the tube is located in the tube sheet which is made with carbon steel and produces a large E.C. signal (magnetic influence) ; cracks can appear at this end within tubes showing important metallurgical noise. The resulting signal is then a summation of four different factors.

If the examination is performed with a single frequency system, the analysis of complex signals requires highly specialized people. This increases the risk of human error and further reduces the reliability of the interpretation. Other requirements are :
- the training of skilled analysts which is long and costly
- good analysts must work in shifts during each inspection to reduce reactor downtime
- no automation can help the analyst efficiently
- random signals originated both by the metallurgical noises and by the wobble of the probes rule out any attempt to study, for the purpose of further recognition, all the EC complex patterns.
  Examinations performed with a multifrequency system offer the following advantages :
- the risk of error is reduced if a skilled analyst can compare the signals obtained for the same defects at different frequencies
- the work of the analyst can now be simplified by using Dr Libby's (1) methods as illustrated hereafter ; this is particularly true for the problem of random noises
- this will allow, in the next future, to reduce notably by automation the human intervention without using the pattern recognition technique which is costly and time consuming.

One of the possible applications of Dr. Libby's principles is the mixing process. Consider e.g.

and b). After mixing (rotation of the patterns, modification of their shape and substraction), the ratio of the signal of the hole with the signal of the TSP (fig.1c) is increased by a factor 15 (23 dB). Fig.1 d to 1 e show the same operation for a TSP without superimposed defect obtained during an in service inspection (ISI). Using the same technique, the random metallurgical noises are at the present reduced by a factor 13 (22 dB).
It must be pointed out that this separating efficiency is reached during the ISI. This facilitates the interpretation of the results as illustraded by the purity of the signal shown in fig.1f which corresponds to an external flaw of 80 % penetration depth observed in a SG tube after three years of operation.

Furthermore, this signal treatment realized by our method does not result in excessive loss of sensitivity : our present detection limits after treatment during ISI work for flaws in straigth parts are the followings :
- smallest external artificial hole analysed :
  2,5 mm diameter   8 % penetration depht
  1,7 mm   "   11 %   "   "
  1   mm   "   25 %   "   "
  0,5 mm   "   60 %   "   "
- smallest external real defect analysed :
  < 5 % penetration depht
For example, fig.1 g shows a 25 % deep external hole of 2,5 mm diameter.

3. EQUIPMENTS AND METHODS USED IN BELGIUM FROM THE BEGINNING TILL TODAY

3.1 Starting with the experience gained in condensor examination with an E.C. single frequency (single track) system the utilities requested to perform the E.C. inspection of the PWR steam generators (SG).
The commissioning inspection of the first three industrial nuclear power plants built in Belgium were carried out with either single frequency-single track, or single frequency - two tracks E.C. equipments.
The concern for SG integrity of the other power plants showed that a more efficient detection system should be used, so that, for every new defect of whatsoever character, the evolution of the SG reliability could be studied.

3.2 The next three paragraphs introduce the equipments used for the first ISI (in service inspection) with the multifrequency system, the problems encountered with their present solutions and a description of the process of the Eddy Current system (methods and equipments). Note that all the improvements were introduced between each inspection and immediately tested in irradiated steam generators. This research and development work had four objectives :
. reduce the irradiation dose of the operators
. shorten the downtime of the plant
. improve the quality and reproducibility of the E.C. signals
. lower the cost of the inspections

3.3 E.C. system used for the first ISI.
For the understanding, the system is divided in two parts :

. The mechanical units : (see figure 2)
  - pusher-puller, remote manipulator and probes were set up inside the water box and in front of the man hole.
  - these items are contaminated.
. The electronic units : (see figure 3)
  - multifrequency E.C., pusher-puller control board, strip chart and magnetic recorder and manipulator control board were installed in a low irradiation field zone (distance: 10 meters of the SG maximum).
  - these items are not contaminated.

The sequence of operation of these units was relatively intricate :
- the probes were calibrated before the inspection by using a ASME type standard tube. These calibrations included each frequency channel and the mixing-modules. The mixing-modules were necessarily incorporated in the E.C. measuring equipment because of the technical limitations (to poor quality of the signals after reproduction).
- the probe was pushed through the complete U-bend tube with compressed air
- the operator counted the tube plates and stopped the pusher when the probe reached the end of the tube sheet on the other side of the water box (an attempt at automatic detection was not reliable enough)
- the same switch started the chart and magnetic recorders and the pulling back of the probe with or without D.C. magnetization
- counting the tube support plates, the operator stopped the puller when the probe reached the guide tube between the manipulator and the pusher-puller
- later, the chart was analysed and the special signals were identified when possible on a memory oscilloscope
- due to the operators' fatigue, inspection slowly progressed. Errors appeared in the settings of the E.C. measuring equipment and the number of probe failures was very high, giving rise to problems described hereafter.

3.4 Problems_encountered_with_the_previous_method.
Main difficulties are of four types :
- mechanical :
  . the absence of automatic control of the pusher-puller produces many probe failures. These failures slow down the rate of tube inspection and increase irradiation doses and inspection cost
  . the time needed for every probe replacement or repair is too long, because of the location of the pusher-puller in front of the main hole where, as a consequence of the security procedures, a work of a few minutes amount to at least one hour
  . the manipulator shows an insufficient resistance to the adverse environment of the water box (temperature, moisture, fall of water droplets, boric acid,...) and also to the shocks during its set up.
  . the use of compressed air to assist the pushing of the probe appears increasingly necessary at each inspection to avoid rapid damages of the probes.
- electronic :
  . the signal to noise ratio of the multifrequency unit and the bridge-balance stability are insufficient
  . the low signal to noise ratio of the magnetic recorder affects the pattern of the E.C.

signals, decreasing its reproducibility
  . the mixing outputs are affected by the noise and the bridge drift. But, it soon appears that a single definitive mixing is unsuccessful during ISI. The standard tube is never representative enough of all the tubes of the SG even if they satisfy the manufacturing specifications and the corrosion of the TSP affects too much the E.C. patterns
- auxiliary :
  . the absence of audio-visual aids increases uselessly the irradiation doses because any crankiness of the manipulator or of the pusher-puller could only be detected by direct observation.
- method :
  . as previously mentioned, the advantage of the signal treatment is not useful for many inspected tubes. The impossibility to correct any setting error requires the re-inspection of many tubes.

3.5 Main_present_solutions.
The main present solutions are listed below and have all been tested in several inspections :
- mechanical :
  . an automatic pusher-puller control board was soon introduced with the following features:
  . speed checking, automatic control during pushing and pulling with detection of end of course ; in the same time, an air-tight tube guide was developped to allow the use of compressed air. These improvements diminish the probe failures by a factor of 4 and consequently lower irradiation doses and inspection cost
  . the location of the pusher-puller is shifted at another level (distant of 4 meters of the man hole) suppressing the time needed for the special security procedures. This has been possible by the introduction of the air-tight fixture and the improvement of the E.C. measuring equipment
  . the maintenance of the manipulator is improved, and each fundamental part is replaced at each new inspection. The time allowed for the training is lengthened by a factor of three. These simple modifications suppressed the troubles and reduced the irradiation doses by a factor of five.
- electronic :
  . the electronic of the E.C. equipment is improved to an actual signal to noise ratio of 72 dB. The distance between the pusher-puller and the E.C. system is brought from a maximum of 10 meters to 18 meters ; the bridge-drift is cancelled by 95 %
  . the tape recorder is changed from a frequency modulation design to a modified digital encoding system with a signal to noise ratio of 72 dB.
- auxiliary :
  . one camera is set up inside the steam generator with remote movement and zoom capabilities
  . one camera is fixed in front of the pusher-puller with remote movement capability
  . lighting supplied by a set of batteries secures the operations inside the water box during a power shortage (as happened during an inspection in February 1977, with an operator inside the water box)
  . audio communication is used between the con-

trol units, the pusher-puller unit and the SG

- method :
  - owing to the signal to noise ratio of the E.C. measuring device and tape recorder, the data collection and the data analysis are split in two units. The data collection unit records the demodulated pure E.C. signals differential and absolute. The data analysis unit performs the mixing calibrations for each necessary signal treatment, for the chart recorder and the E.C. pattern identification by the operator.

3.6 Present E.C. multifrequency system – Data acquisition and data analysis units.

Each equipment is described in appendix 1.
The system is divided in two parts.

- Data acquisition : (located inside the containment) (see figure 4).
  - the probes are calibrated before inspection for each pure frequency only (about five minutes for one probe)
  - when the manipulator is positioned in front of a tube to inspect the operator pushes one single button which starts the following process : (the manipulator is inhibited)
    - the pusher assisted by compressed air introduces the probe inside the tube. A speed checking system stops the pusher if the probe is slowered for any reason
    - the control board detects either the end of the tube or a pre-programmed distance, stops the pusher, switches on the tape recorder, starts the puller and digitally records the tube number, the date, the hour and the position of the probe inside the tube at each moment
    - between the tube sheet and the tube guide, standard defects are introduced. Their signals are recorded at each inspected tube giving each record a reference calibration
    - when the probe reaches the guide tube, the puller and the recorder stop. A signal is then sent to the manipulator to allow a change in position.
  - the speed of inspection is 50 cm/s.
- Data analysis : (located in the administrative building) (see figure 5).
  - every four hours or less, the analyst receives the last recorded tape
  - each demodulated pure E.C. signals is calibrated with the reference signals and possible settings error are corrected
  - the procedure for the signal treatment indicates each specific mixing calibration. Using the recirculating memory (which freezes up the signals resulting of the inspection of up to four meters of a tube – thus simulating a continuously measuring of this part of the tube) these calibrations are performed in about two minutes. The recirculating memory is also used to calibrate each track of the chart recorder (about 2 minutes); then the tubes are chart recorded at four times the speed of inspection. Each record is labeled with the tube number. If the reference signals exhibit some setting errors, these are corrected before continuing the printing process
  - the charts are analysed and the suspicious E.C. patterns are identified using the recirculating memory on an oscilloscope. Due to the speed of the signal treatment, some

special mixing are available to identify complex patterns
  - the recorded date and time allow automatic search of a tube inside a tape at high speed
  - the time between the completion of recording and data analysis is less than two hours. This feature allows to mark a tube to be plugged without introduction of a man in a waterbox using the air-tight fixture of the manipulator. This is presently done by projecting through the tube a dart attached to a nylon line. Tubes are then marked by paint from outside of the waterbox.

The above advantages require an efficient integration of the tasks of data acquisition and data analysis.

4. SOME ASPECTS OF THE ORGANIZATION

4.1 Introduction.

The utility bases the S.G. inspection sheduling on the previous results, the safety guides and the possible defects of the SG observed in other operating power plants. The SG examination delays core loading and thus must be shortened to a minimum.
On the other hand motivation for plugging defect tubes lies in the possibility of maintaining steam capacity and minimizing repairs.
To shorten the SG inspection in the future, a maximum amount of datas is collected to further improve the method.

4.2 Typical organization of an E.C. inspection.

The tasks in the E.C. inspection procedure are :
1. preparing the mechanical and electronic units and training the EC operators
2. training the installation crew in a mock-up
3. installing the data acquisition units in the containment (except the manipulator)
4. installing the data analysis unit in the administrative building
5. installing the manipulator and starting the data collection
6. analysing the E.C. tapes
7. marking defective tubes for plugging for each manipulator position
8. SG returns to service

The installation crew is provided by a specialized company providing workers with the necessary medical surveillance

Task 1 is carried out at the power plant for the contaminated units (pusher-puller, SG camera, manipulator). The calibration of the probes and the checking of the electronic units are carried out at the laboratory. Each E.C. operator is trained at the laboratory using a simulation of the in-service inspection conditions. This taks is performed one week before the inspection.

The training of the installation crew, task 2, starts with the preparation of the mock-up. A video-film is presented to the workers before practical training. This training goes on for three days and is managed by an experienced operator.

The staff is selected and the installation of the data acquisition begins as soon as the con-

77

tainment is accessible. This task 3 is performed in two hours.

The task 4, installation of the data analysis unit in the administrative building is carried out at the same time and needs half an our (including final checking).

The start of inspection depends on the accessibility of the water box. As soon as the authorization is granted by the utility, the cover over the primary connections (hot and cold legs), the SG camera and the manipulator are set up in the water box (task 5). This operation takes between one and two hours, depending on the safety procedures for the SG. To speed up the installation of the manipulator, one tube is marked from outside with paint and checked by the camera. These operations are performed by two workers (one of them going into the SG). Afterwards, the final checking of the position of the manipulator is video-tape recorded for further examination if necessary. During the inspection, each tube is investigated following the process previously described ; three shifts of eight hours ensure a non stop inspection.

A recorded tape is transmitted to the analysis units (task 6) every fourth hour. When all the tubes within the manipulator reach have been inspected, the data acquisition unit waits for the interpretation. During this period, some tubes are inspected twice following the analyst's advise and the tube marking equipment is prepared.

Task 7 is carried out if necessary ; the manipulator is set up in another position or another SG.

At the end of the inspection period, the utility gives its authorization to leave the SG for further service. This last task (8) takes two hours for the equipment inside the containment and half an hour for the analysis units. Table 1 summarizes these tasks link up inside the inspection period.

4.4 The normal inspection period is two to five days depending on the number of SG to inspect and the eventual special measurements. The E.C. crew is composed of :
- 1 person responsible of the inspection
- per shift 1 person responsible for the data acquisition for one to three units
- per shift 1 E.C. operator, per unit
- per shift 1 analyst for one to three data acquisition units
- 1 electronic technicien
- 1 mechanical technicien

The SG installation crew is composed of :
- 1 to 3 men per shift, depending on the numbers of SG to inspect.

Table 2 indicates the actual irradiation doses for the inspection of one SG (2880 tubes measured on a partial and 260 on the total lenght) ; the time of inspection was 3 days including tasks 5, 6, 7, 8 with the addition of defective tube marking and special measurements on tubes.

5.   Future trends.
The Eddy Current multifrequency inspection of PWR steam generators is considered the best method for SG inspection, even though this corresponds to a higher equipment complexity and a higher invest-

ment.
This is traded against reducing reactor downtime due to SG examination, which pays off, rapidly.
The next development step is aimed at reducing the human factor in the data analysis. Indeed, some new equipments as the recirculating memory have shortened the calibration time by a factor of 20. The currently used analysis unit demonstrates :
1°) the possibility to separate the data acquisition from the data analyse.
2°) that analysing multifrequency signals can be as fast as single frequency signals.
3°) the information obtained with the multifrequency approach requires one single probe passage.
Future works should focus on developing probes better fitted to investigate defects in SG. A more rugged mechanical design would also permit a measurement speed to match that of the electronics.
Finally, the design of SG manipulators (XY fixtures, polar fixtures, finger walkers) has to be pursued. During our inspections, each type of manipulator has been tested in irradiated SG. Each system has its advantages and drawbacks, but the stress is on with more reliable systems which also means the simplest design with fewer components.
The Belgian utilities favour such a research and development policy since it has already proved advantageous in the existing power plants.

APPENDIX 1

Description of the equipments composing the multifrequency Eddy Current system for SG inspection

1.   Data acquisition unit (figure 5)

1.1 E.C. multifrequency unit : This equipment has already been described. It should be pointed out that only the pure demodulated frequencies (3 differential modules : 100 kHz, 240 kHz, 500 kHz and one absolute module : 100 kHz) are recorded at this data acquisition unit.

1.2 E.C. probes : These probes are differential type coils with a D.C. magnetization coil to suppress the effect of magnetic oxides. Each probe is calibrated in the laboratory using a standard tube. The calibration values are furnished with the probe during the inspection, so that no in-situ calibration is necessary.

1.3 Pusher-puller and associated control board. This unit pushes or pulles the probe through the tube. Each probe is stored on a cartridge to allow a rapid change. The control board has already been described. One should point out some special features : height of the probe inside the tube is continuously indicated in centimeters, indication of broken probe, speed checking security, automatic process with detection of the end of the tube in both directions. A special link avoids any displacement of the manipulator during the push-pull of the probe. The tape recorder and the digital information of the tube number are processed by this unit.

1.4 Tape recorder and associated digital encoding unit.
These two units are with the tape itself, the essential link between the data acquisition and the data analysis. The quality and reproductibility of such a design principle was the funda-

mental technical evolution to allow the separation of acquisition and signal treatment (manual or computerized). The informations recorded at each track are :
- the two channels X, Y of each frequency module (including the absolute), which means a total of eight channels
- tube number and one digit of general information (probe change, new manipulator position, etc...) are introduced on a little calculator-type keyboard or directly from the manipulator control board
- date, hour, minute, seconds continuously recorded are used as a pointer during analysis for high speed searching of the probe position inside the tube during its movement.

1.5 Manipulator and associated control board.
These units depend on the type of manipulator used (XY fixture - polar fixture - finger walker) Each type has been used during these 21 SG inspections and has shown advantages and drawbacks during their use. The essential remark is that the failures number are a direct function of the parts-count of the unit. The time to install the system in the water box has no importance in comparison of the time lost for one single failure. For these units, the only important features are reliability and speed of change from one tube to another.

1.6 Audio visual aids.
These have already been described. The special features to note are that each camera has deplacement, focusing and zoom capabilities. Their introduction in E.C. system has immediatly reduced the time of inspection by 10 to 20 %. Their contribution to the identification of the failures has been determined. Finally, the video tape recorder provides a permanent document of each manipulator position, tube marking and tube plugging operations.

2. Data analysis unit (figure 6).

2.1 Tape recorder and associated digital decoding unit.
Each recorded data is decoded and transmitted either in analog or digital way. The digital outputs are send to the recirculating memory (see 2.2). The analog outputs are sent to the E.C. analyser. These eight analog channels are issued either from the tape either from the recirculating memory. The tube number and other digital informations are used by a microprocessor unit to be printed on the chart recorder or for further digitized treatment.

2.2 Recirculating memory.
This unit has three objectives ; a random access memory stores the eight channels and outputs them to the digital decoder on a continuous way (as if the probe was permanently scanning the same part of the tube). A variable window allows the cancellation of unwanted parts of the record and the selection of the signals to submit on the signal treatment. The second objective is to

record on programmable read only memories the standard or real defects as if they were continuously scanned by the probe. This feature allows a convenient process to identify complex E.C. patterns. Further, the last objective is to link the analyse unit to a computerized process using the standard IEEE 488 communication system for easy interfacing.
The introduction of this equipment has shortened the time of calibration (particularly important for the multifrequency E.C. systems due to the mixing-settings) by a factor of 20 at least. Presently, the calibration for a new probe including special signal treatment and chart recorder setting is five minutes maximum.

2.3 E.C. analysis and display unit
Receiving the eight analog signals, this unit allows the correction of eventual setting errors during the data collection. The standard defects recorded with each tube are used as reference for these calibrations. Further, the mixing modules are used as previously described for the E.C. signal extraction. Two outputs are provided for the strip chart recorder and for an eventual tape recorder. Each pure demodulated frequency signal and mixed output is investigated using a large screen memory oscilloscope.

2.4 Chart recorder and associated tube number printer.
The previously calibrated signals are recorded on chart at four to eight times the speed of inspection. This means that a four hours tape is printed in half an hour to one hour. A needle-printer records the tube identification on the side of the chart during the recording of the analog channels.

2.5 Audio aids
Each tube number is recorded on the audio channel of the tape recorder in case of failure of the digital tube-number equipment.

3. Remarks
Since the separation of the data collection and the data analysis, all the setting errors during the inspection were corrected and there has been no need to mesure a tube twice. Further, particular signal treatments were realized on tapes recorded during previous inspections. This satisfies one of the fundamental objectives of E.C. inspection: the study of the evolution of the SG integrity for every new defect of whatsoever character.

BIBLIOGRAPHY
1. Introduction to electromagnetic non destructive test methods.
   Hugo L. LIBBY - WILEY-INTERSCIENCE, 1971.

2. System summary of a Westinghouse pressurized water reactor nuclear power plant.
   George MASCHE.
   PWR System Division - 1971 - WESTINGHOUSE

Table 1 (✷) Operation in the water box to set up the manipulator.

Task	J-8	J-7	J-6	J-5	J-4	J-3	J-2	J-1	J	J+1	J+2	J+3	J+4	J+5
1		←				→								
2						←			→					
3								←→						
4								←→						
5									(✷)		(✷)		(✷)	(✷)
6									←					→
7											IF needed		IF needed	
8														

(✷) At each change of the manipulator position.

Table 2

	Cumulated irradiation doses
EC crew	1,5 R
SG installation crew	3 R

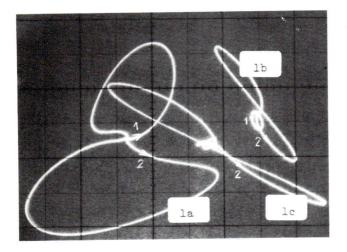
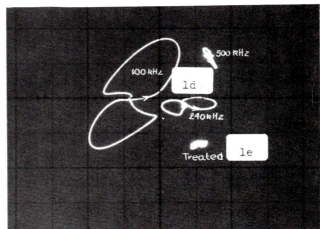
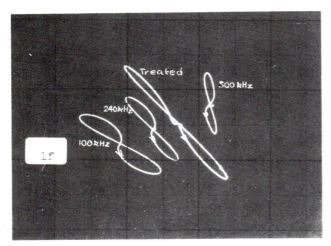
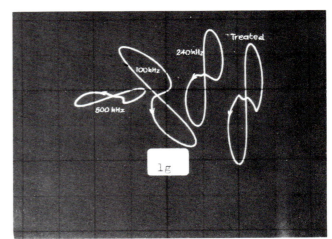

Fig. 1

Fig. 2

Fig. 3

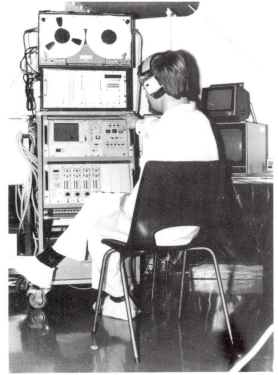

Fig. 4

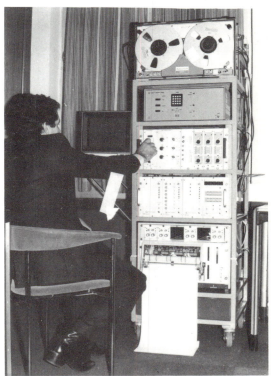

Fig. 5

81

C31/79

DEVELOPMENT OF NDT TECHNIQUES FOR THE INSPECTION OF WSGHWR PRESSURE TUBES

B. S. GRAY, BSc, P. J. HIGHMORE, PhD, J. R. RUDLIN, PhD, and A. G. COOPER
UKAEA, Risley Nuclear Power Development Laboratories, Risley, Warrington, Cheshire

The MS of this paper was received at the Institution on 1 December 1978 and accepted for publication on 2 February 1979

SYNOPSIS   The fuel for the Steam Generating Heavy Water Reactor at Winfrith Heath is contained in vertical Zircaloy pressure tubes and is cooled by boiling light water. This paper describes the development of NDT techniques for the inservice examination of the pressure tubes to provide continuing assurance of the absence of axial crack-like defects.

The resultant equipment has to operate in water-filled tubes in the presence of the radiation field due to the irradiated fuel elements in adjacent tubes. Also, a layer of surface oxide on the inside of the tubes has been found to significantly affect the behaviour of a prototype inspection device. To provide adequate sensitivity in these conditions, without the occurrence of unnecessary spurious indications, a combination of techniques has been developed. This involves the use of ultrasonics in both pulse-echo and 'pitch and catch' mode together with a single frequency eddy current technique.

Laboratory work using artificial defects is described and also how the development programme was modified to accommodate the results of in-reactor tests using a prototype device.

Reference is also made to the development of CCTV equipment to provide a supplementary visual examination.

INTRODUCTION

1.   The fuel for the 100MWE Steam Generating Heavy Water Reactor at Winfrith Heath is contained in 104 vertical Zircaloy pressure tubes and is cooled by boiling light water (ref. 1). The pressure tubes were carefully inspected (ref. 2) when the reactor was constructed but the possibility of generating defects during normal operation has been recognised. This paper describes the development of NDT techniques to provide assurance of the continued absence of axial crack-like defects.

INSPECTION REQUIREMENTS

2.   The pressure tubes are manufactured from Zirconium alloy (mostly Zircaloy 2) with a bore of 130mm, thickness 5.0mm and an installed length of approximately 4600mm. As a result of creep and growth during reactor operation, the inspection technique must be able to cope with small increases in diameter whose magnitude varies along the length of the pressure tube.

3.   A channel has to be inspected after the removal of its irradiated fuel element but without, necessarily, the removal of fuel from adjacent channels. Hence, as reported by Black and Perry (ref. 3) the conditions in pressure tubes during shutdown with which inspection equipment must be compatible are:

- fully flooded with flowing water at 20 to 50°C

- a peak gamma dose rate of about 10^6 rems/h

- a peak neutron dose rate of about 3×10^4 rems/h

- corrosion of the internal and external pressure tube surfaces

4.   The volumetric inspection development has been aimed at the reliable detection of axial defects with a through wall thickness of 0.5mm and a length greater than 25mm. The objective is to provide an inspection technique capable of incorporation in an automated inspection system with the ability to inspect a channel of WSGHWR in about 1½ hours.

5.   In addition a CCTV system is required to permit a visual examination of surface conditions.

CHOICE OF TECHNIQUE

6.   The initial development work to meet this inspection requirement has been described by Cowburn (ref. 4). He reported on the initial selection of a pitch-and-catch ultrasonic technique with pulse-echo ultrasonics being considered as a possible supplementary technique. Both techniques utilised angled shear waves with a frequency of 5MHz and a beam angle of 50° to the normal in the metal. Preliminary reactor trials were carried out in 1976 and revealed the potential significance of a layer of oxide present on the inner surface of pressure tubes. Development of an eddy current technique was started to improve the inspection capability for bore defects. As described in the following, the recommended inspection procedure now involves a concurrent use of all three techniques.

DEVELOPMENT OF VOLUMETRIC INSPECTION TECHNIQUES

Test specimens

7. Since the in-reactor trials have yielded no confirmed defect indications, technique development has been based on laboratory produced defects. To date the latter have included machined slots with a width of about ¼mm and depths in the range 1/3 to 4mm, and fatigue cracks.

8. Fatigue specimens were prepared from pieces of Zircaloy plate 60mm long x 14mm wide x 16mm thick. Crack starter slots 2mm deep were machined across the centre of the 14mm x 60mm face before fatiguing the samples in 3 point bend. After the starter slots had been machined away the back surface was machined to leave a final specimen with a thickness of 5mm. Similar sized calibration specimens were produced containing machined slots with nominal depths of ½, 1 and 1½mms.

9. No samples of reactor corroded pressure tube are available for laboratory use and studies of the effect of oxidation have had to be based on the use of laboratory oxidised material.

10. Thus, following consideration of the 1976 and 1977 in-reactor trials a number of oxidation treatments were used to produce samples whose ultrasonic behaviour was examined. Significant variations in behaviour were noted between specimens and a judgement was made that one particular treatment best simulated the effect of in-reactor conditions. This procedure, which involved autoclaving in steam at 550°C for 19 hours with fast heating and cooling, was subsequently used for all tests involving oxidation effects.

11. Owing to the size limitations of the available autoclaves the oxidised specimens discussed in this paper were in the form of quadrants and, to avoid experimental difficulties due to the distortion which occurs if the oxide does not cover all surfaces, most results were obtained with specimens oxidised on both inner and outer tube surfaces. In general, it is anticipated that this will result in a lower signal to noise ratio than in the reactor where corrosion is expected to be largely confined to the inside of the pressure tube. Thus the laboratory data should be pessimistic in this respect.

12. An autoclave which has the capacity to oxidise complete ring specimens up to 50mm long is now available and this will permit future experimentation with undistorted samples oxidised on the inner surface only. Also, future studies will include oxidised fatigue specimens to supplement the existing results from clean specimens.

Experimental equipment

13. Several pieces of apparatus were used to obtain the results discussed in this paper. On a laboratory scale, apparatus has been used in which the probes (either ultrasonic or eddy current) move over independently supported specimens, and a jig is available to permit stressing of the fatigue crack specimens during tests. For tests in the reactor the probes are on a holder with wheels running on the inner surface of the pressure tube.

14. A prototype device which is largely manually operated has been used for most of the in-reactor measurements and its detailed design, particularly the probe arrangements, has varied from year to year as techniques developed. Figure 1 illustrates the principal features of the inspection head which is normally scanned round the tube at axial intervals of 25mm. A laboratory version of this device has also been used during the work and reference is made later to a few results obtained during the initial commissioning of the semi-automated device described by Black and Perry (ref. 3).

The pitch and catch ultrasonic technique

15. The basic details of this technique in the form developed for WSGHWR are summarised in Fig. 2. Current development has been based on the use of 5mm diameter probes and the development instrumentation has been based on a commercial flaw detector with a modified amplifier to provide an approximately logarithmic output scale. Data display has involved either a high speed recorder or a graphics terminal incorporated in a minicomputer facility.

16. For simple machined slots in clean tubing the technique offers good sensitivity for reliable detection together with a simultaneous sizing capability. Moreover it is a 'fail-safe' technique in that equipment malfunction should modify an existing signal. Laboratory tests have confirmed that the magnitude of the reduction in backwall echo due to a machined slot in a clean tube increases monotonically with slot depth up to at least 80% penetration. The relative sensitivity to large and small defects can be modified by changes to the linearity of the flaw detector response but Fig. 3 illustrates the sensitivity achieved using typical settings of our instrumentation. The difference in sensitivity for internal and external surface breaking defects is clearly shown but the defect types can be distinguished as indicated.

17. The in-reactor trials in 1976, 1977 and 1978 demonstrated two effects of major importance to the use of this technique for routine inspection purposes. Firstly the magnitude of the backwall echo varies along the pressure tube and secondly an oxide nodule on the inner surface can produce a signal similar to that expected from a short surface breaking crack. Also of significance was a general increase in noise level due to the presence of surface oxide and, in 1977 and 1978, evidence that the oxide was causing drag on the inspection device which could result in movement of the probes away from the tube surface.

18. With regard to the 'spurious' indications, Fig. 4 shows three sets of data obtained from one region of a particular pressure tube. Fig. 4 a) and b) compare the results obtained in 1976 with a first set of data for 1977. Allowing for a change in system gain, there is excellent agreement between the results which are equivalent to those which would be produced by a machined slot with a length of 4mm and a depth of about 0.7mm. However Fig. 4 c) demonstrates

that they are not due to a genuine defect by showing the results of a repeat examination after chemical cleaning of the pressure tube. The indication has disappeared and this is believed to result from the crushing of a softened oxide nodule by the wheels of the inspection device. Other supporting evidence for this interpretation is discussed in the sections on other techniques. The number of such indications when using a 25mm scanning raster has been found to vary widely from tube to tube.

19. It has also been observed in the reactor that the signal/noise ratio deteriorates at the level of the fuel element grids and this is thought to be due to variations in oxide thickness and surface roughness.

20. Subsequently, laboratory tests have been carried out on slots machined through corroded tube and on tube samples which were oxidised after introducing the artificial defects. Nominal slot depths of $\frac{1}{2}$, 1 and 1$\frac{1}{2}$mm were used for this work and some of the results are shown in Fig. 5 which is a computer generated diagram from the data obtained with a series of scans at 1mm intervals. Fig. 5 demonstrates that the defects are satisfactorily detected but there is no longer a clear relationship between signal amplitude and defect depth. Further work is required to establish the improvement in performance to be obtained in the laboratory by the use of specimens with an uncorroded external surface.

21. Other laboratory tests have involved fatigue cracks and the detectability has varied. One particular specimen yielded no observable change in backwall echo. In the other specimens comparison of crack profiles deduced ultrasonically, using the calibration data for machined slots, with the depths measured after brittle fracture indicated that the depths were systematically under estimated by between 20 and 50%. Although this is presumably due to the cracks being partially transparent to ultrasound, no effect has been detected when the external closure stress was deliberately varied up to 6MPa.

The pulse echo ultrasonic technique

22. Use of pulse-echo for the detection of surface breaking defects, using the angled transducers fitted for a pitch and catch examination, relies on the corner effect with inner wall defect detection involving a full skip technique as shown in Fig 6.

23. By comparison with pitch and catch, the pulse echo system has been found to be more sensitive for the detection of small defects but the signal amplitude saturates more rapidly as the size increases. Also, the programme of work reported in this paper has indicated a greater reliability for the detection of fatigue cracks, with all cracks in our laboratory specimens being detected. However their depth was again underestimated when comparing signal amplitudes with calibration data from machined slots.

24. In reactor, no response was detected from the oxide nodule referred to in Fig. 4, and elsewhere only minimal reflection at the front face has been detected in a tube yielding numerous spurious indications for pitch and catch examination.

The eddy current technique

25. In view of the potential limitations of the ultrasonic techniques and to provide long-term assurance of the ability to detect significant defects which might be caused by fuel handling, development of a single frequency eddy current technique was begun after the first reactor trials.

26. A commercial instrument incorporating a vector display has been utilised after modification to improve the frequency stability of the oscillator. Probe development included an investigation of the effect of variations in dimensions, number of turns and core material, temperature, lift-off, operating frequency and lead length. A frequency of 500kHz was selected as a suitable compromise between the requirements for good sensitivity and the practical limitations set by the need to operate with 25m of cable connecting the probe(s) to the instrumentation.

27. The probes used are wound in the laboratory and consist of about 50 turns of 48 SWG insulated copper wire about a ferrite core with a complete probe being about 2$\frac{1}{2}$mm in diameter. The probe used for the 1978 reactor trials is illustrated in Fig. 7 which also indicates the design used in 1977.

28. Absolute and differential modes were compared in the 1977 reactor trials but the absolute mode has been used for most of the laboratory investigations with hard copy records being made of both bridge outputs in addition to observing the vector display. In general the bridge has been adjusted so that variations in probe 'lift-off' only affect one of the outputs and the 'non-lift-off' output used for defect detection.

29. Fig. 8 illustrates the sensitivity achieved at 2mm lift-off for clean machined slots indicating a detection threshold at about $\frac{1}{2}$mm.

30. As anticipated the presence of oxide has not directly affected the signal/noise ratio or sensitivity of the technique. Thus no response was obtained from the region yielding the ultrasonic response illustrated in Fig. 4. An unanticipated achievement of the eddy current technique has been an ability to assist with the characterisation of defects. This has arisen from the spatial variation observed for certain defects and from the fact that any instantaneous ECT observation yields two pieces of data.

31. Fig. 9, for example, compares 'non-lift-off' data for traverses over the oxidised specimens with slots machined before and after oxidation. The results clearly show the enhanced metal loss in the former specimens.

32. Fig. 10 contrasts the vector displays obtained for a clean fatigue crack with clean machined slots. The different character of the defects is again revealed, in this case as a difference in the phase. It is thought that

this arises from the differing separation of the crack faces and theoretical studies are in progress in an attempt to explain the effect.

33. Examination of the laboratory fatigue crack specimens gave successful detection in all cases but, as might be expected from Fig. 10, they cannot be directly sized using calibration data for machined slots. In practice, when using the 'non-lift-off' data a weighting factor of about 1.5 must be applied to the depth derived from the normal calibration curve.

34. In the reactor situation, with the type of inspection device shown in Figs. 1 & 7, the oxide layer has an important indirect effect on the ECT measurement through its effect on the separation between the probe and the metal surface. Such variations in lift-off produce variations in sensitivity and further work is required to allow routine allowance for this effect, perhaps from analysis of the signal including lift-off effects.

35. Further work is also in progress to assess the importance of material variability on the sensitivity of the technique, while further consideration is required on the relative merits of absolute and differential probe operation for different types of possible defect.

Discussion of volumetric inspection

36. When used in combination the three techniques described provide a reliable procedure for detecting significant axial crack-like defects which are orientated normal to the tube surface and may also be used to determine the defect length. The eddy current technique offers good prospects of characterising internal wall defects and providing a useful estimate of their through wall thickness.

37. These conclusions are based on laboratory tests since no defects have been confirmed during reactor proving trials of the techniques. Further work is in progress to extend the laboratory results to inclined slots.

38. An important practical merit of the combined techniques is the ability to eliminate indications due to oxide nodules without the delays necessary for more detailed surveys of a pressure tube. (Scans are normally carried out at intervals of 20 or 25mm). The combination also retains the high sensitivity and hence greater reliability of detection of the ultrasonic techniques for the smallest defects of interest while recognising the current inability of the ultrasonic techniques to characterise and size defects in the presence of oxide. The deficiencies of ultrasonic amplitude measurement for the determination of defect depth are in general agreement with the conclusions of Coote (ref. 5) in connection with his studies for Zirconium-Niobium tubing in CANDU type reactors. The nature of the oxide layer in WSGHWR has prevented the use of the sizing technique described by Coote involving detailed measurement of the spatial variation of pitch and catch signals.

DEVELOPMENT OF CCTV

39. The provision of a means for visual inspection of a component will always be desired by reactor operators even though the presence of an oxide layer may conceal fine cracks. In practice CCTV pictures obtained during the development programme contributed significantly to the understanding of the behaviour of the prototype inspection device. For instance the CCTV pictures clearly show the wheel tracks from the volumetric inspection device over most of the tube and have revealed oxide discontinuities in the areas yielding the worst signal to noise ratios during pitch and catch ultrasonic inspection.

40. The CCTV equipment used for pressure tube inspection has comprised conventional commercial cameras mounted in special housings with the signal recorded on videotape. The major problems have been associated with the provision of appropriate lighting in a relatively confined space and with variable surface conditions.

41. An attempt to solve the lighting problems in 1976 by using a silicon diode detector instead of a standard videcon was nullified by a large increase in electrical noise. Subsequently a rotating lighting system has been demonstrated which gave good illumination over a small area for examination at x10 magnification. Work is in progress to permit a larger area to be examined at lower magnification with a view to reducing the time required to look at a whole tube.

DISCUSSION

42. Techniques have been developed in the laboratory and tried out in WSGHWR which appear capable of meeting the requirement for the reliable detection of any axial crack like defects which exceed ½mm in depth and are more than 25mm long.

43. Since the laboratory work has utilised test specimens which were oxidised on both surfaces, it is anticipated that the actual sensitivity of the ultrasonic technique in the reactor will be better than that estimated from the laboratory work. While it is conceivable that growth of the oxide layer could reduce the sensitivity of the ultrasonic examination techniques, the eddy current technique is expected to provide continuing assurance of the ability to detect possible defects. If necessary, by modifying the coil design and lowering the operating frequency an eddy current technique could probably be developed to detect large defects which did not penetrate the inner surface of the tube.

44. No fundamental problems have been identified in providing CCTV facilities for surface inspection within a pressure tube although further work is required to provide an optimum lighting system for work at low magnifications.

45. The only significant irradiation damage observed during the reactor trials described here has been the browning of conventional glass components in the CCTV system and steps have been taken to replace them with radiation resistant equivalents. So far, in spite of

several days inreactor use, the ultrasonic and eddy current transducers have not revealed evidence of deleterious effects.

CONCLUSIONS

46. Techniques have been developed for volumetric inspection of WSGHWR pressure tubes in order to detect significant axial defects although some further work is required to establish the detailed response of defects with particular orientations and profiles eg. gouges. The concurrent use of ultrasonic and eddy current techniques is recommended to provide a reliable inspection system with some ability to characterise defects in the presence of significant surface oxide. The presence of the surface oxide layer means that the detailed engineering design of the probe holder and location system is important for the performance of the final inspection system.

47. The radiation environment does not lead to major problems in providing a CCTV camera for use in a pressure tube, but space limitations have led to difficulties in providing for adequately uniform illumination for examination at low magnification.

ACKNOWLEDGEMENTS

48. The authors would like to acknowledge the willing assistance of their colleagues at RNL

and AEEW in performing the work described in this paper.

49. The permission of the Managing Director of the UKAEA, Northern Division, to publish this paper is gratefully acknowledged.

REFERENCES

1. CARTWRIGHT H. The design of the Steam Generating Heavy Water Reactor. 1967. Proceedings of IAEA Symposium on Heavy Water Power Reactors (Vienna).

2. HANSTOCK R.F., LUMB R.F. and WALKER D.C.B. Ultrasonic inspection of tubes. Ultrasonics 1964, Volume 2 pp109-119.

3. BLACK W.S.A. and PERRY A. WSGHWR pressure tube inspection equipment. Presented to IEE, colloquium on 'Advances in remote inspection techniques'. 25 May 1978, London.

4. COWBURN K.J. NDT methods for pressure tube reactor inspection. CSNI Report 14. Proceedings of Specialist Meeting on the Ultrasonic Inspection of Reactor Components. September 1976 (Daresbury), 1978.

5. COOTE R.I. Ultrasonic assessment of crack size in CANDU pressure tubes. CSNI Report 14. Proceedings of Specialist Meeting on the Ultrasonic Inspection of Reactor Components - September 1976 (Daresbury), 1978.

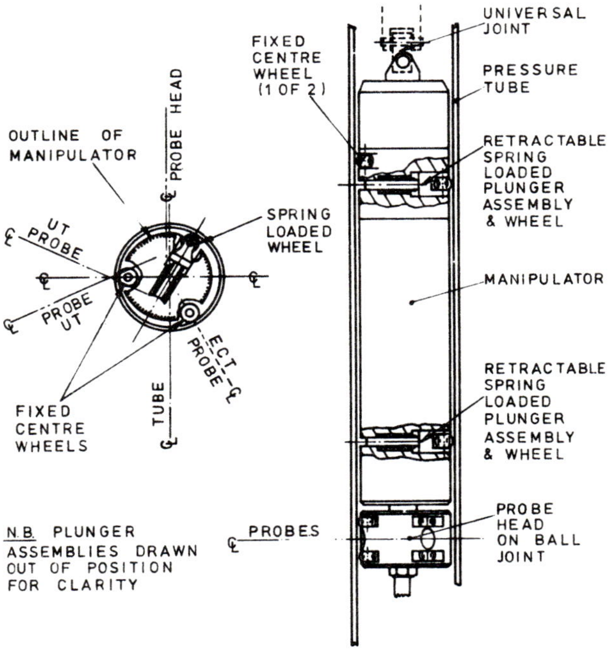

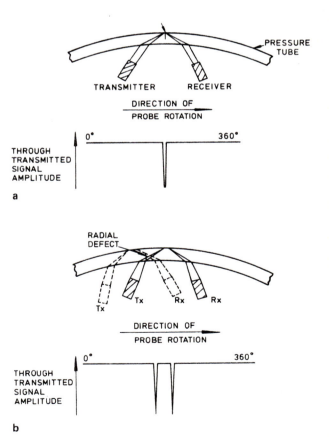

Fig. 1 Diagram of prototype in-reactor device

Fig. 2 Recording indications obtained by the ultrasonic pitch-and-catch method

a External radial defect
b Internal radial defect

© I Mech E 1979

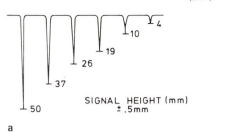

2.45 1.95 1.45 1.1 0.75 0.33 DEFECT DEPTH
(mm)

10
4
19
26
37
50

SIGNAL HEIGHT (mm)
± .5mm

a

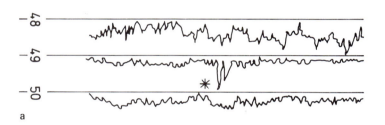

2.4 1.9 1.5 1.08 0.73 0.3 DEFECT DEPTH
(mm)

11
8
5.5 2.5
15
19

SIGNAL HEIGHT (mm)

b

Fig. 3 Typical pitch-and-catch recordings

a External defects (machined slot)
b Internal defects (machined slot— calibration
date 7 April 1977)

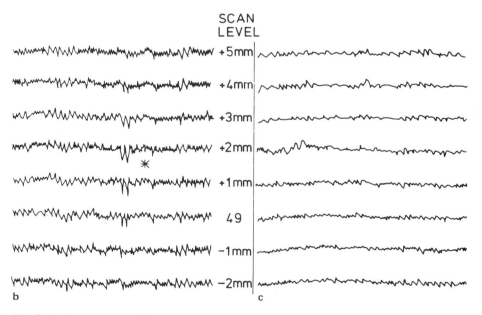

48

49

50

a

SCAN
LEVEL

+5mm

+4mm

+3mm

+2mm

+1mm

49

−1mm

−2mm

b   c

✳ APPARENT

DEFECT INDICATION

Fig. 4 Indications recorded in pitch-and-catch around an oxide nodule

a Spring 1976
b 1977 (pre-clean inspection
c 1977 (post-clean inspection)

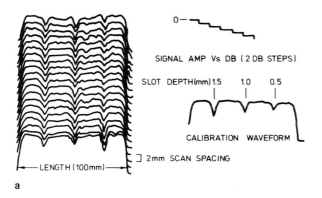

SIGNAL AMP Vs DB (2 DB STEPS)

SLOT DEPTH(mm)1.5   1.0   0.5

CALIBRATION WAVEFORM

2mm SCAN SPACING

LENGTH (100mm)

a

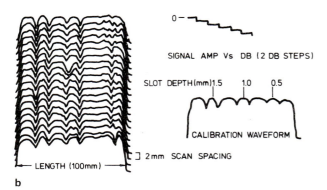

SIGNAL AMP Vs DB (2 DB STEPS)

SLOT DEPTH(mm)1.5   1.0   0.5

CALIBRATION WAVEFORM

2mm SCAN SPACING

LENGTH (100mm)

b

Fig. 5   Laboratory pitch-and-catch data for oxidized tubing
   a Slots machined through oxidation (outside surface)
   b Machined slots under oxidation (inside surface)

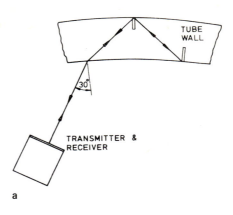

a

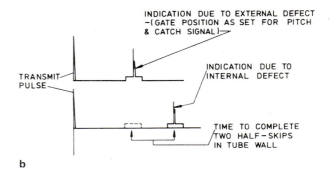

b

Fig. 6   Diagrams illustrating pulse-echo method in angled shear
   a Ultrasonic beam path
   b Flaw detector (A scan) display

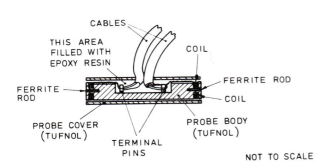

a

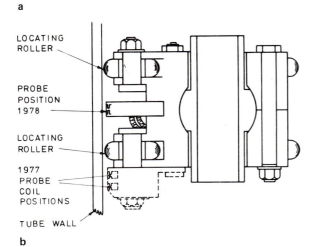

b

Fig. 7   In-reactor eddy current probe
   a Probe design — 1978
   b Detail of probe mounting arrangements

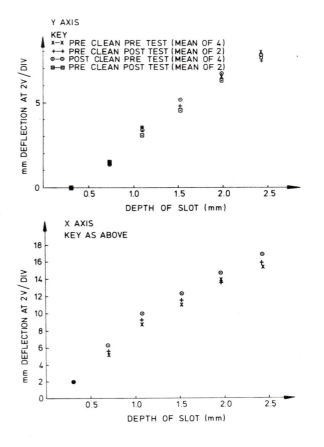

Fig. 8   Calibrations of ECT system for reactor trals, 1977

89

TYPICAL SIGNALS FROM SLOTS CUT THROUGH CORROSION

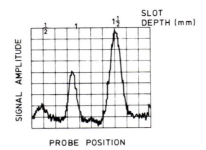

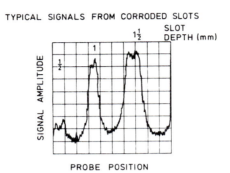

Fig. 9   Eddy current inspection of corroded pressure tube

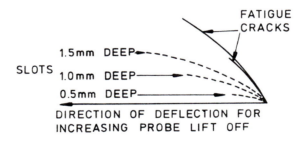

Fig. 10   Vector display of eddy current signals obtained from
some machined slots and fatigue cracks

C32/79

THE DEVELOPMENT OF A REMOTE GAUGING AND INSPECTION CAPABILITY FOR FUEL CHANNELS IN CANDU REACTORS

M. P. DOLBEY, PhD and O. A. KUPCIS, PhD
Research Division, Metallurgy Section, Ontario Hydro, Toronto, Canada

The MS of this paper was received at the Institution on 23 August 1978 and accepted for publication on 18 September 1978

SYNOPSIS   Equipment under development for the inspection and gauging of pressure tubes in CANDU (Canadian Deuterium Uranium) type reactors is described. A brief overview of the mechanical scanning system is presented followed by a detailed description of the measurement and data processing systems for the gauging of diameter and wall thickness, volumetric inspection of the tube wall and gauging of the annular gap between the pressure tube and the calandria tube. Experience of testing ultrasonic transducers in very high (10^6 Roentgens/hour)(R/h) radiation fields is reviewed.

INTRODUCTION

1.   In 1975, Ontario Hydro, Research Division, undertook the development of a new generation of equipment for inspecting and gauging pressure tubes in CANDU type reactors. Pressure tubes, so named because of the high pressure and temperature of the primary heat transport system, are made of a zirconium alloy and have a length of 6.30 metres, a wall thickness of 4.1 millimetres and an inside diameter of 103 millimetres. Each end of the tube is attached to a stainless steel end-fitting that is 2.5 metres long. It locates the tube within the calandria and provides inlet and outlet for the heavy water coolant and for loading fuel. The pressure tube is surrounded by a thin wall zirconium alloy calandria tube. The 9 mm annular gap between the two tubes is maintained at intervals by garter spring spacers. The whole assembly, known as a fuel channel, is illustrated in Figure 1.

2.   With the history of pressure tube inspection at Ontario Hydro's Pickering Nuclear Generating Station (ref. 1) and the increasing demands of in-service gauging and inspection of the pressure tube and rolled joints, a programme was started in 1975 to develop a system which can operate in a channel without dewatering it, be fully remote controlled and require little or no radiation exposure to personnel. The objective is to develop a drive mechanism that can be attached to and operated by the fuelling machine and its control system, which is capable of moving an inspection head through the defuelled channel in a controlled scan pattern while the reactor is shut down.

3.   The development is referred to as the CIGAR (Channel Inspection and Gauging Apparatus for Reactors) project. The prototype unit is being designed to be used at Ontario Hydro's Bruce A nuclear generating station and is expected to be operational in late 1980.

OBJECTIVES OF THE DESIGN

Inspection objectives

4.   The inspection and gauging system must be capable of fulfilling the mandatory gauging requirements established by the Canadian government's regulatory agency, the Atomic Energy Control Board. For Bruce GS A, it is required to measure the diameter of 3 pressure tubes in each reactor every 5 years. This is to assure that the diametral creep is within design limits. There are also requirements to measure parameters that will indicate pressure tube sag and surface roughness.

5.   The system should measure parameters that will assist in predicting the need for pressure tube maintenance or replacement. The measurement of wall thickness will indicate thinning due to diametral creep or corrosion. The measurement of the minimum gap between the pressure tube and the calandria tube indicates the amount of sag or curvature in the pressure tube.

6.   The system should measure parameters that will provide improved information for the design of future fuel channels. Historically, Chalk River Nuclear Laboratories have performed measurements on pressure tubes in their test reactors to provide this information. However, because the neutron flux in power reactors is higher than in the smaller test reactors, radiation-induced changes are accelerated and the effects of large cumulative doses are observed more quickly. It is clearly in our interest to have the best information available for the design of our future reactors.

7.   Finally, the system should make measurements to assure tube integrity from the standpoint of unknown or unexpected generic problems. It is not economically feasible to routinely check all or even a large sample of channels for highly unlikely isolated events that could lead to pressure tube failure. However, a volumetric inspection of a few tubes might show up unforeseen problems common to all tubes in time to take remedial action.

Operational objectives

8.   The prototype equipment is being designed for Bruce GS A, as its fuel handling system is similar to future stations such as Bruce GS B and Darlington GS A. The reactor will be shut down and on maintenance cooling during the inspection.

The channel to be inspected will be defuelled but will remain filled with primary heat transport coolant (heavy water). A full inspection and gauging of the pressure tube (excluding defuelling) will require a maximum of 3 hours and no man-rem will be expended during a normal inspection.

9. The design goal for the minimum operating time between maintenance or replacement of the inspection head measurement transducers is 100 hours. The severe environmental conditions found in the reactor core during inspection make this a challenging goal. A summary of these conditions is presented in Table 1.

Table 1

Environmental conditions in a pressure tube during inspection

Medium	Heavy water	pH 10
Temperature	100°C max (normally 30 to 50°C)	
Pressure	1.5 MPa max ($\simeq$200 psi)	
Radiation	gamma field	10^6 R/hr
	neutron flux	10^9 neutrons/(cm^2.s)

MECHANICAL DESIGN CONCEPT

10. The system being developed for the inspection and gauging of pressure tubes is illustrated schematically in Figure 2. The mechanical equipment required can be considered in two packages; the components that operate in the fuel channel and the external drive mechanism that manipulates the inspection head in the channel.

In channel components

11. After the reactor has been shut down for an inspection, the fuelling machine will remove all the fuel from the selected channel. It will remove the standard closure plug and replace it with the channel entry module, an interlocking set of components approximately 90 cm long that is illustrated in Figure 3. A modified closure plug seals the end of the channel and is penetrated by a combined mechanical-electrical drive rod connector. A sleeve attached to the back of the closure plug supports a carrier that holds the inspection head and is used to transport it through the larger diameter end fitting to the pressure tube. The package is locked together by a number of mechanical locking mechanisms that can only be released by insertion and manipulation of the external drive rod.

12. The inspection head, shown in Figure 4 consists of measurement transducers, centering mechanisms, drive rod load decoupler and support, and the drive rod connector. Ultrasonic techniques are being employed for diameter and wall thickness gauging and for volumetric inspection. An eddy current device is being developed to measure the gap between the pressure tube and the calandria tube. The centering mechanisms maintain the rotational axis of the inspection head coincident with the axis of the tube to within one per cent (1%) as required for accurate measurement of wall diameter. The drive rod load decoupler and support prevent the weight and centering of the drive rod from affecting the inspection head centering. The drive rod connector consists of a

substantial mechanical connection 38 mm diameter that can be remotely locked or unlocked from the drive rod at specific stations. It contains a watertight electrical connection for 12 coaxial cables that are used to connect the transducers on the inspection head to the measurement instrumentation on the fuelling machine.

External inspection drive mechanism

13. The inspection drive mechanism is designed to be attached to a fuelling machine. Power for the inspection head motion is obtained by clutching into various fuelling machine drives which will perform idle motions while the inspection head is operating. The scanning motion is controlled automatically by the fuelling machine computers in the station control room.

14. The drive mechanism will be mounted on the fuelling machine in the central service area and is transported to the reactor vault through a tunnel that runs under the station. When the machine is in the vault, it can be positioned to any fuel channel. After locating a channel that has been defuelled for inspection, the fuelling machine would install the inspection module. After disengaging the channel, the fuelling machine would be moved a precise amount to align the end of the inspection drive mechanism with the center of the channel to be inspected. By a series of axial and rotational motions, the drive rod would lock onto the inspection head, free it from the closure, transport it through the end fitting in the carrier and lock the carrier into place at the end of the endfitting. The inspection head is then free to be scanned through the pressure tube to perform the inspection. The normal scan motion consists of 25 mm axial steps followed by a 400° rotation in alternate directions. A much finer scan spacing can be employed over small areas if detailed inspection is required. The drive is capable of traversing the head through the full length of the pressure tube, a distance of 10 m from the entry to the channel.

MEASUREMENT & DATA GATHERING SYSTEMS

General description

15. The proposed measurement systems would consist of an ultrasonic facility for diameter and wall thickness gauging, an ultrasonic flaw detection system, and an eddy current system for measuring the gap between the pressure tube and the calandria tube.

16. Local monitoring of the ultrasonic and eddy current data will not be practical, as it would require personnel situated close to the reactor face while the apparatus was driven through the fuel channel by the fuelling machine. Instead, the information and control signals will be transmitted to the common equipment room (adjacent to the station control room) on an existing television cable. At this location, a data processing and recording unit will merge the position data from the fuelling machine encoders with the inspection and gauging data, and will store the appropriate information. Some gauging and inspection electronics will be necessary on the fuelling machine head and multiplexing will be employed for transmission of the amplified ultrasonic signals

to the monitoring station.

17. The existing television cable consists of three 75 ohm coaxials and numerous single conductors arranged in seven groups. The cable length is of the order of 350 m. Wave velocity for the cable is about 2×10^8 m/s, which will result in a go-return signal taking almost 2 microseconds. Laboratory experiments show that it is unlikely that it will be possible to interpose the cable directly between a transducer and its pulser-receiver unit. However, transmission of the amplified receiver signal was acceptable, with a transmission loss of 3 to 5 decibels.

The gauging system

18. The use of ultrasonic techniques for both wall thickness and diameter gauging of tubes is well established. The system employs two diametrically opposed transducers at a known gauge length. By measuring the time required for a wave front of ultrasound to reflect back from the inside and outside walls of the tube, and knowing the velocity of sound in the water and the tube, the diameter and wall thickness of the tube can be measured. However, the velocity of sound in a medium varies with temperature. A third transducer measures and compensates for variations in the velocity of sound by reflecting ultrasound off a reflector that is a constant distance from the transducer.

19. The measurement of wall thickness depends on the velocity of ultrasound in the wall material. A question arose as to whether irradiation damage of the pressure tube material would change the material properties sufficiently to cause a change in the velocity of sound. A study was carried out by Atomic Energy of Canada Limited on samples of pressure tube removed from an operating reactor and on unirradiated specimens (ref. 2). The irradiated samples had been in service at full power for approximately 1400 hrs during which they were exposed to an average flux of 2×10^{13} neutrons/(cm^2.s). The study concluded that this amount of irradiation had no measurable effect on the velocity of sound in the Zirconium-2 1/2% Niobium pressure tube material.

20. The specifications for the gauging system are given in Table 2.

Table 2

Gauging system specifications

Inside diameter	Range 102 mm to 107 mm Resolution ± 0.02 mm Accuracy ± 0.05 mm Data Req'd Max, Min and Average diameter to be recorded at 25 mm intervals along tube.
Wall thickness	Range 3.40 mm to 5.60 mm Resolution ± 0.01 mm Accuracy ± 0.03 mm Data Req'd Max, Min and Average wall thickness to be recorded at 25 mm intervals along tube.

21. A block diagram of the proposed gauging measurement system is shown in Figure 5. In this system data processing would be done on the fuelling machine with only digital information

being passed back, on command, to the monitoring station. At the control station, only numerical information would be displayed rather than, for example, A-scan display.

22. The controller would initiate the taking of the measurements dependent on radial position and then expect, in return, 16 bits of information. This would allow 8 bits of information for both diameter and thickness measurements, permitting accuracy to be within 1 part in 256. (The data will be transmitted in the form of a tolerance from a standard dimension.)

23. The gauging system will be pulsed at a repetition rate compatible with the local system. The required dimensions of diameter and thickness will be generated continuously with an 8 pulse running average being maintained. These averages will be continuously compared with maxima and minima in four storage registers. If appropriate, the values stored in the registers will be replaced by current ones. Axial and radial position information will also be stored. The running averages will also be summed during each circumferential scan to produce an overall average diameter.

24. The gauging system would utilize one of the available coaxial conductors and a few of the single conductors.

The volumetric inspection system

25. Volumetric inspection will be accomplished using 45° shear wave pulse-echo ultrasonic techniques. The system will consist of transducers oriented to detect radial-axial indications and radial-circumferential indications. The radial-axial subsystem will consist of two line focus transducers 25 mm long, for detection and two spot focus transducers for flaw sizing. The radial-circumferential subsystem will consist of two groups of transducers, with five transducers per group, for detection and two spot focus transducers for sizing.

26. The mode of operation will be to make an initial pass through the tube employing circumferential scans at 25 mm intervals during which the detection transducers are active and the sizing transducers are dormant. (Diameter and wall thickness measurements will also be made during this pass.) The location of all indications with amplitude greater than an established reference will be noted. A second pass will then be made to perform a detailed inspection of indications detected during the initial pass. During this pass the spot focus devices will be used, while the line focus and group assemblies of transducers will be dormant.

27. The specifications for the volumetric inspection system are given in Table 3.

Table 3

Volumetric inspection system specifications

Volumetric Inspection	Range 100% coverage Resolution: Ability to detect a reflector equivalent to a spark eroded notch 0.075 mm deep x 6.25 mm long. Data Req'd: Location and peak amplitude of all indications greater than reference.

28. A block diagram of the volumetric inspection system is presented in Figure 6. A single coaxial conductor will be used for transmission of the four analog data groups. The single conductors will be used for control and status information.

29. Each transducer, or transducer group, will have its own rudimentary pulser-receiver unit, without displays, gates, alarms etc. Each pulser will be under central control through the pulse sequencer. The control center trigger will produce a 2-bit code which in turn will cause the pulse sequencer to initiate action by one of the four pulsers. The received signal will then be amplified and transmitted to the control center. Spurious reflected signals detected by any of the other receivers will be blocked and not transmitted.

30. At the control center the received analog data will be directed to one of four processor-displays. The processor-displays will be standard ultrasonic instrumentation but will not require a pulser section. It is expected that repetition of the order of 1000 pulse/s can be achieved, if the electronics are the only constraint.

31. Radial and axial position displays will be made at the control center.

32. Automatic signal processing equipment at the control center will digitize the amplitude of any in-gate signal to produce a 4-bit data set (accurate to 1 part in 16). A digital display of position, in-gate signal amplitude, transducer identification and alarm (with count) will be provided for automatic flaw signal assessment.

33. A cathode ray tube (CRT) display will be provided to indicate flaw positions on an isometric display of the pressure tube.

34. The quantity and type of data to be stored permanently during detailed scanning of suspect indications is currently under investigation. Besides the standard methods of analog chart storage and digital storage of signal amplitude and gate position, the possibility of total digitization of the A-scan response is being considered. The amount of data generated by this technique is very large and storage would be feasible for only a limited number of detailed inspections. The advantage of the method is that it provides full information for post inspection analysis. The exact configuration of the data handling and storage system has not yet been finalized.

Pressure tube-calandria tube annulus gap

35. A system is being developed to measure the minimum annulus gap between the pressure tube and the calandria tube using eddy current techniques. Because of gravitational loads on the horizontal tubes it can be assumed that the minimum gap will occur at bottom dead center of the tubes. The measured variable, complex impedance of a coil, is affected by many parameters such as pressure tube wall thickness, resistivity and temperature, the gap between the coil and the pressure tube (liftoff) and the gap between the pressure tube and the calandria tube. Computer programs have been developed to assess the effects of these variables on the change in impedance for different coil configurations and impressed frequencies.

36. In the proposed system the inspection head will carry one coil lightly spring loaded against the inside surface of the pressure tube. After the gauging and volumetric scan is completed, the eddy current probe will be oriented to bottom dead center of the pressure tube and traversed through the tube. Complex impedance measurements (amplitude and phase) for three different impressed frequencies will be made at 100 mm intervals. The six resulting values will be processed by a classifier to estimate into which of three ranges the measured gap falls (see Table 4). A multiple linear regression technique is then employed to estimate the actual minimum annular gap within the accuracies specified in Table 4.

Table 4

Tube annular gap gauging specification

Annular gap   Range: 0 to 15 mm
Accuracy:
170% to 40% nominal gap ± 1.0 mm
40% to 15% nominal gap ± 0.4 mm
15% to 0% nominal gap ± 0.1 mm
Data Req'd: measurements at 100 mm intervals along length of tube.

37. A schematic of the annular gap measurement system is presented in Figure 7. The eddy current coil on the inspection head will be connected to a constant current wave generator and the impedance analyzer mounted on the fuelling machine head by one of the coaxial cables in the drive rod. One of the coaxial cables and a number of single conductors in the television cable will connect this equipment to the control center. At each measurement station in the tube, the controller will sequentially initiate measurements at each test frequency and will expect, in return, 16 bits of information for each measurement. This would allow 8 bits of information for both amplitude and phase measurements permitting accuracy to be within 1 part in 256, which is deemed adequate for the application. The data processing system will accumulate the data, perform the classification and linear regression analysis and store the results with axial tube position information.

38. A prototype system is currently being manufactured and testing will begin shortly.

ULTRASONIC TRANSDUCER PROCUREMENT

38. The difficulty of procuring ultrasonic transducers to withstand the environmental conditions in the fuel channel was considerably underestimated and this problem has not yet been fully resolved. Transducer life of 100 h under these conditions (summarized in Table 1) has been chosen as a design goal.

Transducer testing to date

39. Considerable testing of transducers has been carried out by Atomic Energy of Canada's Chalk River Nuclear Laboratory both in gamma cells and in reactor during shutdown conditions (ref. 3). A total of 17 transducers were tested with very variable results. A number of conclusions were drawn from this study.

40. A significant number of test failures were

due to problems with the test procedure or equipment and could not be attributed to the environmental conditions.

41. Most standard commercially available transducers will not withstand the temperature requirement. Normal transducers subject to high temperature may fail suddenly due to an open circuit caused by excessive differential expansion of components, or their response may gradually deteriorate due to partial or complete delamination of the active element (crystal) from the damping material.

42. Tests of transducers specially built to withstand high temperature appear to be unaffected by total doses of radiation two to three times the design goal. A number of failures of this type of transducer characterized by sudden open circuits, have occurred, but they appear to be due to thermal effects rather than radiation induced effects. It is hoped that improved manufacturing will solve this problem. A new set of high temperature transducers has been procured from the same manufacturer who was alerted to the reliability problem. Further tests are currently under way.

43. It has been observed, and is generally acknowledged by manufacturers, that high temperature transducers have poorer ultrasonic characteristics than standard transducers.

Future transducer testing

44. Due to the importance of procuring reliable transducers that will withstand the test environment, a more flexible and less expensive method of testing transducers has been instituted. Temperature and radiation testing are being done separately. Temperature testing is performed in the Ontario Hydro Research Laboratory before and after radiation testing. Radiation testing is performed by lowering up to 5 transducers at a time onto a tray of spent fuel in the irradiated fuel bay at Bruce GS A for periods of approximately two weeks. During this period the gamma radiation field decays from 10^6 R/h to 3.5×10^5 R/h providing the transducers with a cumulative dose of approximately 2.2×10^8 Roentgens.

45. It has been suggested that the effects of simultaneous exposure to temperature and radiation may be more detrimental than the sum of the individual effects. While this criticism is accepted, it is believed that the present technique will allow the most promising transducers to be selected for more rigorous testing at a later time.

46. A number of high temperature transducers are being procured from several manufacturers. All transducers will be built to the same specification to allow their ultrasonic and environmental performance to be compared. On the basis of tests of these units, a larger sample of the most successful type of transducer will be procured for further testing.

47. With these actions, we hope to be able to resolve the problem of obtaining transducers for severe environments in the near future.

SUMMARY OF PROGRESS AND FUTURE PLANS

48. Ultrasonic measurement systems for diameter and wall thickness gauging and for volumetric inspection have been developed. However, the availability of radiation and temperature resistant UT transducers has not yet been satisfactorily demonstrated. Current work is expected to resolve this problem.

49. An eddy current device for measuring the minimum gap between the pressure tube and the calandria tube is under development at the Ontario Hydro Research Laboratory.

50. The design of in-channel mechanical components is essentially complete and prototypes of most components have been completed and tested. Some minor changes are expected as the external drive mechanism design is finalized.

51. The preliminary design layouts of the external drive mechanism have been completed and detailed layouts are in preparation.

52. Transmission of data will require multiplexing information over a television cable between the reactor and the station control center. Design of the system is in progress.

53. The broad requirements of the data evaluation and storage system have been decided on and detailed specification of the equipment is in progress.

54. The objective is to have CIGAR, Channel Inspection and Gauging Apparatus for Reactor, operational in late 1980.

REFERENCES

1. KUPCIS, O.A. Nondestructive Inspection of Pressure Tubes at the Pickering Nuclear Generating Station. Proceedings of the Third Conference on Periodic Inspection of Pressurized Components I. Mech. E. Conference Publications 1976-10.

2. SCHANKULA, J.J. An Investigation into the Effects of Irradiation Damage in Zr-2.5 wt% Nb Pressure Tubes on the Velocity of Ultrasound. Atomic Energy of Canada technical Memorandum SP-UT-9. October 1977.

3. COOTE, R.I. and WARD, M.J. Effects of Irradiation During Reactor Shutdown on Ultrasonic Transducers. Atomic Energy of Canada technical Memorandum SP-UT-6, August 1976.

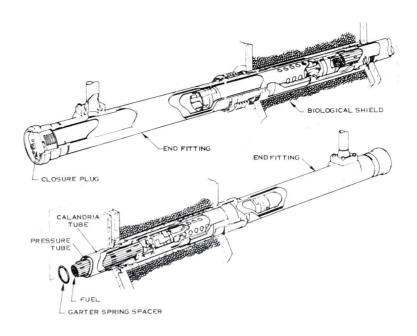

Fig. 1   CANDU reactor fuel channel

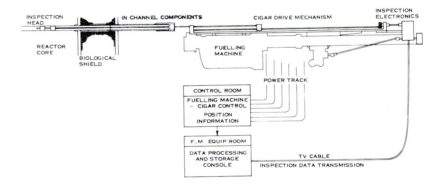

Fig. 2   Schematic of the CIGAR system

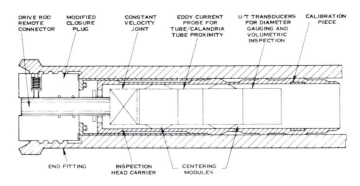

Fig. 3   CIGAR channel entry module

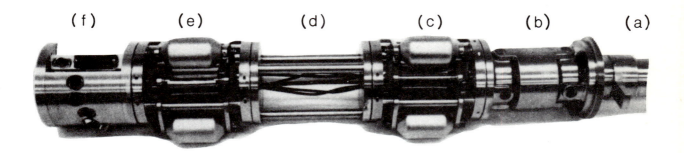

(f)	(e)	(d)	(c)	(b)	(a)

(a) DRIVE ROD CONNECTOR

(b) DRIVE ROD SUPPORT AND UNIVERSAL JOINT

(c) REAR CENTERING MODULE

(d) EDDY CURRENT TRANSDUCER MODULE

(e) FRONT CENTERING MODULE

(f) ULTRASONIC TRANSDUCER MODULE

Fig. 4 CIGAR inspection head

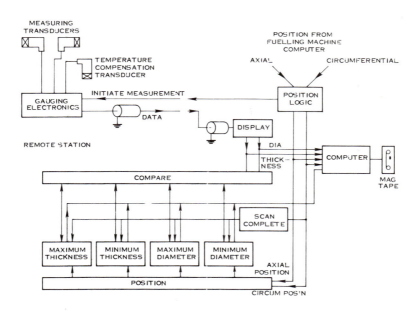

Fig. 5 Schematic of gauging measurement system

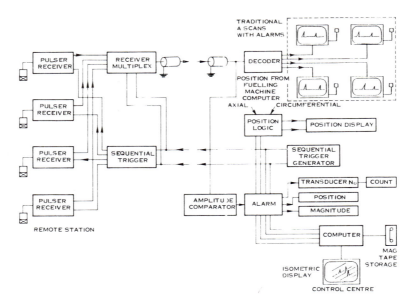

Fig. 6 Schematic of volumetric inspection system

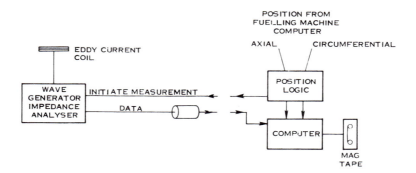

Fig. 7 Schematic of annular gap measurement system

C33/79

IMPROVED ULTRASONIC FLAW-DETECTION AND ANALYSIS TECHNIQUES FOR INSERVICE INSPECTIONS ON PRESSURE VESSELS

Dr.-Ing. H. WÜSTENBERG, and Dipl.-Ing. A. ERHARD,
Bundesanstalt für Material-prüfung (BAM), Berlin
and
Dipl.-Ing. G. ENGL,
Kraftwerk-Union AG (KWU), Erlangen

The MS of this paper was received at the Institution on 30 October 1978 and accepted for publication on 19 December 1978

SYNOPSIS   Experiences with inservice inspections on PWR and BWR have shown that the evaluation of the results needs an improved classification and analysis. Especially the near cladding areas deliver spurious indications lowering the signal significance. Therefore we try to improve the methods for detecting defects in this area by optimized longitudinal wave TR-probes, broad band shear-wave-probes, a tandem-technique with a better focused sensitivity area and a simple signal averaging. Results about the performances of those techniques are reported.   As a special analysing technique linear-acoustic-holography has been adapted to cladded surfaces. A comparison between focusing probes and linear-holography on some artificial defects in cladded test-blocks will be reported.

INTRODUCTION

1.   During the last years in the Federal Republic of Germany a lot of experiences have been collected with the ultrasonic inservice inspection with a multiprobe system including longitudinal wave probes, 70° longitudinal wave T/R probes for the near surface areas, 45° shear wave probes and shear wave probes in tandem arrangements (ref. 1).

2.   During the experiences many problematic areas have been found. In this contribution we will try to present improved techniques concerning three major requirements (Fig. 1):

1. The signal to noise ratio especially inspecting the near cladding area from OD or ID has to be improved. Especially the ultrasonic inspection of the near surface area for cracks starting at the surface is to be regarded. A special problem consists in the fact, that the surface area is often disturbed by geometric reflectors.

2. Using planar probes the broader beams may assure a better reliability of detection, but the significance of the signals received by the inspection system is more difficult to classify. For many reasons in Germany non-focused transducers are applied with similar sound fields used during the inspection in the shop. The severe effects of the cladding on the sound field, especially that of shear waves observed by many authors (ref. 2, 3, 4) let us assume that focused sound field may in some places be systematically deviated so that uninspected blind areas are not reliably enough avoided. But with planar probes one has of course more difficulties with the classification and evaluation of indications.

Classification of criteria other than the amplitude must be investigated in order to avoid, that a too large amount of indications has to be analysed by special techniques.

The problem of the signal significance during an inservice inspection can be explained on the example of the sensitivity setting for a tandem technique (ref. 5). The sensitivity setting on the reference block must be corrected by additions due to the transfer behaviour, due to fluctuations of the transfer behaviour and due to the inhomogeneous sensitivity distribution in the inspected volume zones. All additions may in some cases reach 12 or even 18 dB. Such a conservative sensitivity setting leads to an increase of indications and as a consequence to a reduced signal significance. Especially the amplitude of an indication - for all conventional, manual ultrasonic inspections used as the first information source for classification - looses even more its relation to the importance of a reflector.

3. Techniques for a more precise analysis of indications are needed and should be implemented in an ISI-system.

IMPROVEMENT OF THE SIGNAL TO NOISE RATIO IN THE NEAR CLADDING AREA

3.   As for all coarse grained materials we have different possibilities to improve the signal to noise ratio in that area:

1. The use of short pulses
2. The use of lower frequencies
3. Application of concentrated sensitivity fields, for instance by transmitting receiving techniques or focusing probes

4. The use of longitudinal waves
   (ref. 6)

4. In order to compare different techniques according to the signal to noise ratio in the near cladding area, the test block in Fig. 2 containing two notches of 2 and 4 mm depth starting at the cladding surface, three flat bottom holes under 45° in different distances to the interface and a slot representing large under-cladding cracks has been inspected by 1, 2 and 4 MHz 45° and 40° shear waves, 2 MHz 45° and 40° longitudinal waves in half the skip distance, focused sound fields of 1 and 2 MHz and with a special longitudinal creeping wave probe of 2 MHz.
   The signal to noise ratio of flat bottom holes seems to increase with the shear wave frequency. The signal to noise ratio for the indication of the notches representing an acoustic reflection behaviour as we must assume for cracks decreases with higher frequencies (Fig. 3). Therefore we decided to use 1 MHz shear waves if possible in a range of incidence angles between 40 and 45°. The use of focusing probes may improve for 2 MHz and also for 1 MHz the signal to noise ratio for the detection of cracks, but the difference towards planar 1 MHz shear wave probes is not dramatic.

5. The use of longitudinal waves with an angle of incidence of about 40° – 45° will assure a sufficient signal to noise ratio, but due to the many spurious indications from shear waves, their use must be restricted to special areas. A drawback with longitudinal waves for inspecting the opposite surface in half the skip distance consists in the fact that using the corner effect energy losses by wave mode conversion at both reflecting surfaces (the crack and the complementary surface of the test object) cause a decrease in amplitude.

6. An improvement for the detection of cracks in the cladded area (including cracks beginning at the surface) may be the application of longitudinal creeping wave probes as shown in Fig. 4 (ref. 7, 8). The wedge of this probe has an angle in order to reflect totally the longitudinal waves. In the near field area of the probe the longitudinal wave is propagating as a creeping wave along the surface. This wave is attenuated by the constant irradiation of shear waves. But the signal to noise ratio for surface cracks, which can be reached during an inspection from the ID, is similar to that one of a 1 MHz shear wave probe for an OD-inspection.

7. The signal to noise ratio in the cladded area for the crack detection seems to be optimal using 1 MHz shear wave probes under an angle of 40° to 45°. For an ID-inspection the use of creeping waves represents a good alternative. Inclined longitudinal waves can be applied in areas where the always excited shear waves will not interfere.

8. A special problem has been the sensitivity of the tandem technique especially in the near surface zone due to indications produced by a scattering at the cladding and its interface. In order to reduce this noise we tried to optimize the sensitivity distribution of a tandem arrangement by a suitable choice of the crystal sizes and distances between the probes. Fig. 5 shows the sensitivity distribution of a normal tandem arrangement and of the optimized tandem arrangement. It has been necessary in this case to enlarge the probe further away from the inspected area. The probe in the near defect position operates at the end of his near field. The sensitivity for spurious indications from the cladding area has been decreased by more than 7 dB. During the corresponding base-line inspection the amount of spurious indications from that area could be considerably reduced. The optimization has been calculated by an echodynamic computer model for the tandem technique (ref. 9, 10), which is also used to determine an optimized sensitivity distribution for the tandem technique.

GEOMETRIC REFLECTORS ON CIRCUMFERENTIAL WELDS

9. Fig. 6 demonstrates two possibilities to improve the discrimination between geometric and defect indications during the inspection of circumferential welds. The inspection of such welds with ultrasonic techniques in the primary circuit is in many places restricted by the fact, that geometric indications will hide the defect. One possibility to avoid a heavy noise by geometric indications consists in the use of focused sound fields. During the construction of a probe for this purpose we must regard the fact that the sound beam is used in between the 1/2 and 1 skip distance, that means, we have to take into account a reflection on a curved surface. First results with such probes demonstrate that the good lateral and longitudinal resolution of focused sound fields may enable us to discriminate between defect and geometric reflectors based on an evaluation of the probe position and the time of flight during an automatic inspection.

10. A more simple possibility consists in the use of creeping wave probes (ref. 8). The above mentioned creeping wave probe from Fig. 4 can also be used on ferritic surfaces. For ferritic or cladded surfaces it is necessary to use different probes due to the fact, that the longitudinal wave velocity of the cladding and the stainless steel may differ by more than 250 m/sec. Such probes not only excite longitudinal creeping waves, but also a longitudinal wave beam inclined with 75 to 79° depending on the size of crystal. By an optimal choice of the crystal size it will be possible to have a sensitivity for de-

fects near the coupling surface and also in the inner wall area. The half skip distance can be used to detect defects on the opposite surface. One drawback of this technique is an inhomogeneous sensitivity for different depths, which must be equalized. In order to avoid confusions due to too many indications, e.g. from shear waves, a fixed position of the probe towards the weld is recommended. In that case a fixed gate may pick up from the A-scan presentation of the echogram the areas where defects not produced by the geometry will appear at the screen. Another possibility for discriminating between geometric and defect indications is presented on the right-hand side in Fig. 6. The use of focusing probes delivers well defined indications which can be related to each region according to the probe position and the time of flight of the indication.

11. A third possibility to discriminate between geometric and defect reflectors consists in the use of the linear acoustic holography (ref. 12, 13). Fig. 7 demonstrates the principle of the linear holography. Within an aperture range the real and imaginary part of an echo indication received in a gated zone is measured and stored in a computer. Correcting the measured hologram (that means the real and imaginary parts of the echo indication) within the aperture by a quadratic phase function we focus on a radius r and an observation angle α_0. The corrected hologram can be transformed by the fast fourier transformation, and by that procedure we reconstruct the sound field intensity around a circle of radius r of the middle-point of the aperture. If we apply this to a circumferential weld with a root and a cup producing spurious indications, we may nevertheless discriminate a defect. This is demonstrated in Fig. 8. The differentiation between the three sources of indications only based on the A-scan presentation is nearly impossible.

12. We have to take into account geometric reflectors not only for thinner walled tubes, but also in some areas of the pressure vessel, for instance the coverhead of a PWR or the bottom of a BWR. For such areas the discrimination between geometric and defect indications based on the dynamic behaviour of the time of flight for different indications is possible. First experiences on test blocks and during base-line inspections will be available in a short time.

AVERAGING TECHNIQUES

13. Averaging techniques will be used not only to improve the signal to noise ratio as for instance is now discussed and realized for coarsed grained materials (ref. 14, 15, 16), but also in order to reduce the amount of datas, which must be stored. In the Federal Republic most experiences with base-line and inservice

inspections have been collected with a system using an averaging of all echo-amplitude datas of each different functions of the multiprobe system. The averaging is carried out in a volume element limited by the displacement between two inspection traces, the gate width and especially for the tandem technique the depth of the sensitivity zones. By the use of such a technique the signal to noise ratio for larger reflectors will be increased, whereas the signal to noise ratio for small indications is not too strongly reduced.

14. A more advantageous technique for improving the signal to noise ratio will be a spacial averaging of different A-scans within the same volume elements. But this requires the use of very quick averaging devices, which are actually under development. A result of a spacial averaging of the A-scan is demonstrated in Fig. 9. We obtained this result on a cast stainless steel specimen. The echo indication is produced by a 2 MHz T/R-probe for 45^0 long-waves on a 3 mm $\emptyset$ side drilled hole. The averager (built by Melvin Linzer at the NBS in Washington) (ref. 17) samples the HF A-scans at a 40 MHz rate with a 4 bit accuracy. 2^{12} A-scans have been used during a probe movement parallel to the axis of the side drilled hole with about 10 mm total displacement.

CLASSIFICATION AND ANALYSIS

15. In order to improve the signal significance, not only averaging techniques, but also the evaluation of the dynamic behaviour of an indication during the displacement of the probe system and the spectral distribution of a pulse and its dynamic variation can be used. Investigations about a classification based on this dynamic behaviour of the spectrum are actually on the way. First results indicate that the differentiation between small spherical reflectors and larger reflectors like a slag inclusion or a crack is possible. It is expected that especially spurious indications produced by the structural noise at positions with an unexpected high sensitivity can be suppressed by the evaluation of the "spectral dynamic".

16. Whereas the development of better classification criterias and techniques is not sufficiently progressed for the time being, the performances of analysing techniques - describing defect sizes in length and depth or other quantitative measures - have been strongly improved.

17. Fig. 10 shows the scan of a defectuous area with a focusing probe with direct coupling. Length and depth extensions of this defective area could be derived from the scanning procedure and afterwards verified by opening. If suitable focal spot sizes are used, an accuracy of the defect sizing sufficient for

fracture -mechanical estimates is possible.

18. For an inservice inspection it seems, that the linear acoustic holography offers the same potential to analyse already detected reflectors as focusing probes. But a linear holographic scanning procedure is easier to match to different conditions and cases to be expected at a pressure vessel. Focusing probes must be adapted to the sound path and to the geometry of the object. A linear acoustical holography can be added without too much modifications to a conventional inservice inspection system and activated, if necessary. The principle of linear acoustic holography (ref. 11, 12) is explained in Fig. 7.

19. Fig. 11 shows two examples of test reflectors scanned through a cladding. The frequency of the shear waves used has been 1 MHz. Right-hand side in the figure a holographic reconstruction of a scan in a tandem arrangement of two flat bottom holes is presented. During the scanning the gate has to be positioned by a computer. Left-hand side a scanning in a single probe arrangement with about 50° angle of incidence and 1 MHz shear wave is presented.

20. Although the physical resolution for both cases reaches not more than 3 resp. 4.5 mm, the two reflectors could be discriminated and the total defective area clearly defined. Those experiences prove that the analysis of a defective area through the cladding is possible with the techniques actually available.

CONCLUSIONS

21. The optimization of detection capability of an ultrasonic inservice inspection system in the past has increased the number of indications and lowered the signal significance. Therefore improved means for enhancing the signal to noise ratio and the classification possibilities of ultrasonic indications had to be developed. In our contribution we tried to demonstrate, what kind of improvements are possible and already developed. The problem is not yet solved sufficiently, but the collected experiences and the approaches under investigation or discussion have shown, that the major difficulties with the performance of the ultrasonic inspection can be solved.

REFERENCES

1. ENGL G., MULLER G., SEIGER H., WUESTENBERG H. Inspection techniques and equipment based on German requirements on ISI. Proceedings "Specialist meeting on the ultrasonic inspection of reactor components", Daresbury, Engl., Sept. 27-29, 1976.

2. KOLB K., WOELFEL M. Zur US-Prüfung von Kernreaktor-Druckbehältern. Mat. Prfg. 17 (1975) Nr. 10 Okt.

3. LAUTZENHEISER C.E., AHLBERG G. Design and metallurgical problems associated with preservice and inservice inspection of nuclear systems. Mat. Eval. 30 (1972) 5, p. 99-102.

4. WUESTENBERG H., SCHULZ E. Investigations concerning the influence of austenitic claddings on the sound fields of ultrasonic probes. VII. IC of NDT Warszawa June 4-8, 1973, H-03.

5. ENGL G. Basisprüfung des Reaktordruckbehälters; Tandem, Impulsecho und Durchschallung. Seminar ZfP in der Kernreaktortechnik-Ultraschallprüfung, IzfP, 9.3.1978, Saarbrücken.

6. WUESTENBERG H., NEUMANN E. Improved probe techniques at the ultrasonic inspection of coarse grained materials. 6th Water reactor safety research information meeting im NBS-Gaithersburg/Md. USA, 6.-9.11.1978.

7. ERHARD A., WUESTENBERG H., MUNDRY E. 90°- Longitudinalwellen-Winkelprüfkopf zum Nachweis oberflächennaher Risse. DGZfP-Conf. and Exhibition on NDT, Mainz, 24.-26.4.1978.

8. WUESTENBERG H., ERHARD A., KUTZNER J. Detection and analysis of near-surface-cracks by ultrasound. Proceedings "1th International Symposium on ultrasonic materials characterization" June 7-9, 1978, NBS Gaithersburg/Md., USA.

9. Dreidimensionale Echodynamik für Einkopf-, Tandem- und SEL-Prüftechnik mit Integration über die Fehlerfläche. Bericht der Fachgruppe 6.2 Zerstörungsfreie Materialprüfung, Laboratorium 6.21 "Mechanische und thermische Prüfverfahren" der BAM, Berlin, 7.Dez. 1977. BERTUS N., WUESTENBERG H.

10. WUESTENBERG H., KUTZNER J., ENGL G. Dependence of echo amplitude on defect orientation in ultrasonic examinations. 8th World Conf. on NDT 6.-11.9.1976, Cannes/France.

11. KUTZNER J., WUESTENBERG H., MOEHRLE W., SCHULZ E. Zonenaufteilung, Empfindlichkeitseinstellung und Prüfkopfhalterung bei der manuellen Ultraschallprüfung mit Tandemverfahren. Materialprüfg. 17 (1975) Nr. 7 Juli, S. 246-250.

12. KUTZNER J., WUESTENBERG H., KAPS U. Acoustical holography with numerical reconstruction as a tool for the analysis of ultrasonic indication. 8th World Conf. on NDT 6.-11.9.1976, Cannes/France.

13. WUESTENBERG H., MUNDRY E., KUTZNER J. Experience with flaw size estimation by ultrasonic holography with numerical reconstruction. ASM, ASTM, ASNT, ANS International Conf. Nondestructive Eval. in

the Nuclear Industry 13-15 Febr. 1978, Salt Lake City/Ut., USA.

14. GOEBBELS K., von KLOTH. Signalanhebung bei der Prüfung grobkörniger Werkstoffe. Seminar "ZfP in der Kernreaktortechnik" des IzfP, Saarbrücken, 8./9. März 78

15. GOEBBELS K., HOELLER P. Quantitative determination of grain size and detection of inhomogeneities in steel by ultrasonic backscattering methods. First Int. Symposium on materials characterization, June 7-9, 1978, NBS Gaithersburg/Md., USA.

16. NABEL E., JUST T. Diskussionsbeitrag zur Verbesserung des Signal-Rausch-Verhältnisses bei der US-Prüfung grobkörniger Werkstoffe. Neuere Verfahren zur Analyse von Ultraschall-Befunden. Vorträge der Fachausschüsse Sonderprüfverfahren und Ultraschall-Prüfverfahren, 24.2.1977, Berlin. (DGZfP)

17. LINZER M., SHIDELER R.S., PARKS S.1. Ultrafast signal averaging and pulse compression techniques for sensitivity enhancement. First International Symposium on ultrasonic materials characterization, June 7-9, 1978, NBS Gaithersburg/Md., USA.

Detection of Near Surface Defects:
Noise due to the Cladding Crack at the Surface Geometric Indications
Classification of Indications
Analysis of Indications

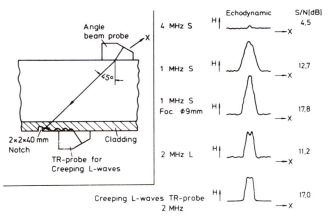

Fig. 1 Problematic areas for an ISI with ultrasound

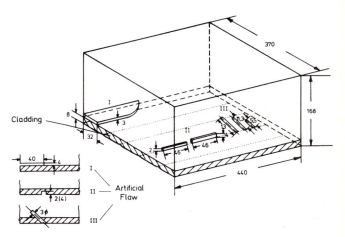

Fig. 2 Testblock for the inspection of the cladded area

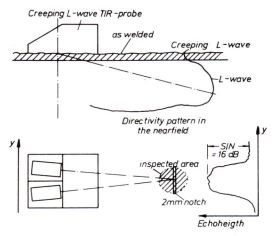

Fig. 3 Surface crack detection influenced by a cladding

Fig. 4 Use of creeping waves in the cladded area with TR-probes

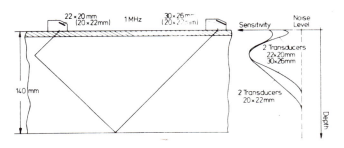

Fig. 5 Reduction of the noise from the cladding for a tandem technique

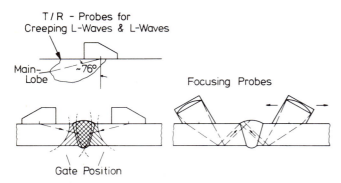

Fig. 6 UT-methods with improved discrimination between defects and geometry

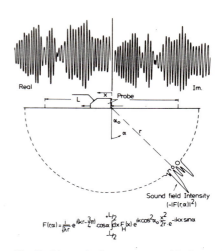

Fig. 7 Numerical reconstructed holography

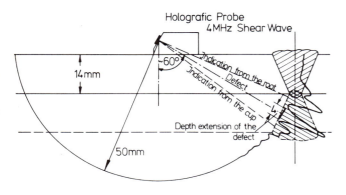

Fig. 8 Discrimination between geometry indications and defects by the linear-acoustic-holography

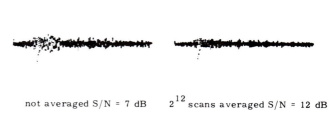

T/R-L-Wave Probe: 45° 2 MHZ 3 mm ⌀ SDH 27 mm Depth

not averaged S/N = 7 dB 2^{12} scans averaged S/N = 12 dB

Fig. 9 Averaging of A-scans during probe movement

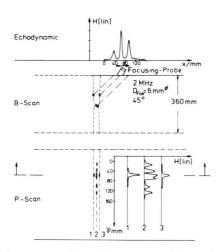

Fig. 10 Echodynamic, B- and C-scan presentation of weld defects by focusing probes

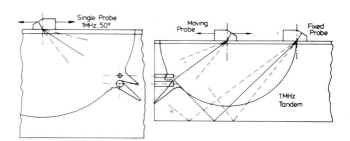

Fig. 11 Linar-acoustical-holography on a cladding component

C35/79

MAPPING RESIDUAL STRESS BY ULTRASONIC TOMOGRAPHY

B. P. HILDEBRAND and T. P. HARRINGTON
Battelle, Pacific Northwest Laboratories

The MS of this paper was received at the Institution on 19 December 1978 and accepted for publication on 12 January 1979

SYNOPSIS   It is known that internal stress concentrations can give rise to microcracks which then grow when the structure is subjected to external forces. It has also been found that the velocity of sound is altered as it propagates through a region of stress. In this paper we discuss a technique called Computer-Assisted Tomography (CAT) and describe an application that provides pictures of stress fields. We report the results of both simulated and experimental models used to evaluate the technique. We conclude that the CAT approach has great potential for locating and mapping residual stress in metals.

INTRODUCTION

1. One of the outstanding problems of non-destructive testing is the location and measurement of areas of residual stress. In the manufacture of large structures, such as pressure vessels, a number of very large welds are required. The heat-affected zone surrounding the welds will contain residual stresses due to uneven cooling rates. In practice, these stresses are relieved by heating the entire vessel to some appropriate temperature and then carefully controlling the cooling rate. This standard procedure has not always been successful in relieving the residual stress that is induced by the welding process.(Ref. 1) Structural designers realize this, and try to compensate for the possibility of stress concentrations by overdesigning. This leads to penalties in terms of both added weight and cost. Consequently, there is high motivation for finding and delineating such concentrations.

2. Currently there is no satisfactory test for measuring the success of stress-relief procedures. Thus, it may happen that high residual-stress regions exist in the pressure vessel. If these regions occur in critical areas (such as nozzle-to-vessel joins), cracks may develop, which would require subsequent plant shut-downs and expensive repair. If it were possible to detect and map residual stress, local stress relief could be applied, and future problems could be avoided.

3. Standard Nondestructive Testing (NDT) examinations consist of radiography and ultrasonic pulse-echo. Neither of these techniques can reveal the presence of residual stress; the former because it shows only density variations, and the latter because stress regions are not sharply defined; hence, they do not reflect much sound. It is known, however, that the velocity of sound in a solid is affected by stress. This phenomenon is a third order effect, and has been used primarily to determine the Lame and Murnaghan elastic constants for various materials.

4. A number of studies have shown that it is possible to measure velocities with sufficient accuracy to detect and calculate residual stress.(Ref. 2-4) Typically, for steel, the stress-acoustic constant is approximately 0.197 nsec/cm/6895 kPa. This constant means that a 6895-kPa increment in stress produces a 0.5 nsec change in travel time through a 2.54 cm thick specimen. Since it is relatively easy to measure time-of-flight changes to this accuracy, the sensitivity of the proposed method should be approximately 6895 kPa.

5. At the present time, a number of researchers have made velocity measurements for the computation of stress. This data, however, has been obtained in the form of spot measurements or profiles. No one has yet attempted to take a number of profiles at different angles in order to reconstruct a cross-section of a velocity anomaly, and hence a cross-section of the residual stress. The purpose of this paper is to discuss our efforts to develop a method for mapping stress anomalies utilizing velocity information and an image reconstruction technique known as Tomography.

6. A Tomogram could be described as a picture of a slice. Over the past several years CAT scanners have revolutionized the field of diagnostic medicine. This technique uses X-rays to obtain visible, thin slices through any section of the human body. We are applying the same reconstruction algorithms used in medical applications, (except for the substitution of ultrasound for the X-rays), to generate velocity maps of cross-sections of thick metal sections.

7. In the following section we briefly discuss the principles of Tomography, and describe the Algebraic Reconstruction Technique (ART) we are using to construct the velocity profile. Section 3 describes several computer simulations for testing the ART algorithm. In Section 3 we also discuss the effect of varying several parameters and describe the effect they have on the quality of the reconstruction. In

Section 4 we describe an experimental apparatus, which we have developed in our laboratory, for obtaining time-of-flight measurements. Section 4 also includes a discussion of reconstructions generated from these experimental data, and a description of the calibration procedure we are using in order to relate the velocity of sound to stress. In our conclusion (Section 5), we comment on the validity of this method for mapping residual stress and also describe a prototype instrument that we are building to test this technique.

PRINCIPLES OF IMAGE RECONSTRUCTION IN COMPUTED TOMOGRAPHY

8. As we noted in the introduction, Tomography has had its greatest impact to date in the field of diagnostic medicine. A good source of references on medical Tomography can be found in a paper by Brooks and Dechiro.(Ref. 5) The principles of image reconstruction using Tomography have also been applied in a number of other disciplines: Electron Microscopy, Radio Astronomy, and Nondestructive Testing.(Ref. 6-11)

9. To clarify the discussion that follows, a brief digression into the principles and terminology of Tomography will be necessary. Consider Figure 1a, which is a particular cross-section of a general three-dimensional object. Define the velocity distribution of the object as $v(x,y,z)$. Assume an ultrasound pulse, transmitted from location t, traverses the object along path ℓ to a point r where it is received. The time-of-flight along path ℓ is a function of the velocity distribution along the path, and is given as the line integral

$$t_\Theta(\rho) = \int_\ell \frac{ds}{v}$$

where ρ and Θ define the path ℓ from t to r. If the variable ρ is allowed to vary continuously over the object at a constant angle , a one-dimensional projection of the velocity distribution will be obtained.

10. The reconstruction problem is to estimate the velocity of the cross-section $v(x,y,z_0)$ from the projection values. Clearly, this can only be done if the projection values are available for a large number of angles Θ_i. Also, in actual practice the projection values are available only at a discrete number of the ρ_j for each angle Θ_i. Thus, the reconstruction is an approximation to the actual velocity field, with resolution being a function of the spacing between rays $\Delta\rho_j$ and the degrees between angles $\Delta\Theta_i$.

11. In theory, the velocity field can be reconstructed quite easily from the projection information. To see this, visualize a square grid superimposed over the object of Figure 1a as shown in Figure 1b. If we assume that the velocity within each square (pixel) is constant, then the total travel time through the grid becomes the sum of the travel times in each of the pixels. Each ray that passes through the velocity field will intersect various pixels. The length of the ray segment in each pixel can be determined quite easily by using geometry. If a ray of length l_k lies in a cell with velocity v_k, the travel time through the cell

is l_k/v_k. Thus, the total time-of-flight can be expressed as a linear equation whose unknowns are the velocities v_k. By using a number of independent rays that is equal to the number of grid cells, a linear system of algebraic equations can be generated and solved for the unknown velocities.

12. The difficulty with solving for the unknown velocities by inverting a system of algebraic equations is that, even for small grid systems, the size of the resulting matrix is prohibitive. For example, consider a grid system of 20 by 20 pixels. This means that there are 400 unknown velocities to solve for. If we take 20 independent views with 20 rays per view the result is 400 equations in 400 unknowns, or a 400 by 400 matrix to invert. Furthermore, the matrix will be sparse, since each ray will usually intersect less than 30 pixels.

13. Other reconstruction algorithms are available that overcome the limitations imposed by the matrix inversion method, some of these alternatives are:

(1) Convolution, which is based on the work of Johann Radon (who first solved the equations governing image reconstruction in 1917).(Ref. 12-13)

(2) Fourier Transform, which is the spatial frequency version of convolution.(Ref. 14)

(3) Back Projection, which was used in the first attempts to produce Tomograms of living patients

(4) Iterative Techniques, which are probably the most widely used reconstruction methods.

ITERATIVE RECONSTRUCTION ALGORITHMS

14. All the iterative algorithms start with an "initial guess" as to what the image looks like. Typically, the average value of the projections is divided among all the pixels that make up the image. The algorithms then adjust the pixel values to bring them into better agreement with the measured projections. These algorithms are iterative in that they continually sequence through the set of projection data, updating the pixel values until a stopping criteria is met. An example of a stopping criteria would be a measurement of the change that has been made to the image during an iteration. If the change is less than some prespecified minimum, the iteration is stopped.

15. The three most popular iterative reconstruction algorithms currently in use are: the Iterative Least Square Technique (ILST) of Goitein, (Ref. 15) the Simultaneous Iterative Reconstruction Technique (SIRT) introduced by Gilbert, (Ref. 16) and the Algebraic Reconstruction Technique (ART) discovered by Gordon, et.al.(Ref. 7, 17) Our preliminary investigation of both ART and SIRT indicates that ART produces better reconstructions than SIRT, given the amount and nature of the data that is available.

16. You will remember from the discussion earlier in this section that the time-of-flight

for a given projection can be approximated as a sum of the time-of-flights in individual pixels. Briefly, the ART algorithm works as follows: for each ray in the data set, ART compares the experimentally obtained time-of-flight with the time-of-flight calculated using the current pixel values. If they are different, ART updates the velocity in each of the pixels as a function of the length of the ray path in the pixel. After the pixels along the ray have been modified, the time-of-flight calculated from the pixel values matches that obtained experimentally. ART then cycles through the remainder of the rays in the data set, repeating the process described above. However, as each new ray is processed, the value of previously updated pixels will be changed. This is the reason for the need to iterate. It has been shown, however, that the image improves after each iteration and converges to the best solution in a least squares sense.(Ref. 17)

17. The majority of medical CAT scanners in use today employ one of two geometries for obtaining profiles. These two geometries are usually referred to as fan beam and parallel beam. Figure 2 points out the differences between these two approaches. In the parallel beam method a set of measurements are obtained by scanning the source and detector linearly past the patient. The entire scanner assembly is then rotated by a fixed amount and the scan is repeated. In the fan beam approach, the source rays are formed into a fan of narrow beams that encompass the patient. The rays are received simultaneously by an array of detectors. For this geometry, the source and detector array are also rotated about the patient. Since the fan beam method does not require a linear translation of the source and detector, it is capable of much superior performance.

SIMULATED DATA

18. To test the feasibility of using time-of-flight information along with the ART algorithm to reconstruct velocity fields, we have simulated several different images on the computer. As mentioned in the previous section, the time-of-flight for a particular ray crossing a velocity field is given by the line integral

$$t_\Theta(\rho) = \int_\ell \frac{ds}{v}$$

To obtain simulated time-of-flights on the computer, this integral can be discretized to:

$$t_{\Theta_i, \rho_j} = \sum_{k=1}^{K} \ell_{kij}/v_k$$

where ℓ_{kij} is the path length of the ray defined by Θ_i and ρ_j through the cell k. And K is the set of all cells which intersect the line ℓ.

19. To visually compare the reconstructions with the simulated velocity fields that were used to generate the data, we construct both isometric views and black-and-white gray scale images of the reconstruction. We will use both of these techniques to display results in this paper. To obtain a more qualitative idea of how the reconstructions compare with the original image, we calculate several error parameters at each iteration of the ART algorithm.

The discrepancy is defined as:

$$\delta^q = \left\{ \frac{\sum\limits_{k} (V_k^q - V_k)^2}{\sum\limits_{k} (V_k - \overline{V})^2} \right\}^{1/2}$$

This is a normalized Euclidian norm where V_k is the velocity value of the kth pixel of the test image, $\overline{V}$ is the average velocity of the test image, and V is the velocity of the kth pixel after the qth iteration. This equation shows that the discrepancy is the ratio of the root-mean-square error to the standard deviation of the test picture. This measure was suggested by Gilbert,[16] and has since been used by Herman, et.al.[17] and Colsher.(Ref. 16, 17, 8)

20. The second error parameter we calculate is the mean relative (or average) error which is defined as:

$$E_{av}^q = \frac{\sum\limits_{k} |V_k^q - V_k|}{\sum\limits_{k} V_k}$$

This has also been used by Sweeney and Colsher.(Ref. 18, 8)

The final error function that we calculate is the residual, which is defined by:

$$R^q = \left\{ \sum_i \sum_j \left[P(\Theta_i, \rho_j) - P^q(\Theta_i, \rho_j) \right]^2 \right\}^{1/2}$$

where $P(\Theta_i, \rho_j)$ is the actual measured time-of-flight for the ray path defined by Θ_i and ρ_j and $P^q(\Theta_i, \rho_j)$ is the computed time-of-flight for the same path using the pixel values after the qth iteration. This measure indicates the degree to which the reconstructed image satisfies the measured time-of-flight data. Notice also that this is the only error measure of the three that is valid for experimentally obtained data. This criterion has been employed by (Colsher.Ref. 8)

21. An obvious deficiency of the first two error criteria is that the test image must be discretized. Several of the velocity fields that we used to generate our simulated data are circular and thus difficult to discretize. This problem tends to negate the usefulness of these two criteria.

22. All of our reconstructions were performed on a PDP 11/70 minicomputer. It is not valid to discuss execution time, since this computer operates in a multi-user environment. On the average, however, each iteration of the ART algorithm required approximately 30 sec of wall clock time and 5 sec of computer time. All of the isometric and gray scale images presented in this report were post-processed after the ART reconstruction program was completed.

23. In our initial investigation of the ART algorithm we simulated parallel beam geometry, as described in the previous section. Figure 3a represents a velocity field having two islands of 2% velocity increase in a uniform region. Figure 3b shows the reconstruction when time-of-flight profiles are taken over a

full 180° field of view. Notice that the tops of the two islands in this reconstruction are fairly flat and the steep sides approximate the test image very well. In Figure 3c a 90° field of view was available to the reconstruction algorithm. In this image the tops of the islands are somewhat irregular and the sides have less slope than the 180° field of view reconstructions. The 90° data base, however, still provides an adequate reconstruction of the test image. Figure 3d shows the result of using a 45° data base. In this reconstruction, the two islands have been smeared into the uniform velocity region. The tops of the islands are rough, and the height does not represent a 2% velocity difference. The walls of the islands in this reconstruction also have a very shallow slope. The reason for the poor reconstruction with a 45° field of view is that the majority of the rays pass through one of the two islands. As discussed in the last section, the velocity differences are divided among all the pixels in the path of the ray. Since there are few rays which pass completely outside the two velocity islands, the smearing effect is not counteracted.

24. A major goal of our present work is to develop procedures for mapping residual stress in thick metal sections (typically 10 to 20 cm). In many practical situations, it may not be feasible to obtain time-of-flight profiles by either of the geometries discussed in the previous section. For instance, it may be physically impossible to position a detector inside a pressure vessel.

25. To overcome this difficulty, we have been investigating the feasibility of locating both the transmitter and the receiver on the same side of the metal section. Figure 4a and 4b illustrate the differences between this geometry and those discussed in the last section. Figure 4a illustrates the case in which both sides are accessible. The only way in which different angular profiles can be taken is to launch the waves at different angles by tilting the source. Theoretically, it is possible to launch waves over a $\pm 90°$ field. However, note that the receiver would need to be moved farther away with increasing angle. Hence, a practically obtainable field of view with this method is $\pm 45°$.

26. Figure 4 illustrates the dilemma produced by the requirement for single surface inspection, and the solution. The back surface of the section is used as a reflector, with the receiver placed to receive the reflected signal. Note, however, that the reconstruction will now include a mirror image as well as the object itself. Also note that in both cases the total geometric path length changes as a function of angle of view. Both of these peculiarities must be taken into account by the reconstruction algorithm.

EXPERIMENTAL DATA

27. To obtain actual data to test our proposed reconstruction method, we ran several different experiments. Our first experiment was designed to test the accuracy of the equipment we were planning to use. For this experiment we used the parallel beam geometry to collect time-of-flight data from objects submerged in a water bath. In our second experiment we also used parallel beam geometry but made measurements on a section of type 1018 mild steel, in which we embedded an oversized pin to induce stress. In our current work we are employing a variation of the reflected beam geometry to analyze the physical characteristics of the reconstruction.

28. The equipment we are using to measure time-of-flight includes:

    (1) Velonex Model 570 Pulser
    (2) Metrotek MP 215 Pulser
    (3) Metrotek MR 101 Receiver-Amplifier
    (4) Metrotek MG 703 Time Interval Gate
    (5) Nortec No. VM 16-2.25-0 Broadband Unfocused Ultrasonic Transducer
    (6) Hewlett Packard 5345A Time Interval Counter
    (7) Hewlett Packard 9825A Desk Calculator/Computer.

29. The Velonex 570 Pulser generates a 0.5 sec, 600V pulse with a repetition rate of 2000 Hz. This pulse starts the Time Interval Counter and energizes a Nortec V-Z-16-2.25-0 2.54-cm, wide-band transducer. A Nortec R-Z-0.025-10-0, 0.064-cm, wide-band receiver senses the pulse, a sense amplifier amplifies and thresholds it and sends it to the Time Interval Counter. The counter, a Hewlett Packard 5345A, measures the time between the start pulse and the received pulse by counting internal clock pulses generated in the interval. The instrument can be set to average over intervals from 10 nsec to 20,000 sec; its worstcase accuracy is ± 0.7 nsec and its worstcase resolution is ± 11 psec. With a 1 sec averaging time we have consistently seen a repeatability better than ± 0.3 nsec.

30. A Hewlett Packard 9825A Calculator was added to the equipment list prior to our most recent experiment. This calculator allows us to semiautomate the data gathering process; it has been programmed to control a set of stepping motors that are used to scan the transducer assembly past the object being examined. The calculator also has the capability to plot the data as it is being taken. This gives us visible assurance that the measurements being made are correct. Finally, the calculator has a permanent storage capability in the form of removable magnetic tape cartridge. In our current configuration, the time-of-flight measurements are recorded on the magnetic tape while the experiment is in progress. After all the data has been collected, it is read off the magnetic tape and transmitted over a communications line to a PDP 11/70 MiniComputer, where the reconstruction is performed.

FIRST EXPERIMENT

31. In our first experiment we wished to test the feasibility of the measurement system. To accomplish this, we developed a model consisting of a rubber glove filled with a water-alcohol mixture. The ratio of water to alcohol was intended to provide a 2% increase in the velocity of sound over the surrounding water. We found this model to have several advantages:

(1) Ease of implementation
(2) Easy adjustment of the velocity of sound in the sample to any desired value in the range of interest
(3) A well-defined shape for reconstruction testing

32. Three disadvantages of this model became apparent as we moved toward smaller velocity differentials. First, a variation with temperature in the density of the ambient fluid gave rise to an observable, slowly varying velocity change which must be accounted for in the data analysis. Fortunately, because the thermal mass of the water bath is sufficiently large, data correction is not difficult.

33. Second, the rubber glove sample is sufficiently pliant to make it sympathetic to small turbulences brought about by mechanical movement of the scanning apparatus within the tank. This is overcome by allowing a short settling-time between positioning of the apparatus and data collection.

34. Figure 5 shows the measurement system used for this early set of experiments. The transducer assembly is scanned past the model with time-of-flight measurements taken at discrete intervals. The object is then rotated on the turn-table through a fixed angle and the transducers are again scanned past the model.

35. Figure 6 shows a reconstruction obtained when the two fingers of the rubber glove are filled with a water-alcohol mixture corresponding to a 2% velocity increase. The data base for this reconstruction consists of 19 scans taken at 5° increments from 0° to 90°. Each scan consists of 20 rays taken at 0.0254 cm intervals over a span encompassing the two fingers.

SECOND EXPERIMENT

36. In our second experiment, we developed a model that we feel provides a realistic test in steel as well as a controlled stress field of known shape. As shown in Figure 7a, a slab of type 1018 mild steel 5.08 x 1.27 x 7.62 cm was used as the base metal. A 0.635-cm hole was drilled as shown. A pin, of diameter 0.639 cm, was also fabricated from the same steel stock. The pin was cooled in liquid nitrogen (-190^oC), the block was heated to 1000^oC and the pin was inserted in the hole. After the model stabilized to room temperature, the sides were ground to remove scale and to assure constant thickness. The model was then mounted on the turntable and scanned. In this case, due to refraction, the angle through which the model was rotated was nonlinear. That is, the angular increments were calculated to achieve 5° increments in the steel. Fifty-one measurements were made per scan over a 2.29 cm span centered on the pin.

37. Rough computations show that the pin is subjected to about 6.895×10^8 kPa compressive stress, which must be balanced in the surrounding metal with a distributed tensile stress. Figure 7b shows a reconstruction of the velocity field in the steel section in the vicinity of the pin. The reconstruction shows that, inside the pin itself, the velocity has decreased indicating compressive stress. The region surrounding the pin has a higher velocity than the base line, indicating an increase in tensile stress.

Third Experiment

38. In this experiment we tested the reflected beam ART algorithm. We again used a water-alcohol mixture in a water bath to simulate changes in velocity due to stress. In these experiments, however, the transmitter and receiver are positioned on the same side of the object. A mirror centered between the two transducers (and on the opposite side of the rubber glove) is used to simulate reflection off the back wall of a metal section.

39. The Hewlett Packard Calculator was included in the data acquisition loop to accomplish the control, data storage, and quality assurance tasks mentioned earlier in this section. Using the calculator decreases the time needed to obtain a complete data set by almost 50% over our earlier manual mode. This is important because, as was discussed earlier, the temperature of the water bath changes during the day and this causes changes in the velocity of sound in water.

40. For this experiment we took 23 different views that included angles from 10° to 120° in steps of 5°. At each angle, 31 time-of-flight measurements were taken as the transducer assembly was scanned past the object. The distance between measurements was 0.732 cm.

41. Figure 8a shows a 3-D plot of the reconstruction obtained from this data. It definitely indicates a sharp velocity change in the region of the finger. The computer calculated a maximum velocity change for this reconstruction of 2.7%, which is close to our measurement of actual velocity in the glove. Figure 8b shows the same reconstruction, using a gray-scale plot to indicate velocity change.

Calibration

42. In order to relate velocity measurements to stress, we have initiated a calibration experiment. The purpose of this experiment is to obtain velocity measurements as a function of stress, both tensile and compression. To accomplish this, we fabricated tensile and compression specimens from 6.45-cm, A516-74A, grade 70, pressure vessel steel plate. These were placed in an MTS 810 material-test-system machine located in our laboratory.

43. Time-of-flight measurements were made with a system consisting of a Metrotek MP 215 High-energy Pulser (driving a wide-band transducer), and a Metrotek MR 101 Receiver and Metrotek MG 703 Interval Gate (driving an HP 5345A Time Interval Counter). The transducer was coupled directly to one side of the specimen and stress was applied. The MG 703 was set to gate out the first and second echos from the back surface of the specimen. This eliminates any possible error due to the coupling thickness. We have found this measurement to be repeatable to 0.1 nsec. Transverse strain was measured with a Lion Precision Corporation Metri-Gap 300-3 Capacitive Micrometer to a precision of 25 μm. A thermocouple was also

attached to the specimen to keep track of the temperature as the specimen was stressed.

44. The results of these tests are summarized in Figures 9 and 10. Curves 1, 2 and 3 of Figure 9 are derived from one tensile specimen and curve 4 is derived from another. Curve 1 represents a test run well below the yield point, curve 2 is a test run of the same specimen taken past the yield limit and curve 3 is a final run taken to failure. This set of curves is extremely interesting since it seems to indicate that the specimen retains a memory of its last test if taken beyond yield. Curves 1 and 2 follow the expected path of a linear velocity increase with stress. However, after stressing the specimen into yield, as was done in curve 2, the behavior of the velocity (curve 3) is that of a specimen under compressive load. It appears as though a net residual compressive stress is present in the sample. The tension introduced by the machine must first overcome the residual compressive stress before the velocity again increases. This residual stress appears to be about 2.07×10^5 kPa. Note also, that the yield strength of the material increases from about 3.10×10^5 kPa to 4.55×10^5 kPa due to work hardening (as is well known). Curve 4 is the result of a test on a second identical specimen, performed to corroborate the stress acoustic constant derived from the first specimen. This factor, which is the slope of the curve, is 7.33×10^{-4}%/ 6.9×10^3 kPa/2.54 cm.

45. In Figure 9 the curves are drawn only for the stress region before yield is reached. This is because the readings become erratic past this point. For this material the velocity change appears to flatten out. However, this may be due to the well known behavior of materials in tension such as dislocation and slip fractures in the material. In compression, a much smoother behavior is revealed, as shown in Figure 10. In this case, linear velocity decrease (with about the same stress acoustic constant) occurs up to the yield point. After the material begins to yield, the velocity decreases in a highly non-linear manner with the slope increasing precipitately. This, too, is expected from theory.

CONCLUSION

46. We have discussed an application of Computer-Assisted Tomography (CAT) for locating and mapping regions of residual stress. The simulations and experiments described have demonstrated that velocity anomalies of 2% can be quite easily resolved. In work not reported here we have also experimentally mapped velocity anomalies as low as 0.2% and feel that 0.05% is technically feasible. These velocities translate to a sensitivity of 6895 kPa in a 2.54-cm-thick region.

47. We feel that the reflected beam geometry represents an important advance in the development of methods for inspecting structures which do not physically lend themselves to either the parallel beam or the fan beam geometries. Thus, the reflected beam geometry could have important applications in the area of in-service inspection.

48. We are currently designing a prototype instrument for measuring stress in an online production environment. The instrument, as we envision it, will consist of an array of transducers placed in contact with the metal surface. Two elements of the array would be selected to form a pitchcatch arrangement. One is used to transmit a pulse and the other to receive it. The choice of elements defines the angle of the ray to be measured. After a time-of-flight measurement has been made, the selected set of two elements are electronically moved over by one element and a second measurement made. In this way, a whole profile can be made very quickly.

49. After one profile is made, the separation between the two elements is changed by selecting two different transducers and the process is repeated, giving another profile at a different angle. All of the necessary data can thus be taken in a few seconds or less, depending upon the accuracy desired.

50. The time-of-flight measurements will be fed directly into the memory of a micro-processor. When a complete set of data is available, the computer executes the ART algorithm and displays the resulting reconstruction on a CRT display. Using a micro-processor, it appears possible to obtain reconstructions in less than a minute. If velocity anomalies appear to be present in the reconstructions, local stress relief procedures could be applied.

This research is sponsored by the Electric Power Research Institute under Contract RP504-2.

REFERENCES

1. Gott, K.E., "Residual Stresses in a Weldment of Pressure Vessel Steel," AB Atomenergi Report, NyKoping, Sweden (1978).

2. Hsu, N.H. "Acoustical Birefringence and the Use of Ultrasonic Waves for Experimental Stress Analysis," Experimental Mechanics, May 1974.

3. Smith, R.T. "Stress-Induced Anisotropy in Solids--The Acousto-Elastic Effect," Ultrasonics, July-September 1963.

4. Noronha P.J. and Wert, J.J. "An Ultrasonic Technique for the Measurement of Residual Stress," J. of Testing and Evaluation, March 1975, Vol. 3.

5. Brooks, R.A. and DiChiro, G. "Theory of Image Reconstruction in Computed Tomography," Radiology, December 1975, Vol. 117, pp 561-572.

6. DeRosier, D. J. and Klug, A. "Reconstruction of three-dimensional structures from electron micrographs," Nature (London), January 1968 No. 217, pp 130-134, 13.

7. Gordon, R., Bender, R. and Herman, G. T. "Algebraic Reconstruction Techniques (ART) for Three-Dimensional Electron Microscopy and X-ray Photography." J. Theoretical Biology, December 1970, Vol. 29, pp 471-481.

8. Colsher, J.G. "Iterative Three-Dimensional Image Reconstruction from Projections: Applications in Electron Microscopy" (Ph.D. Thesis) University of California/Livermore, UCRL-52179, December 1976.

9. Bracewell, R.N. and Riddle, A.C. "Inversion of Fan-beam Scans in Radio Astronomy", The Astrophysics J., November 1967 Vol. 150, pp 427-434.

10. Kruger, R.P. and Cannon. T.M. "The Application of Computed Tomography, Boundary Detection, and Shaded Graphics Reconstruction to Industrial Inspection" Materials Evaluation, April 1978 Vol 36 No. 5.

11. Falconer, D.G. and Gates, D.C. "Reactor-Component Inspection with Computed Tomography" EPRI NP-213, Project 610-1, July 1976.

12. Radon, J. "Uber die Bestimmung von Funktionen durch ihre Integralwerte langs gewisser Mannigfaltigkeiten" Ber. Akad. Wiss. (Leipzig) Math. Phys. Klasse Vol 69, pp 262-277, 1917.

13. Rammachandran, G.N. and Lakshminarayanan, A.V. "Three-dimensional reconstruction from radiographs and electron micrographs: Applications of convolutions instead of Fourier transforms", Proceedings of National Academy of Science U.S.A. September 1971, Vol 68, pp 2236-2240, .

14. Mersereau, R.M. and Oppenheim, A.V. "Digital Reconstruction of Multidimensional Signals from their Projections" Proceedings of the IEEE, October 1974, Vol 62 No. 10, pp 1319-1338.

15. Goitein, M. "Three-dimensional density from a series of two-dimensional projections," Nucl Instr Meth, June 1972, Vol 101, pp 509-518.

16. Gilbert, P. "Iterative methods for the Three-dimensional Reconstruction of an Object from Projections" J. Theoretical Biology, July 1972, Vol 36, pp 105-117.

17. Herman, G.T., Lent, A. and Rowland, S.W. "ART: Mathematics and Applications A report on the Mathematical Foundations and the Applicability to Real Data of the Algebraic Reconstruction Techniques" J. Theoretical Biology, 1973, Vol 42, pp 1-32.

18. Sweeney, D.W. "Interferometric Measurement of Three-dimensional Temperature Fields" (Ph.D. Thesis), University of Michigan/Ann Arbor, 1972.

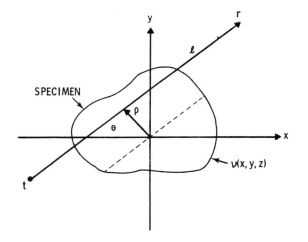

a

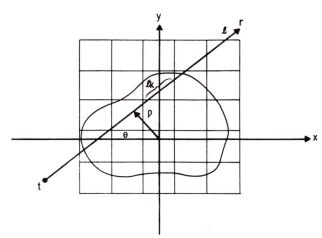

b

Fig. 1

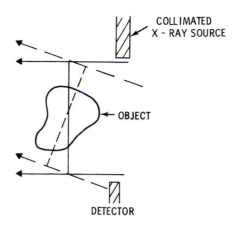

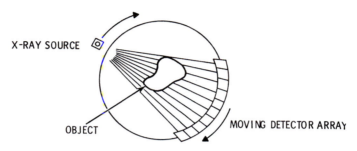

Fig. 2

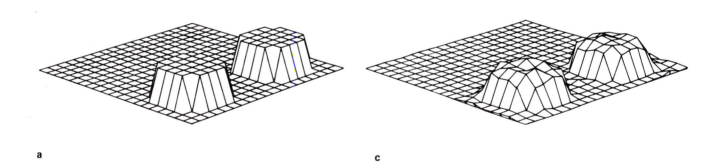

a

c

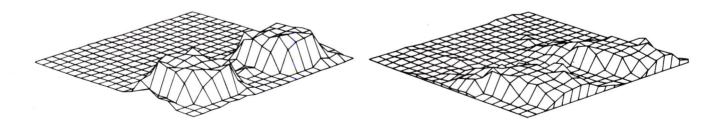

b

d

Fig. 3

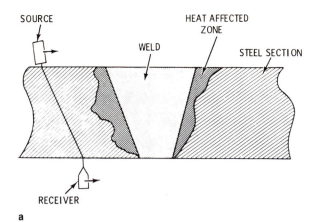

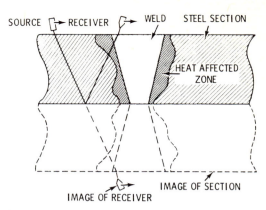

a

b

Fig. 4 a Scanning thick sections by transmission
         b Scanning thick sections by reflection

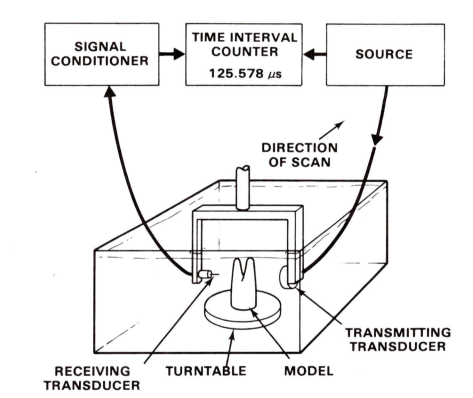

Fig. 5

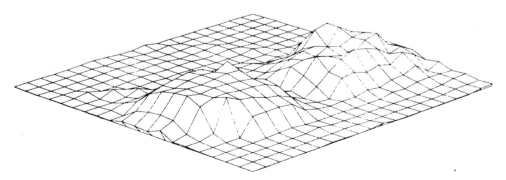

Fig. 6

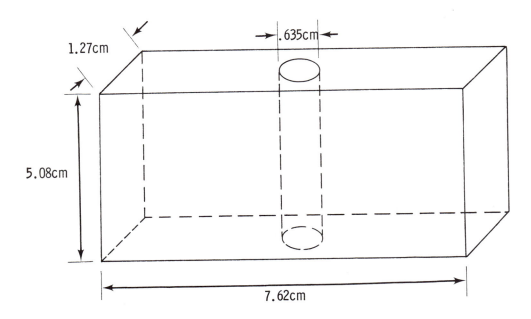

1.27cm

.635cm

5.08cm

7.62cm

a

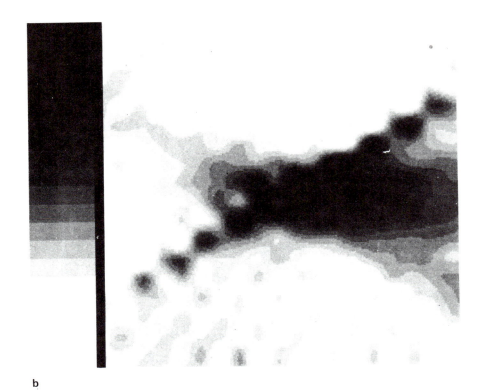

b

Fig. 7

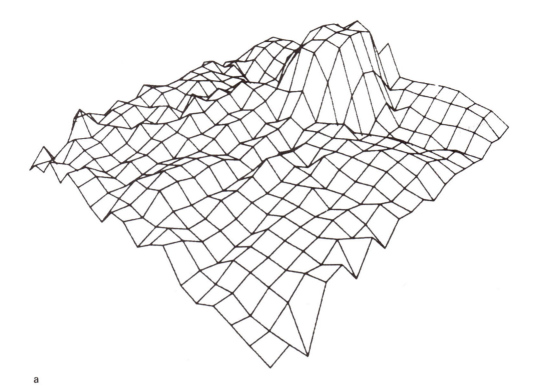

a

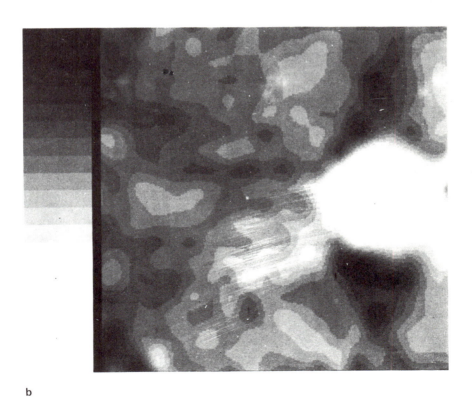

b

Fig. 8

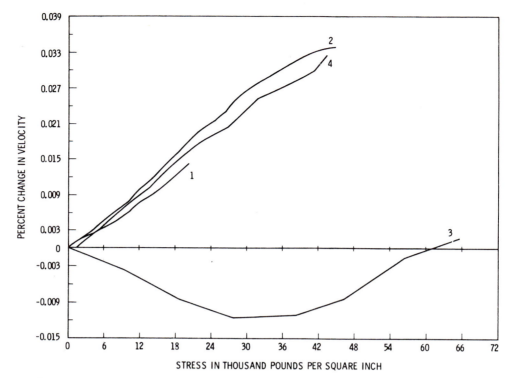

Fig. 9

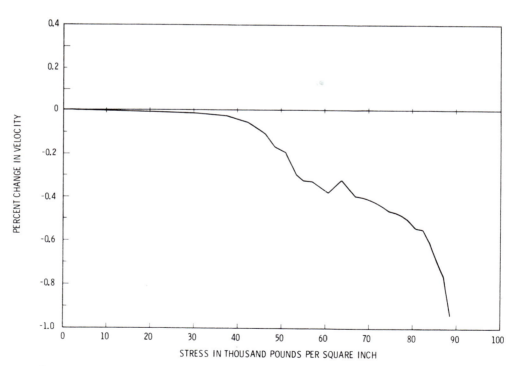

Fig. 10

AUTOMATIC DATA RECORDING DURING MANUAL ULTRASONIC INSPECTION

T. L. ALLEN, BS, R. F. TRIGILIO, BS, A. R. WHITING, BS
Southwest Research Institute, San Antonio, Texas

The MS of this paper was received at the Institution on 18 October 1978 and accepted for publication on 2 November 1978

SYNOPSIS   Conventional manual ultrasonic results are entirely dependent upon the training, skill, and attention to detail of the operator and the details and accuracy of the recorded data. A system has been developed that eliminates most of the manual ultrasonic inspection problems. This system automatically records the position of the search unit with respect to the weld and the associated ultrasonic data. This record is on magnetic tape, and the data is analyzed by a computer.

INTRODUCTION

1.   The major problems associated with conventional manual ultrasonic inspection are that the results are entirely dependent upon the training, skill, and attention to details of the operator and the details and accuracy of the recorded data. These problems increase when the operator works long hours day after day; fatigue will certainly lower operator attentiveness and lower accuracy of recorded data. Mechanized scanning systems reduce/eliminate these problems, but, due to their complexity, are usually utilized only for production runs. An exception is the nuclear power industry where mechanized equipment is utilized due to the high radiation field and/or in order to acquire more reliable data.

2.   Southwest Research Institute (SwRI) made an in-depth study of ultrasonic testing (UT) practices in order to identify areas that could be improved through development of new methods or equipment. Based on this study, it was decided to restrict the investigations to improvements which could be obtained without changing the basic UT method now in widespread use. In making this decision, SwRI did not ignore the desirability of improvements to the basic UT methods nor the probability that such improvements would be forthcoming at some future date. The final decision was to design, fabricate, and test a system which would automatically record the position of the search unit together with the ultrasonic data. Such a system, the search unit tracking and recording system (SUTARS), is shown schematically in Fig. 1.

3.   The criteria which were established for the system included:

(1)   Search unit position in the direction away from the weld (W direction), along the weld (L direction), and angulation (skew) with respect to the weld must be measured remotely. There would be no mechanical connection between the search unit and the component being inspected. This position data would be recorded automatically.

(2)   The ultrasonic data acquired at each search unit position would be recorded automatically.

(3)   The entire A-scan trace within variable gate limits would be recorded and available for computer analysis.

(4)   All data would be recorded on magnetic tape for maximum system portability.

(5)   Computer printout of the data would provide information identical to that recorded by manual ultrasonic methods plus other analysis tools such as B-scan and C-scan plots.

4.   All these criteria were met during development of the first prototype units early in 1978.

DEVELOPMENT

5.   The most significant concept to develop, of course, was the search unit tracking method. Based on the criteria that there be no mechanical connection between the search unit and the component being inspected, the only methods for location of the search unit were material borne elastic waves, airborne sound, or light. All three methods were tried during initial experiments. It was quickly found that material borne elastic waves were not feasible on thin material such as is found in most piping systems. However, both the airborne sound and light methods of location were found to be feasible. The airborne sound method was selected as the one for further development based on the greater amount of time that would be required to fully develop an acceptable light method unit. A location system using light (laser) is potentially more accurate and could be used in explosive atmospheres. Development in this area is continuing.

6.   It was recognized that an airborne sound tracking system was potentially susceptible to interference from ambient noise sources and that the locating accuracy could be degraded by changes in the velocity of sound in air and by Doppler shift resulting from air currents. For this application, however, none of these potential

problems interfered to a perceptible degree. The operating frequency of the tracking system is so high that the only potentially troublesome noise source would be from a small orifice under high pressure. Sound velocity variations due to atmospheric effect are easily accounted for by calibration procedures; and, under all but the most unusual circumstances, the air current velocity encountered in the immediate vicinity during inspections are so low that errors related to Doppler shifts are imperceptible.

SUTARS DATA ACQUISITION

7.   SUTARS consists of two major subsystems: Data Acquisition and Data Processing. In addition, data acquisition is accomplished through several subordinate subsystems:

• Search Unit Position Data Subsystem

• Ultrasonic Data Subsystem

• Parameter Entry Subsystem

• Data Recording Subsystem

Search unit position data subsystem

8.   In SUTARS, position data is acquired by measuring the time it takes a sound wave to travel from a point of unknown position to two points of known position. Fig. 2 illustrates the method. Two sound pulsers are mounted on adjacent corners of a fixture that is attached to the ultrasonic transducer. A set of sound receivers is mounted in a fixed array in known relation to the weld under inspection. The sound wave generated by one pulser is sensed by several of the receivers in the array. The sound propagation times from the pulser to the two nearest receivers is measured by SUTARS and stored in local memory. When called for by the Data Management Subsystem, the values of these two propagation times are transmitted to the Recording Subsystem where they are recorded on magnetic tape. Later, the computer reads these values from the magnetic tape and converts them into position coordinates of the sound pulser by a method of triangulation. The position coordinates of the second sound pulser are similarly determined. Because the geometric relationship of the sound pulses with each other and with the ultrasonic transducer is known, the position coordinates of the transducer and its skew angle relative to the weldment centerline can be calculated. One set of position data consists of the four time values described above plus two numbers that identify the receiver pairs used in the measurement.

9.   Also, a microprocessor was developed to calculate the search unit position in real-time, displaying this data on the front of the unit. The data are also used to generate the area coverage plot.

10.   In order to accurately locate the search unit with respect to the weld, the microprocessor and the computer must know the relationship between the sensor belt and the weld. This is accomplished by:

(1)   Placing the sensor belt near and approximately parallel to the weld.

(2)   Aligning the first microphone with the zero position of the weld or any other known arbitrary position.

(3)   Placing the search unit on the center line of the weld opposite the first microphone and actuating a switch on the face of the control unit.

(4)   Repeating this procedure at each microphone location along the weld.

11.   Experience has shown that an operator can install the sensor belt and establish location within a very few minutes even under adverse conditions.

12.   Electronically, the ultrasonic data collection function is entirely separate from the search unit locating function. This was incorporated into the design so that changes in one function could be made without disturbing the other function. Thus, for example, a change in ultrasonic data collection from the video form to the RF wave form could be made with no impact on search unit location. The main SUTARS control panel with legend is shown in Figs. 3a and 3b, respectively.

The ultrasonic subsystem

13.   The Ultrasonic Subsystem consists primarily of a standard Sonic MK I ultrasonic instrument that has been modified in a minor way to interface with the SUTARS components. Other elements within the Ultrasonic Subsystem include a Time Controlled Gain Subsystem and an Analog to Digital Converter. The fundamental purpose of the Ultrasonic Subsystem is to acquire those signals that exceed a specified recording level and to pass digitized representations of these signals to the data recorder.

14.   A Time Gate Generator is included in the Ultrasonic Subsystem in order to reduce the volume of data recorded to that which is useful for the inspection being made. The Time Gate Generator produces a time gate that is variable in width and position. These variables are adjustable over a wide range of values so that only the signals of interest are enclosed within its boundaries. It is only these signals that are recorded.

15.   A Time Controlled Gain feature is included in the Ultrasonic Subsystem in order to enhance the quality of the signals acquired and to make all signals from equal-size reflectors appear to be of equal amplitude regardless of their time distance from the transducer. In many cases, the required variation of gain is not a simple function of time because of the complex geometry and metallurgical characteristics of the material being inspected. Therefore, in SUTARS, six parameters are adjusted to produce an approximating gain versus time function. These parameters are adjusted with the aid of six artifical signals generated within SUTARS. Initially, the amplitudes of these six signals are adjusted to match the amplitudes of signals derived from a calibration block. This is done with the Time Controlled Gain turned off. After the six signal amplitudes are adjusted, the Time Controlled Gain function is turned on and its six parameters adjusted so that the six signals, which are uniformly spaced over the time span of interest, produce equal outputs from the ultrasonic instrument.

16.   SUTARS allows ultrasonic data collection in two modes:

Mode 1 - Collection of the entire A-scan trace. This is accomplished by recording the A-scan trace within the gate by sampling and recording the amplitude of this trace at 0.3 microsecond intervals. The A-scan trace thus appears as a series of dots separated by approximately 0.5 mm metal path in steel. An interesting feature of this mode is that no matter what amplitude threshold recording level is selected to trigger the selection, once this level has been exceeded anywhere along the trace, the entire trace is recorded.

Mode 2 - Recording of signal amplitude and time. This recording will be made only for signals which are within the gate and are above the selected amplitude threshold recording level. Two adjacent signals must be separated by at least 1.2 microseconds or 2 mm metal path in steel in order to be recorded as two signals. In other words, if two adjacent signals are separated by less than 2 mm metal path, there will be a single record showing the maximum amplitude. This is consistent with the ±2 mm locating capability for the search unit position.

17.   In each mode, the operator has three options:

Option 1 - The gate where signals will be recorded can be varied from almost zero to essentially full screen. In addition, the computer can select smaller gates within the recording gate. This means that, in effect, one can analyze almost an infinite number of gates.

Option 2 - The amplitude threshold recording level may be adjusted from zero amplitude to any higher amplitude. In practice, this level is determined by the applicable Code and, as a practical measure, should be set above the ultrasonic noise (grass) level.

Option 3 - The data sampling rate may be adjusted from 10/sec to 60/sec. A choice of sampling rate should be selected based on the estimated speed of search unit movement and the distance that can be tolerated between data collection points.

The parameter entry subsystem

18.   SUTARS includes a Parameter Entry Subsystem. This subsystem provides the means by which operating data may be manually entered into the system memory for control of the instrument and for recording on tape. It is through this subsystem, for example, that the position and width adjustments of the Time Gate Generator are made. Numeric values representing these adjustments are retained in the SUTARS memory and are periodically recorded on tape under control of the Data Management Subsystem. Later the computer reads these values and makes use of them during the processing of the inspection data. Other parameters which are manually entered into the SUTARS memory are: Operator Identification, Data Sheet Number, Signal Calibration Amplitude, Shoe Angle, Metal Path Length, Sector Number, Pipe Diameter, Sensor Spacing, and a Comment Code. The Comment Code is a numeric value that indicates to the computer what portion of the inspection is being recorded, or it may be used to otherwise annotate the inspection data.

19.   All data entered into SUTARS memory are displayed on the front panel at the time of entry and may be recalled for display at any time thereafter by operator selection. Data entered into SUTARS memory are not lost when the system power is turned off.

The data recording subsystem

20.   The current SUTARS Data Recording Subsystem includes a magnetic tape recorder of the cartridge type. One cartridge tape holds approximately 20 million bits of data. It is estimated that SUTARS operating at the maximum data rate would fill one cartridge in 17 minutes. Under most inspection conditions a tape will last more than one hour. Additional interface circuitry includes two memory banks for double buffering the data received for recording. This permits the data to be received in random bursts while recording is accomplished at a uniform rate over fixed block lengths.

21.   The circuitry in the SUTARS is such that the Data Recording Subsystem may be located several hundred feet from the inspection site, thus making the system more portable and reducing radiation exposure for the Data Recording Subsystem operator.

DATA PROCESSING SUBSYSTEM

22.   The computer software program for data processing is built around a Data General NOVA computer and peripherals as shown in Fig. 4. The steps in data analysis begin with transfer of data from the magnetic tape cartridge to a 9-track tape. This provides for consolidation of data and more rapid input to the computer. The computer operator then selects the gate and minimum amplitude level to be analyzed. The data processing cycle is shown schematically in Fig. 5.

23.   Data Processing is accomplished by essentially recreating the volume of material under examination. This volume, referred to as a sector, consists of the distance along the weld from the first receiving microphone to the sixth (typically 600 mm), the maximum scanning distance away from the weld (typically 200 mm), and the thickness of the material. The software recreates this volume as a mass of cells, each 2.5 mm on a side. Reflector information is processed for each cell which exceeds the selected processing threshold, and this is correlated to the recorded search unit position. The program then displays these data in a variety of output plots. These plots include:

(1)   Parameter listing (shown in Fig. 6). This sheet lists the essential variables of the inspection.

(2)   The inspection coverage plot (shown in Fig. 7). Each X represents a square equal to 0.9 times search unit size, thus assuring coverage overlap sufficient for minimum Regulatory requirements. In the plot shown in Fig. 7, the lack of an X in some areas indicates the search unit was not recorded to have been at that location.

(3)   A listing of position and ultrasonic data similar to that taken during manual inspection (shown in Fig. 8). Manual plotting of this data would yield the same results as manual plotting of manually collected data.

© I Mech E 1979

(4) An analysis in accordance with the rules of ASME Section XI (shown in Fig. 9).

(5) B-scan and C-scan plots (shown in Fig. 10). Each of the small squares in this plot represents a cell two millimeters on a side. This plot also provides a number assigned to each indication.

REPRODUCIBILITY

24. One of the major benefits expected from SUTARS, in addition to more rapid inspection and reduction in radiation exposure in nuclear plants, is that of improved data reproducibility. A study was performed by Southwest Research Institute in June and July of 1978 to correlate SUTARS inspection data with manual inspection data and to determine the reproducibility from one SUTARS inspection to another. In order to accomplish this, seven individual examination tests were performed on a SUTARS evaluation block which contained eight machined reflectors. The seven tests were as follows:

Test 1: Performed using standard manual ultrasonic examination techniques.

Test 2: Performed using SUTARS and water couplant.

Test 3: Identical to Test 2.

Test 4: Identical to Test 2 with a different SUTARS operator.

Test 5: SUTARS recalibrated and Test 4 repeated.

Test 6: SUTARS sensor belt repositioned and Test 5 repeated.

Test 7: SUTARS Test 6 with couplant changed to glycerin.

25. After all tests were completed, the data processing printouts from the SUTARS examinations were compared with each other to determine reproducibility using this system. The data from a SUTARS examination were also plotted along with manual ultrasonic data from Test 1 on a drawing of the SUTARS evaluation block for comparison between manual and SUTARS data.

26. Fig. 11 shows SUTARS data versus manual data as plotted during a portion of this study. As can be seen in the Figure, a high degree of correlation was obtained. During these tests, reproducibility from one SUTARS inspection to another was considered to be much better than had ever been obtained from manual inspection.

CONCLUSION

27. The search unit tracking and recording system has proven to be a reliable, effective method of performing ultrasonic examinations during preservice and inservice inspections. Field tests indicate significant saving in time to perform a high technology task in a difficult work environment. Furthermore, data processing provides significant improvement in records and flexibility for analysis. Reproducibility has been demonstrated to be outstanding.

REFERENCES

1. SUTARS reproducibility study. Quality Assurance Systems and Engineering Division, Southwest Research Institute, San Antonio, Texas, 1978.

2. RENSMEYER M. E. and GROTHUES H.L. Automated data processing for ASME Section XI requirements using the SUTAR system. Quality Assurance Systems and Engineering Division, Southwest Research Institute, San Antonio, Texas, 1978.

3. ALLEN T. L.; GROTHUES H. L.; and JACKSON J. L., Automatic data recording for manual ultrasonic examinations. Quality Assurance Systems and Engineering Division, Southwest Research Institute, San Antonio, Texas, 1978.

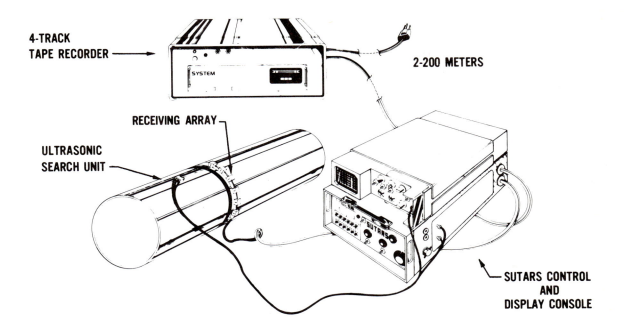

4-TRACK TAPE RECORDER

2-200 METERS

RECEIVING ARRAY

ULTRASONIC SEARCH UNIT

SYSTEM

SUTARS

SUTARS CONTROL AND DISPLAY CONSOLE

Fig. 1 Schematic arrangement of SUTARS data acquisition subsystem

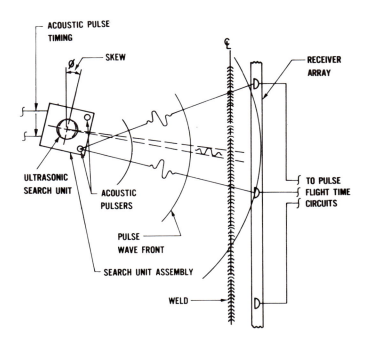

ACOUSTIC PULSE TIMING

SKEW

Ø

RECEIVER ARRAY

ULTRASONIC SEARCH UNIT

ACOUSTIC PULSERS

PULSE WAVE FRONT

SEARCH UNIT ASSEMBLY

TO PULSE FLIGHT TIME CIRCUITS

WELD

Fig. 2 SUTARS search unit location method

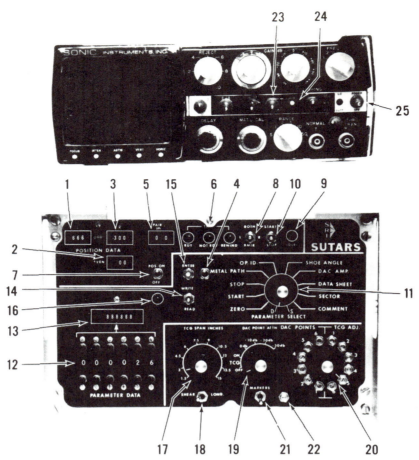

Fig. 3a   SUTARS control unit and ultrasonic instrument

1. 3-digit LED indicates the actual "L" position of the search unit in inches and tenth of inches or in millimeters.

2. 2-digit LED indicates the skew angle of search unit.

3. 3-digit LED indicates the actual "W" position of the search unit in inches and tenths of inches or in millimeters.

4. INCH/MM Switch - determines mode of "L" and "W" readout for (1) and (3).

5. Pair of 1-digit LEDs. Left number indicates space occupied by left spark gap. Right number indicates space occupied by right spark gap. There are 5 spaces. Space 1 is between receivers 1 and 2. Space 2 is between receivers 2 and 3, etc.

6. Three light emitting diode (LED) indicators.

   RDY - Tape recorder ready to record.

   NOT RDY - Tape recorder not ready to record.

   Rewind - Tape recorder rewinding.

7. OFF-ON switch to control power to display items 1 thru 6 (may be used to conserve battery power).

8. 3-position switch: When set left, only left sound source is active; when set right, only right sound source active; center, both sound sources active.

9. LED indicator - lights up when data acquisition subsystem is idle.

10. 3-position switch: Momentary push "up" starts data acquisition subsystem (idle light goes out).

    Momentary push "down" stops and rests data acquisition subsystem (idle light comes on).

11. Parameter Select: Identifies and allocates a space in memory for the data that is going to be entered into 12.

12. Parameter Data: Data is entered through the switches and will be recorded into memory if the system is in the WRITE mode. See steps 14 through 16.

13. Numerical Readout: Displays the data contained in memory for the parameter that has been selected by 11.

14. Read-Write: Controls memory access mode.
    (a) In the READ position the value of the parameter selected by 11 is displayed by 13. Control 12 is inactive in the READ mode.
    (b) The WRITE position arms the ENTER switch 15 and causes indicator 16 to blink intermittently, thereby warning the operator that the actuation of 15 will cause the information contained in 12 to be written into memory at a location determined by 11.

15. Data will be written to memory if control 14 is positioned to the WRITE position and then control 15 is actuated.

16. This indicator will blink intermittently when control 14 is positioned to WRITE. This warns the operator that actuation of control 15 will cause the WRITE mode to be selected. When the WRITE mode is achieved, indicator 16 will illuminate steadily. The functional arrangement of controls 14 and 15 and indicator 16 is designed to prevent un-intentional entry of data.

    Controls 17, 18, 19 and 20 are the operating controls for an advanced Time Controlled Gain feature (TCG). This feature was designed specifically for SUTAR, and its use will provide an accurate and rapid calibration of TCG operation.

21. MARKERS Switch: Displays 3 marker gates on scope.

22. DATA THRESHOLD Adjustment: Allows operator to adjust Data Recording Threshold.

23. CRT DISPLAY Switch: Changes the CRT DISPLAY between UT presentation and sector coverage presentation.

24. CRT ERASE Switch: Erases sector coverage display from CRT.

25. POWER/OFF Switch.

Fig. 3b   Legend call-out for SUTARS

Fig. 4   SUTARS data processing subsystem

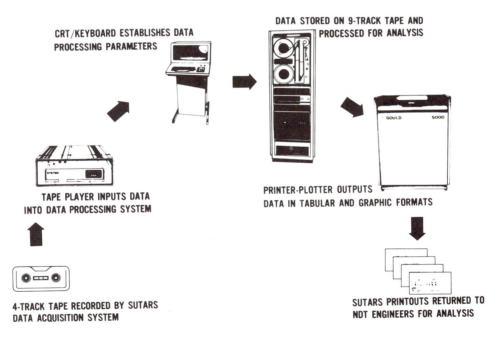

CRT/KEYBOARD ESTABLISHES DATA
PROCESSING PARAMETERS

DATA STORED ON 9-TRACK TAPE AND
PROCESSED FOR ANALYSIS

TAPE PLAYER INPUTS DATA
INTO DATA PROCESSING SYSTEM

PRINTER-PLOTTER OUTPUTS
DATA IN TABULAR AND GRAPHIC FORMATS

4-TRACK TAPE RECORDED BY SUTARS
DATA ACQUISITION SYSTEM

SUTARS PRINTOUTS RETURNED TO
NDT ENGINEERS FOR ANALYSIS

Fig. 5   SUTARS data processing configuration

```
SITE: FILE #2, TAPE #2 (1200 FT)
PLANT: SWRI - LAB
WELD I.D: T-SECT, ODD SECTS (2ND)
OPERATOR I.D: 4180
PROCEDURE NO: S00-50
CALIBRATION SHEET NO: 192483
DATA SHEET NO: 140005

INSTRUMENT I.D:        1
DELAY(1) =      5
DELAY(2) =      9
DELAY(3) =     47

INSPECTION ANGLE:    45 DEGREES
THRESHOLD LEVEL: 50 % DAC
TRANSFER DATA THRESHOLD LEVEL:   0.% DAC
AMPLITUDE LEVEL (DAC):     8.00

OBSERVED METAL PATH:   3.00

WELD REF. POINTS:
    POS.     LENGTH(L)      WIDTH(W)
     1         0.00          0.00
     2         5.00          0.00
     3        10.00          0.00
     4        15.00          0.00
     5        20.00          0.00
     6        25.00          0.00
     7        30.00          0.00
     8        35.00          0.00
     9        40.00          0.00
    10        45.00          0.00
    11        50.00          0.00
    12        55.00          0.00
    13        60.00          0.00
    14         1.50          0.00

NO. OF SECTORS ON WELD:   3

SECTOR I.D. NO:   1
SECTOR LENGTH: 25.73 IN.,     WIDTH: 10.53 IN.
SEARCH-UNIT CELL COVERAGE SIZE: 0.45 IN.
WALL THICKNESS:   0.98 IN.
```

Fig. 6 SUTARS parameter listing

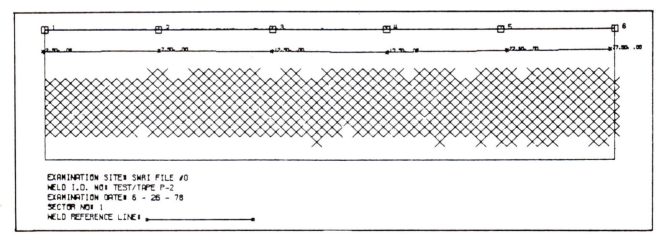

Fig. 7 SUTARS inspection coverage plot

SITE: SWRI FILE #0

WELD: TEST/TAPE P-2

CALIB. SHEET NO: CP-6

DATA SHEET NO: 062302

WALL THICKNESS: 2.0000

EXAMINATION DATE: 6-26-78

EXAMINER I.D. 4653

ANGLE: 46

PROCEDURE NO: 800-50

AMPLITUDE LEVEL: 100.0% DAC

SECTOR NO	INDIC. NO.	XDUCER L1	XDUCER MAX.AMP.L.	XDUCER L2	XDUCER W1	XDUCER MAX.AMP W	XDUCER W2	METAL PATH @ W1	METAL PATH @ MAX.AMP.W.	METAL PATH @ W2	MAX.AMP (% DAC)
1	1	1.7	1.7	3.3	1.6	2.4	2.4	2.70	3.30	3.30	144.8
1	2	4.8	4.9	6.6	2.1	2.4	2.9	2.80	3.00	3.30	144.8
1	3	8.1	8.2	9.8	1.6	2.0	2.5	2.80	3.00	3.30	144.3
1	4	11.7	11.7	13.5	2.3	2.3	2.3	3.00	3.00	3.00	144.8
1	5	15.3	15.3	17.0	2.1	2.1	2.1	3.10	3.10	3.10	144.8
1	6	18.6	18.8	20.6	2.5	2.5	2.5	3.10	3.10	3.00	144.8
1	7	22.4	22.9	24.1	1.8	2.2	2.3	2.90	3.10	3.20	144.9
1	8	26.2	27.1	27.3	2.6	2.6	2.6	3.10	3.10	3.10	135.8

Fig. 8 SUTARS data sheet

SITE: SWRI FILE #0

WELD: TEST/TAPE P-2

CALIB. SHEET NO: CP-6

DATA SHEET NO: 062302

WALL THICKNESS: 2.0000

EXAMINATION DATE: 6-26-78

EXAMINER I.D. 4653

ANGLE: 46

PROCEDURE NO: 800-50

AMPLITUDE LEVEL: 100.0% DAC

INDIC. NO	SECTOR NO	PEAK AMPL	PERCENT T.W.	X START	X LENGTH	X MAX	Y START	Y LENGTH	Y MAX	Z START	Z LENGTH	Z MAX
1	1	144.8	25.0	1.57	1.90	1.67	-0.44	0.80	-0.04	1.50	0.50	1.60
2	1	144.8	25.0	4.67	1.90	4.77	0.07	0.70	0.17	1.50	0.50	1.80
3	1	144.8	25.0	7.87	2.00	7.97	-0.42	0.70	-0.22	1.50	0.50	1.80
4	1	144.8	20.0	11.47	2.00	11.47	0.08	0.60	0.19	1.60	0.40	1.90
5	1	144.8	20.0	15.07	1.90	15.17	-0.30	0.50	-0.10	1.60	0.40	1.90
6	1	144.8	25.0	18.57	1.90	18.67	0.20	0.50	0.30	1.50	0.50	1.80
7	1	144.8	20.0	22.27	1.80	22.67	-0.39	0.60	0.01	1.60	0.40	1.70
8	1	135.8	15.0	26.27	1.30	27.26	0.12	0.30	0.33	1.70	0.30	1.80

Fig. 9 SUTARS advanced data sheet

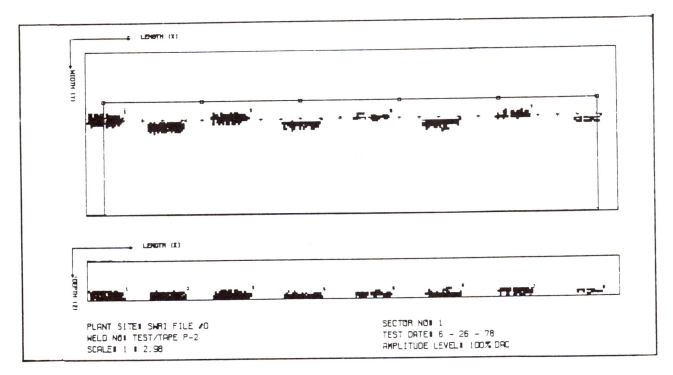

Fig. 10   SUTARS 'B' and 'C'-scan plots

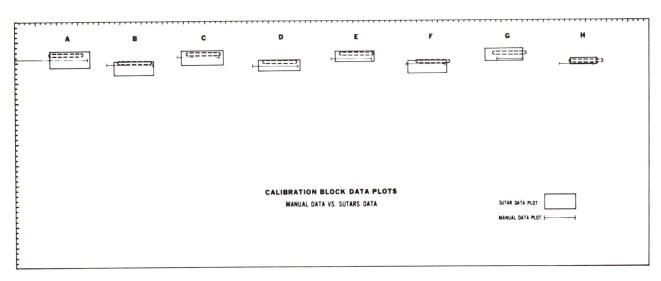

Fig. 11   Manual data versus SUTARS data

C38/79

THE RELATIONSHIP BETWEEN ACOUSTIC EMISSION AND FRACTURE MECHANISMS ILLUSTRATED BY A STUDY OF CLEAVAGE IN MILD STEEL

C. B. SCRUBY, H. N. G. WADLEY and J. E. SINCLAIR
Atomic Energy Research Establishment, Harwell, Didcot, Oxon

The MS of this paper was received at the Institution on 30 November 1978 and accepted for publication on 18 January 1979

SYNOPSIS   A theoretical and experimental approach to the relationship between acoustic emission and fracture mechanisms is presented.   Acoustic emission waveforms are recorded during the cleavage fracture of mild steel, using a calibrated, broad-band detection system.   Source strengths are deduced as time-varying crack volumes using a Green's function derived for an elastic half-space.   A typical source is modelled as a cleavage crack of diameter ~ 50 μm and opening ~ 0.4 μm, which propagated at $\sim 20\%$ shear wave velocity.   The model is in broad agreement with fractographic analysis of the specimens.

INTRODUCTION

1.   The fracture of metals is accompanied by localised strain relaxations which may cause energy to be released as transient elastic waves (acoustic emissions) that propagate away from the fracture source to the test-piece surface.   By using suitable detection methods it should therefore be possible not only to detect and locate fracture events, but also to assess their severity by analysis of the recorded signals. Thus the acoustic emission technique has great potential as a practical nondestructive testing tool for engineering structures.   It has in addition often been suggested that by suitable analysis of the acoustic emission waveform new insights might be gained into the dynamics of microdeformation and fracture.

2.   There has been no shortage of applications for acoustic emission: it has been applied to the monitoring of a wide range of structures from aircraft components to nuclear pressure vessels. Nevertheless, the technique continues to have two major problems associated with it.   Firstly, it suffers from poor reliability: some major flaws emit, while others do not.   Secondly, it has proved difficult to characterise growing defects on the basis of recorded emission measurements alone.

3.   The first problem is crucial: acoustic emission can only be used with confidence if it is known that the critical defects in a given structure are going to emit.   It is thus important to determine which deformation and fracture processes act as sources of emission, and how they are influenced by microstructural variation.   Extensive laboratory experiments have shown that a wide range of defects may generate detectable emission (Ref.1).   In most commercial materials several different types of source may be present, but if specimens are prepared with care it is often possible to isolate a single emission source process.   For example, if pure Al 4 wt% Cu is solution treated, and aged for varying times at about 170°C to produce a range of microstructures (Refs.2,3), it can be shown that the detected emission is associated with the shear of weak Guinier-Preston zones.   The absence of these weak obstacles to dislocation motion in the solid solution, or the presence of very strong obstacles in the peak hardened condition results in almost no emission.   These tests demonstrate that a rapid release of strain energy is required to give detectable acoustic emission. Other systematic tests with aluminium alloys have shown that composition, purity and microstructural variables such as grain size can have a very strong influence on the character of the acoustic emission (Ref.4).

4.   In ferritic steels few systematic studies have been reported.   However, it is clear that the decohesion and fracture of inclusions is an important source of emission in commercial steels (Refs.5,6,7).   The cracking/decohesion of carbides (Refs.8,9) and the escape of dislocations from interstitials either at yield (Ref.10) or during dynamic strain ageing (Ref.8), are further possible sources of emission during deformation.   During fracture a similar complex situation exists.   Brittle crack growth processes such as cleavage (Ref.11), stress corrosion cracking (Ref.12), hydrogen (Ref.13) and temper embrittlement (Ref.14) all generate detectable emission, though the level of activity is dependent upon metallurgical and testing conditions.   However, the normal mode of crack growth in medium and low strength structural steels by slow void coalescence tends to be quiet.   When the same steels are tested in a high strength condition there is evidence that the alternating shear mechanism of fracture does generate emission (Refs.15,16).

5.   During service, emission may be detected from a range of defects, only some of which are critical, and it is important therefore to be able to characterise the emission and assess the severity of the defect source, especially if it is located on a part of the structure which is inaccessible to other NDT techniques.   Workers have attempted to relate amplitude distributions (Ref.17) and frequency spectra (Refs.18,19,20) to source processes, but usually with only very

limited success. The reasons for this are firstly that the ultrasonic pulse carrying the information about the defect must pass through the test-piece to the surface. As it does so there are reflections, mode conversions and attenuation which degrade the pulse so that no simple relationship exists between the surface motion and the source event. Secondly the detection system, and the transducer in particular, may distort the data further by non-linearities and reductions in bandwidth. Finally, much A.E. data is expressed in units, such as "counts", which are not defined in absolute terms, so that results often cannot be more than qualitative. This unfortunately makes it difficult to compare data from different laboratories.

6.   To overcome these difficulties, Wadley and Scruby (Ref.14) used a broadband detection system and showed that differences in recorded waveform could be correlated with changes in crack growth mechanism. The objective of the current Harwell study is to examine in more detail the relationship between acoustic emission waveform and fracture process. In the following section the theoretical basis for the study is outlined, while in the third section the broadband system for waveform measurements is described. For this series of tests one fracture process, cleavage, in mild steel was chosen, and in the fourth section are presented the acoustic emission results when the position of the source was varied within the gauge section of the specimen. Finally in the fifth section the data is discussed and used to produce a model of the emission source process, and this is compared with the results of fractography.

THEORETICAL CONSIDERATIONS

7.   Theoretical studies have not been applied to acoustic emission in a consistent manner until very recently, although the analogous problem of characterising seismological sources (e.g. earthquakes) has received extensive study (Refs.21,22).

8.   The aim of the theoretical studies is to deduce quantitative information about deformation and crack growth processes by analysis of acoustic emission waveforms. This can be tackled in three equally important stages:

(i)   A mathematical description is required of the various deformation or fracture processes of interest. The description can be, for instance, in terms of the time varying distribution of forces which would have to be applied to a perfect body to produce the same elastic disturbance as the process of interest. An alternative, which is more easily related to fracture events, is to model a source in terms of imaginary dislocation distributions. The two schemes are formally related in a simple way through the elastic constants (Ref.23).

(ii)   A transfer function is required to relate given source functions to the surface displacements which will be observed. In space-time problems, such a transfer function is usually called a "Green's function", and expresses the elastic displacement at a given time and position due to a unit impulse of force applied at zero time at a given point. The response to sources extended in space and time can then be obtained by a convolution process. Analytical expressions for elastic Green's functions are however known only for the very simplest of bodies.

(iii)   To work back from observed displacements to a source description poses both a mathematical problem (the inversion of stage (ii), i.e. deconvolution) and an experimental one (the accurate measurement of sufficient displacement wave-forms). Clearly, to determine n independent source parameters as a function of time, we need at least n separate records from the same event.

The Source

9.   Even the simplest of metal fracture processes comprises a complex combination of crack extension and plastic flow, and we have not yet attempted theoretically to model such events. Rather, restricting attention to the growth (or creation) of planar microcracks, it can be deduced that all such processes will appear from a distance to be equivalent to the appearance of a small dislocation loop. For instance, a crack loaded normally (in mode I) will be equivalent to an edge loop whose strength $b\delta A$ (Burgers vector x loop area) is equal to the crack volume. The dislocation loop may in turn be equivalently represented in terms of three orthogonal force dipoles (Ref.23). This type of microcrack model is in general characterised by two constant orientation parameters and three time dependent amplitudes (the crack opening volume and equivalent quantities for two shear modes). For a crack in a half-space, normal displacement measurements at the epicentre can only detect the mode-I opening, and only one orientation, the inclination of the crack to the horizontal, can be distinguished. Thus, a two-parameter (one constant, one time varying) model is appropriate.

The Green's Function

10.   Several authors (Refs.24,25,26) have given solutions for elastic point force problems in a half-space, and corresponding solutions for an infinite plate have recently been computed (Refs. 27,28). More complicated geometries appear as yet to be intractable. For the half-space, the solutions at the epicentre assume closed analytic form, and the spatial derivative of the solution of Willis (Ref.25) were calculated to give surface displacements from force dipoles, quadrupoles, etc. The dipole solutions were combined to form solutions for the dislocation loop models discussed above. Figure 1 shows the result for a horizontal loop (crack) of volume $V = b\delta A\ H(t)$ ("switching on" at t = 0) buried at a depth d. At early times, the delta function part of the longitudinal arrival dominates; its strength is

$$S = \frac{c_2^2}{2\pi d\ c_1^3}\ b\delta A \qquad (1)$$

where C_1 and C_2 are the longitudinal and shear wave speeds, respectively.

Deconvolution

11.   For a given crack orientation, the above procedure yields a solution $U(t)$ for the surface displacement when the source strength (volume) is a unit step function $H(t)$. If a function $Q(t)$ can now be found which, when convolved with $U(t)$,

gives the source function H(t), then the same "operator" Q(t) will also transform any measured surface displacement waveform U(t) into the corresponding source volume $\delta V(t)$. The equation for Q(t) is

$$Q(t) * U(t) = H(t) \qquad (2)$$

12. In the discrete time sampled representation used for the experimental data, this becomes

$$\sum_{m=0}^{n} Q_m U_{n-m} = 1 \text{ for } n = 0,1,2 \ldots \qquad (3)$$

where Q_m is the mth sampled value of Q(t), and similarly for U. This series of equations can readily be solved for each Q_n in turn, starting from $Q_o = 1/U_o$.

13. Finally, from any given sampled displacement record, U_i, the convolution

$$\delta V_n = \sum_{m=0}^{n} Q_m U_{n-m} \qquad (4)$$

gives the source volume history δV_n.

EXPERIMENTAL PROCEDURE

14. It was found to be necessary to develop both a special specimen geometry and a broadband detection system in order to relate acoustic emission waveforms to known fracture events. Careful materials preparation to provide a well characterised acoustic emission source was considered to be equally important.

Specimen Geometry

15. The ideal specimen shape for applying the theory of Section 2 would be a half-space, while testing considerations require the choice of a geometry such as a dumb-bell. The specimen chosen for these tests, the Yobell (Ref.29) (Fig.2) is a compromise between these requirements. First, in order to reduce signal losses by ultrasonic attenuation and geometrical spreading, the source to transducer distance was kept to a minimum ($\sim$18 mm). Secondly, in these tests the waveforms were to be measured at one point, the epicentre, and deformation and fracture were restricted to a 2 mm long, 3 mm diameter, gauge section vertically below the transducer. Thus the specimen transfer function remained approximately constant for all recorded waveforms. Finally, the diameter of the section between source and transducer was made large enough (60 mm) to prevent interference from reflected wave fronts to the first 6 μs of recorded signal.

16. It was realised that the presence of a gauge section could lead to some distortion of the waveform which would vary with source depth. In order to examine this effect, three specimens were therefore tested, with 0.5 mm deep, 60° notches at the top (A), middle (B) and bottom (C) of the gauge section (see Fig.2).

Measurement of Surface Motion

17. In order to calculate the time history of an acoustic emission source, we must accurately measure the motion of the specimen surface over the full frequency spectrum of the source. Typical surface displacements may be $\lesssim 10^{-10}$m and source spectra can extend beyond 10 MHz (Ref. 29). The piezoelectric transducers used for conventional ultrasonics fulfil the sensitivity criterion, but do not have adequate bandwidth, nor are they straightforward to calibrate (Ref. 30). A parallel plate capacitor can be made to satisfy both requirements for use as a detector of acoustic emission (Ref.31).

18. The surface of the specimen forms one plate, and the second circular plate of area A is held a distance x above the surface and at a potential V so that the displacement dx due to the arrival of an elastic wave causes a small change, dq, in the charge stored in the capacitor. The sensitivity is

$$\frac{dq}{dx} = -\frac{\varepsilon AV}{x^2} \qquad (5)$$

where ε is the dielectric constant of the gap between the plates.

19. The spherical wavefronts from the source do not reach all parts of the transducer simultaneously. The consequent time delays lead to loss of phase coherence and limit the effective bandwidth. The bandwidth of the transducer can be shown to be inversely proportional to A (Ref.31). The capacitance transducer used for the current tests had A = 28.3 mm^2, x = 2.7 μm, V = 50 V and ε_{air} = 8.85 x 10^{-12} Fm^{-1}. Thus the sensitivity was 1.72 x 10^{-3} Cm^{-1} surface displacement. The transient response of the transducer has also been measured using a laser interferometer (Ref.30).

20. The transducer was coupled to a wide-band low-noise charge amplifier (Ref.29), and the signal from this was conditioned by a 35 kHz – 45 MHz band-pass filter, further amplified, and recorded with a Biomation 8100 transient recorder (Fig.2). The recorder digitised the acoustic emission waveform at 10 ns intervals with 8-bit precision. The digitised waveform was stored, for later analysis, on a magnetic disc in a PDP8/E computer. For the current experiment, the detection system had a risetime of 20 ns and a sensitivity of 8.5 x 10^{-3} V pm^{-1}. Steps were taken to ensure that the sensitivity remained constant for all three tests. The electrical signal from the amplifier was also monitored with a R.F. power meter to give a qualitative guide to the rate of emission energy release as a function of applied load.

Materials Preparation

21. Cleavage fracture of ferrite generates acoustic emission and is metallurgically one of the best defined fracture processes. It is known generally to involve the fast fracture of single grains (Ref.32). Cleavage is usually only observed at low temperatures, which makes it inconvenient for acoustic emission experiments. However it can sometimes be induced at room temp-

erature using .notched specimens and a rapid
quench rate.

22.   Thus for the experiments reported here,
three notched Yobell specimens were machined from
a bar of commercial mild steel, composition as
shown in Table 1. After annealing at 850°C for
30 minutes, they were quenched in iced brine to
produce a martensite structure with a prior
austenite grain size of 21μm. The specimens were
tested to failure in an Instron 1195 screw-driven
tensile testing machine at a crosshead displacement rate of 0.1 mm min^{-1}. The load and R.F.
power meter output were plotted on a chart
recorder as a function of time. The fracture
appearance of each specimen was examined by
scanning electron microscopy.

EXPERIMENTAL RESULTS

Load and A.E. Power Data

23.   The load and acoustic emission power are
shown as a function of time for all three tests
in Fig.3. At low load, a non-linear load-time
relation was observed due to the high compliance
of the gripping system (which contained electrical insulation). Beyond this region a linear
relation was observed, indicative of only limited
plastic deformation. The first detected
emissions occurred at a load of ∿0.5 kN for each
test, and their rate of generation increased as
final fracture was approached. Specimen B, with
the notch at the centre of the gauge length,
generated more emissions (236) than either
specimen A (108) or C (91). The origin of the
difference may have been the lower stress con-
centration in specimen B, the consequent need for
a larger critical crack length to initiate catastrophic fracture, and hence a greater quantity
of subcritical crackgrowth.

Fracture Appearance

24.   All three specimens underwent a mixed
fracture mode. The fracture faces were covered
with isolated cleavage facets connected by
shallow, large diameter dimples, Fig.4. The area
of a typical cleavage facet was ∿1000 μm^2 and
was generally consistent with cleavage cracks
extending over one or two grains.

Measured Waveforms

25.   Typical acoustic emission waveforms from
each test are shown in Fig.5a; it can be seen
that A resembles the theoretical waveform for a
dislocation loop source (Fig.1) when account is
taken of the removal of the low frequency
component by signal conditioning. It can also be
seen that the δ-function of the calculated L
arrival is replaced by a pulse of finite duration
and height, as a consequence of the finite
spatial extent and lifetime of a real source.

26.   Comparison of typical waveforms from
specimens B and C with specimen A shows that a
second pulse followed the first when the source
was located deep within the gauge. This "echo"
pulse was consistently present during tests B and
C and never during A, and must be associated with
the acoustic response of the gauge section. The
time difference between the two pulses was almost
always larger for C, suggesting that the origin
of the echo was a reflection from the top of the
gauge section where there is a sudden change in

cross-sectional area.

INTERPRETATION

27.   The time dependence of emission activity
(Fig.3) and fractographic results (Fig.4)
strongly suggest that the source of acoustic
emission was the formation of subcritical cleavage cracks, with a range of areas, typically
∿1000 μ m^2.

28.   We now apply the analysis of Section 2 to
determine the acoustic emission source from each
measured waveform by deconvolution (2.3). The
Green's function was assumed to be that for a
half space (2.2), taking the distance from epicentre to notch root as source depth, d. The
source was modelled as an infinitesimal dislocation loop, oriented parallel to the surface
(2.1).

29.   Deconvolution resulted in a source volume
history for each wave-form (Fig.5b). The fall in
calculated volume to zero and negative values at
long times is a result of high-pass filtering at
35 kHz. The small error introduced for short
duration events has been neglected.

30.   From each deconvolved waveform, the maximum volume attained (V_1) was measured as shown
in Fig.6. Physically this should represent the
maximum volume reached by a freshly nucleated
microcrack. In order to facilitate comparison
between the three tests, histograms of volume,
V_1, were plotted (Fig.7a). The extent of the
distribution was restricted by the limited
dynamic range of the 8-bit precision transient
recorder. A percentage of waveforms overloaded
the recorder in each test and could not be used
to calculate source volumes.

31.   A peaked distribution, weighted towards
smaller values of V_1, is observed for all three
specimens. However, the histograms for B and C
also exhibit an increasing proportion of large
volume sources, so that the mean value of V_1
apparently increases with depth (Table 2). This
effect is due to the second, echo peak discussed
above, which gives a volume contribution that
adds on to that of the longitudinal peak. This
serves to emphasise the great care that must be
taken in interpreting acoustic emission sources
when no account is taken of nearby specimen
boundaries.

32.   In order to reduce the effect of echo
pulses a different volume parameter, V_2, was
chosen (Fig.6). The volume value was measured at
the time of the first peak in the transient and
then assuming that the peak was symmetrical this
value was doubled to give V_2. This assumption
leads to inaccuracy in V_2 as a measure of the
final volume of a freshly nucleated microcrack.
In particular it can be seen from Fig.6 that the
longitudinal peak is often unsymmetrical so that
the source apparently takes considerably longer
than time τ to reach maximum volume. The long
trailing edge of the longitudinal peak may
possibly be a feature of the crack growth process
itself, which would suggest that there are significant slow relaxations around the crack tip;
however interference from the circular notch
cannot be discounted. When the histograms of
V_2 are compared (Fig.7b) it is seen that the
three specimens now give the same source volume
distribution within statistical limits. Thus,

neglecting any effect due to the proximity of the circular notch, V_2 can be considered as a system independent source parameter.

33.   A further parameter, the source lifetime τ can also be measured for each emission. The effect of gauge echoes again needed to be reduced and τ was calculated as twice the time for the first pulse to rise from zero to its maximum value (Fig.6). The assumptions implicit here are the same as those for calculating V_2 above. The lifetime histograms for the three tests (Fig. 8) were statistically indistinguishable suggesting that τ is also a system independent parameter, and can be used together with V_2 in attempts to model the emission source.

34.   The mean values, $\bar{V}_2$ and $\bar{\tau}$, for all the recorded transients were calculated from the values from each test (Table 2) and can be used to model an acoustic emission source during these tests. Thus a typical source event was a microcrack whose volume increased by 600 μm^3 ($\bar{V}_2$) in 88 ns ($\bar{\tau}$).

35.   If we assume a horizontal, circular, elastic crack of radius a (Fig.9) then with mode I loading under uniform applied stress σ, the crack faces will open a distance 2$\tilde{b}$ given by (Ref.33)

$$\tilde{b} = \frac{2\,(1-\nu^2)\,\sigma\,a}{E} \qquad (6)$$

where ν is Poisson's ratio, and E is Young's modulus.

36.   The tensile stress in the vicinity of a small microcrack within a notched specimen is difficult to estimate. Clearly, it will vary during the test as the applied load is raised and the uncracked ligament decreases. For the purposes of this interpretation we shall assume $\sigma = 10^9$ Nm^{-2} (an estimated nett-section stress). Using $\nu = 1/3$ and $E = 2.1 \times 10^{11}$ Nm^{-2}, equation (6) yields $\tilde{b} = 8.5 \times 10^{-3}a$. Substituting for $\tilde{b}$ in the formula for the volume of an oblate ellipsoid:

$$V = \frac{4}{3}\pi a^2 \tilde{b} \qquad (7)$$

gives,

$$V = 3.5 \times 10^{-2}\,a^3 \qquad (8)$$

Thus a = 26 μm for the typical source volume of 600 μm^3, and
$\tilde{b} = 8.5 \times 10^{-3}\,a = 0.22\ \mu m$.

It is noted that for given V, a is proportional to $\sigma^{-1/3}$ so that errors in the estimation of σ do not lead to large errors in crack radius. Crack opening however is more sensitive to errors in σ. Allowance for plasticity would increase $\tilde{b}$ and decrease a.

37.   The typical crack area was estimated from fractography as 1000 μm^2 (section 4.2). If this were a circular crack it would have a radius of 18 μm. The calculated value of a = 26 μm is

in remarkably good agreement with this, bearing in mind the assumptions made in the calculation and the scatter in actual cleavage facet areas.

38.   It is, finally, possible to estimate crack speed. It is assumed for this purpose that a typical cleavage crack is initiated at one grain boundary and is arrested when it reaches another, so that its mean speed is $2a/\tau = 590$ ms^{-1}. This value, which ignores acceleration and deceleration, is 18% of the shear wave velocity. However, this estimate must be treated with the same caution as the crack radius, because of the many assumptions which have been made. Brittle fracture velocities for steel of 1370-2000 ms^{-1} have been reported (Ref.34). The velocity estimated above is somewhat less, but no allowance has been made for plasticity which would have the effect of lowering crack velocity.

SUMMARY

39.   This series of tests and the accompanying theoretical studies have shown that careful acoustic emission waveform measurement and careful interpretation can give valuable new insights into the nature of fast fracture events. In particular, it has been possible to deduce the strengths of the emission sources as time-varying crack volumes. From calculated average values of crack volume and lifetime a typical source has been modelled as a cleavage crack of diameter $\sim$50 μm, opening $\sim$0.4 μm and velocity $\sim$20% shear wave velocity.

40.   There remains, however, a need for further work to establish in more detail the relationship between acoustic emission and fracture mechanisms. Much needs to be done to improve experimental accuracy, in addition to increasing the number of measured fracture source parameters. It is furthermore necessary to extend these studies to more general types of testing configuration, such as crack opening geometries. If these prove successful it may then be possible to determine the physical basis for flaw characterisation by waveform analysis on engineering structures.

ACKNOWLEDGEMENTS

    We wish to thank Mr. G. Shrimpton for his practical assistance, and Drs. B.L. Eyre and G.J. Curtis for many helpful discussions during this study. Finally we wish to acknowledge funding during the development of the broadband technique by the Procurement Executive (M.o.D.), through Admiralty Marine Technology Establishment.

REFERENCES

1. Wadley, H.N.G., Scruby, C.B. and Speake, J. Submitted to Int. Met. Rev.

2. Wadley, H.N.G. and Scruby, C.B. Metal Science J., 1978, June, 285-289.

3. Fenici, P., Kiesewetter, N. and Schiller, P. EUR-5550, C 1974-75, 173.

4. Wadley, H.N.G. and Scruby, C.B. Unpublished work.

5. Ono, K., Huang, G. and Hatano, H. 8th World Conf. N.D.T., 1976, Cannes, France.

6. Ono, L. and Lanoy, R. 4th Acoustic Emission Symposium, Sept. 18-20, 1978, Tokyo, Japan.

7. Bentley, M.B. and Birchon, D. Proc. Inst. Acoustics, 1976.

8. Holt, J., Palmer, I.G. and Goddard, D.J. Berichte 20th Symp. der Deutschen Gesell-schaft fur Metallkunde, Munchen, 1974, 24.

9. Ingham, T., Stott, A.L. and Cowan, A. Int. J. Pres. Ves. and Piping, 1974, 2(1), 31.

10. Krasovskii, A.Y., Novikov, N.V., Nadezhoin, G.N., Likatskii, S.I. and Bakalinskaya, N.D., Problemy Prochnosti, 1976, 8(10), 3-7.

11. Radon, J.C. and Pollock, A.A. ASTM STP 559, 1974, 15-30.

12. Chaskelis, H.H., Cullen, W.H. and Krafft, J.M., ASTM STP 559, 1974, 31-44.

13. Dunegan, H.L. and Tetelman, A.S., Eng. Fract. Mech., 1971, 2(4), 387-402.

14. Wadley, H.N.G. and Scruby, C.B. To be published in Acta Met.

15. Clark, G. and Knott, J.F. J. Metal Science, 1977, 11(11), 531-536.

16. Wadley, H.N.G., Scruby, C.B., Furze, D. and Eyre, B.L. To be published in J. Metal Science.

17. Ono, K., Mat. Eval., 1976, Aug., 177-184.

18. Crostack, H.A. Ultrasonics, 1977, Nov., 253-262.

19. Graham, L.J. and Alers, G.A. Ultrasonic Symp. Proc. IEEE, Oct. 4-7, 1972, 18.

20. Ringshall, N.W. and Knott, J.F. 4th Acoustic Emission Symposium, Sept.18-20, 1978, Tokyo, Japan.

21. Gilbert, F. and Dziewonski, A.M., Proc. Roy. Soc., 1975, 278A (1280), 187-269.

22. Stump, B.W. and Johnson, L.R. Bull. Seism. Soc. Amer., 1977, 67(6), 1489.

23. Burridge, R. and Knopoff, L. Bull. Seism. Soc. Amer., 1964, 54, 1875-88.

24. Pekeris, C.L. and Lifson, H. J.A.S.A., 1957, 29(11), 1233.

25. Willis, J.R. Phil. Trans. Roy. Soc. A., 1973, 274, 435.

26. Johnson, L.R. Geophys. J.R. Astr. Soc., 1974, 37, 99-131.

27. Pao, Y.H. "Elastic Waves and Non-Destructive Testing of Materials," 1978, A.S.M.E., A.M.D. 29, 107.

28. Simmons, J. and Willis, J. Unpublished work.

29. Scruby, C.B., Collingwood, J.C. and Wadley, H.N.G. 1978, J. Phys. D., 11, 2359.

30. Speake, J.H., Moss, B.C. and Braedel, T. N.D.T. International, 1978, in press.

31. Scruby, C.B. and Wadley, H.N.G. J. Phys. D., 1978, 11 1487-1494.

32. McMahon, C.J. and Cohen, M. Acta. Met., 1965, 13, June, 591-604.

33. Knott, J.F. "Fundamentals of Fracture Mechanics", 1973, Butterworths, London, 58.

34. Roberts, D.K. and Wells, A.A. Engineering, 1954, Dec., 820-821.

TABLE 1

COMPOSITION OF MILD STEEL

ELEMENT	C	N	O	P	S	Mn	Ni	Cr
CONCENTRATION	0.26	0.007	0.003	0.019	0.025	0.7	0.02	0.01

	Mo	Nb	V	As	Sb	Sn
	<0.005	<0.005	0.001	0.02	<0.005	<0.005

TABLE 2

Specimen No.	depth d/mm	transients recorded	transients overloaded	mean $V_1/\mu m^3$	mean $V_2/\mu m^3$	mean τ/ns
A	17.1	108	22	1120	600	92
B	17.7	236	52	1280	590	86
C	18.4	91	24	1590	635	88

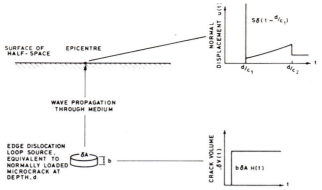

Fig. 1 The dislocation loop model for the source event, and the vertical surface displacement at the epicentre, using the Green's function calculated as described in section 2.2

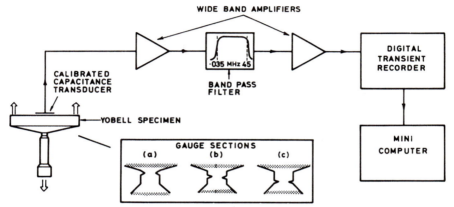

Fig. 2 The YOBELL specimen design, and the broad band detection system employed for the tests. The inset shows the position of the 60° notch for the three specimens tested

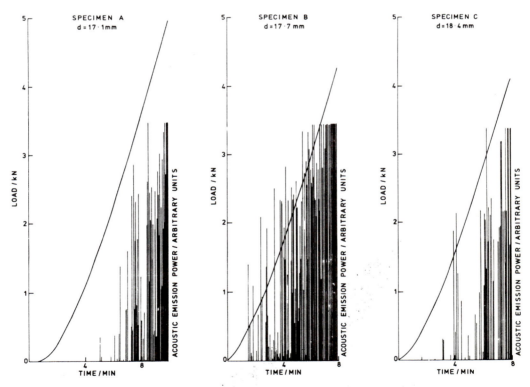

Fig. 3 Showing the load as a function of time for the three notched specimens tested. No units are given for the acoustic emission because the power measurement was uncalibrated. (It is only a qualitative guide to activity.)

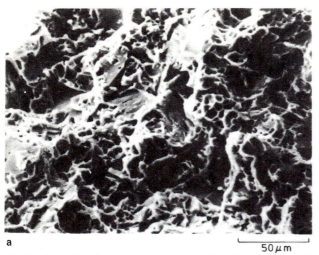
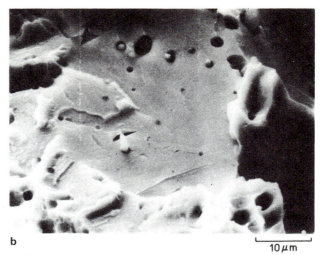

a

b

50 μm

10 μm

Fig. 4 Scanning electron micrographs of the fracture face of specimen C

    a Shows both cleavage and ductile dimple fracture
    b Shows an enlargement of one cleavage facet

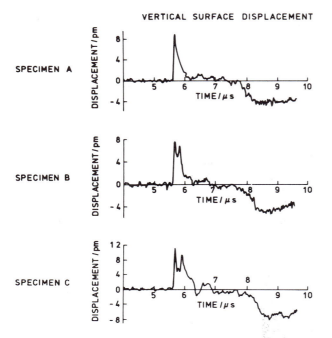
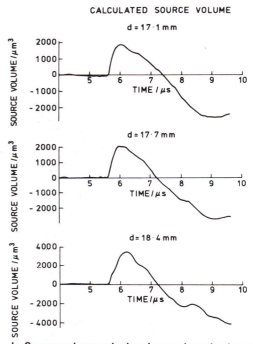

Fig. 5 a Measured surface displacement for a typical emission from each specimen. Note the echo pulse in B and C which is absent in A

          b Source volume calculated assuming a horizontal dislocation loop model at depth d. The volume drops from a maximum and becomes negative because of low frequency filtering

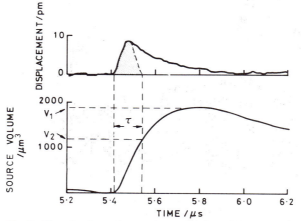

Fig. 6 Showing how the maximum (V_1), and first peak (V_2) volumes and lifetime (τ), used in the histograms of Figs. 7 and 8, are measured from a typical transient

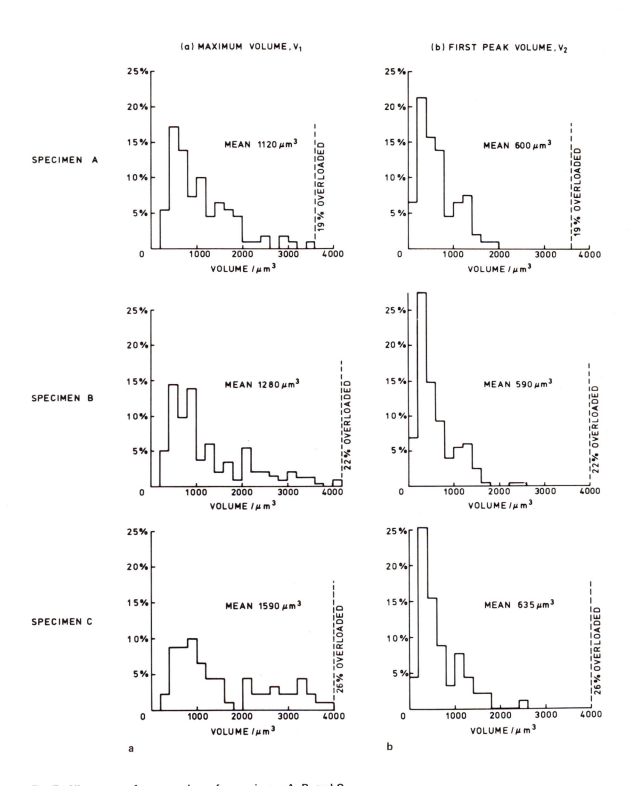

Fig. 7 Histograms of source volume for specimens A, B, and C
    a The mean maximum volume (V_1) increases with source depth due to the contribution of the echo pulses
    b The volume of the first peak (V_2) ignores the echo pulse and the histogram is independent of source depth

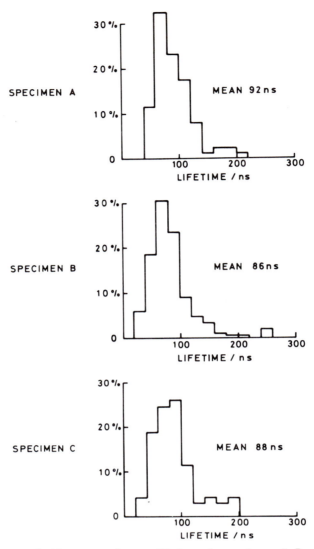

Fig. 8   Histograms of source lifetime τ for specimens A, B, and C, where τ is twice the leading edge risetime (Fig. 6). The mean lifetime is independent of source depth

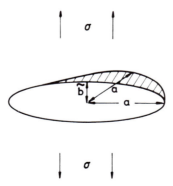

Fig. 9   Ellipsoidal model for elastic crack used to estimate crack opening $\tilde{b}$

C41/79

EXPERIENCES WITH ACOUSTIC EMISSION MONITORING IN NUCLEAR POWER PLANTS

R. GOPAL, BEE, MS, PhD
Westinghouse Pressurized Water Reactor Systems Division, and
J. R. SMITH, BS in EE, MS
Westinghouse Pressurized Water Reactor Systems Division

The MS of this paper was received at the Institution on 4 December 1978 and accepted for publication on 19 January 1979

SYNOPSIS   Acoustic monitoring tests have recently been conducted at a defueled nuclear plant under simulated operating conditions. The test program has not yet been completed, but this paper presents some of the results to date. Tests have included reactor noise measurements, leak detection tests with controlled leaks of primary water, and development of a modified plant startup technique which greatly assists monitoring for crack growth signals during pressurization. Also discussed in the paper is a concrete example of how acoustic leak detection has helped to increase one plant's availability.

INTRODUCTION

1.   Acoustic monitoring is becoming a valuable inservice inspection technique for evaluation of the structural integrity of pressure vessels and other pressure retaining components, and is expected to be especially useful in commercial nuclear power reactors. Substantial monetary benefits can be realized in nuclear plants by increasing their availability, and surveillance systems such as acoustic monitoring can reduce unscheduled downtime through early detection of abnormalities.

2.   In large structures, such as reactor pressure vessels, regions of acoustic emission activity can be located through the use of a moderate number of transducers and some relatively sophisticated system electronics, generally including a minicomputer. The signal processing equipment must measure the relative arrival times of acoustic emission pulses at the sensor sites, and use that information to perform a triangulation operation. Further signal processing can then be included, as desired, to evaluate signal characteristics or grade the activity levels of various source regions.

3.   Leak detection on a reactor, on the other hand, is made feasible because acoustic signals are generated when a pressurized fluid escapes through a metal boundary. For this type of acoustic signal, the source of energy is the fluctuation in the flow momentum of the fluid. In general, the signal from a leak is wideband and continuous, very much like random noise, although at lower frequencies (less than 100 kHz) it is certainly possible to excite fluid flow, cavity, or structural resonances which can greatly influence the frequency characteristics. With proper selection of frequency passband, the signal can easily be detected by piezoelectric transducers attached to the surface of the pressure boundary, and detection at a distance can be quite good, while the background noise is minimized.

4.   Recent acoustic monitoring test programs have definitely emphasized leak detection over acoustic emission (AE) detection, since the simpler and highly reliable nature of leak detection, together with the immediate utility of the results, have brought it much closer to commercial practicability. An acoustic leak monitoring system is sensitive, response time is fast, and a leak source can be located through processing of the relative strength of its signal at different transducer positions by making use of the signal's attenuation with distance from the source.

INDIAN POINT UNIT 1 TEST PROGRAM

5.   Most recently a series of experiments have been conducted at the Indian Point Unit 1 nuclear power plant, with the help of the Consolidated Edison Company of New York and sponsorship of the U.S. Department of Energy. The purpose of the program has been to both further demonstrate acoustic leak detection and location processing capability and to measure acoustic emission generation on the reactor vessel. At this writing Phase 1 of the two-part program has been completed, and the Phase 2 program is underway.

6.   On the Indian Point Unit 1 reactor the reactor vessel and two of the four primary coolant loops were available for testing. A total of 28 acoustic emission sensors were employed in the test program: 14 on the reactor vessel, 10 on one coolant loop, and 4 on a second coolant loop. On the reactor vessel the sensors were primarily intended for acoustic emission flaw growth monitoring, although lack of accessibility severely limited sensor locations to only the top head, top flange, and bottom head areas. The limited access is shown clearly in Figure 1, which indicates the vessel shell sensor locations. No part of the reactor midsection could be reached.

7.   Leak monitoring tests were performed by mounting sensors on the piping of the two coolant loops, with one loop instrumented extensively.

Both loops were monitored to detect natural plant leaks, but the better-instrumented one was also used for attenuation measurements and controlled leak generation tests. For this loop the specific sensor locations and their corresponding channel numbers are illustrated in Figure 2, which is essentially a simplified top view, except that the vertical cold leg piping to the left of channel 24 is shown horizontally to better indicate its dimension.

Indian point acoustic monitoring technique

8. To couple the sensors to the reactor, a moderate force was supplied by a compression spring in each mounting fixture. No ultrasonic couplant was used in Phase 1, since such compounds generally are not permitted in reactor service. Phase 2 testing has employed such couplant, though. The sound-sensitive contact surface of a sensor is less than 3 mm (1/8 in) diameter, however, so that the spring force actually applies a large pressure to the contact face. A dry-mounted sensor is not as sensitive as one coupled via viscous grease, of course; the reduction in sensitivity is nominally a factor of three (9.5 dB).

9. The sensor mounting fixtures themselves were attached either magnetically or with steel bands. A sensor detects acoustic signals by means of a high-temperature piezoceramic, and contains an integral electrical matching network designed to be loaded by a 95-ohm non-resonant transmission line. With this construction the amplifier or preamplifier may be placed as far as 200 m from the transducer. The other end of the transmission line is terminated by the necessary matching impedance in parallel with the step-up transformer of a low-noise amplifier stage.

10. The original test plan called for the amplifiers and other electronics to be located inside the reactor containment, on the operating deck, which would have required cabling of about 50 m (150 ft) to each sensor. However, because of the difficulties associated with containment access, together with the radiation exposure and the possibility of equipment contamination, the decision was made to keep all electronics outside the containment. Cable lengths to the sensors therefore increased to between 150 to 210 m (480 to 700 ft). Cable runs were made continuous from sensors to amplifiers, however, and no noisy containment building penetrations had to be traversed with the cables run through the open air lock. It was then possible to successfully place all amplification outside of the reactor containment, in the motor-generator-set room, together with the other equipment.

11. Figure 3 illustrates the frequency response characteristics of a typical sensor, plotted together with the response of a very-wide-band reference. Test programs several years ago were responsible for determining the midband frequency of 500 kHz, so as to get the sensitive range above most of the mechanical background noise, but still stay below the frequencies where attenuation of the metals becomes large.

12. The operation of the remainder of the acoustic monitoring system is indicated in the block diagram of Figure 4. The system was designed for "on-line" operation; that is, acoustic emission events were located, counted, and displayed as they occurred, and the background rms acoustic noise levels were monitored continuously.

Reactor noise measurements

13. A primary aim of Phase 1 was to determine the feasibility of acoustic monitoring of the Indian Point Unit 1 reactor through plant operating condition background noise. Previously, the major source of interference to reactor acoustic monitoring had been found to be flow noise from the primary coolant pumps. Under operating conditions, the masking effect of the flow noise could severely restrict the range of acoustic emission detector or require that the acoustic emission signals be unrealistically large. With the limited accessibility of the reactor pressure vessel, it is not possible to simply increase the number of monitoring channels to reduce the channel-to-channel separation.

14. The possibility exists that allowance may be made on future plants for attachment of acoustic waveguides in inaccessible areas, but in a retrofit situation the only available locations for reactor vessel AE sensors are at the vessel top and bottom. For leak detection on the coolant loops the situation is much better, since sensors can be more freely positioned. The presence of significant background noise can decrease the available leak sensitivity, however, and necessitate the use of a significantly large number of sensors.

15. On the Indian Point Unit 1 reactor, the reactor vessel and two out of four coolant loops were available for testing. Each loop contained two 1.2 MW primary coolant pumps of the canned-motor type, connected in parallel at the steam generator outlet to discharge into the reactor cold leg piping. Table 1 shows the background noise levels that were measured at three stages of the reactor heatup with all four pumps running. The listed rms levels represent a combination of both acoustic and electronic noise; the electronic noise contribution in all cases was constant at about 100 mV$_{rms}$.

16. The first column of the table shows that upon initial pump startup the noise levels were extremely high, especially on the cold-leg piping. As the heatup progressed towards the 204°C (400°F), 105 bars (1520 psig) conditions of the third column, though, the noise levels decreased greatly. This effect had been expected, since similar results had been obtained in tests on other reactors (Ref. 1). The decrease in noise had been found to be caused by the large increase in pressure, not by temperature escalation.

17. This plant had significantly less noise present on the reactor vessel and hot legs than had been measured at other plants. Because of the very long acoustic paths of both the cold-leg piping and the U-shaped steam generators (see Figure 5), the attenuation of the flow noise was much greater.

18. For example, Table 2 shows the reactor noise levels measured at full power at the Prairie Island Nuclear Plant, a modern Westinghouse plant design with very short coolant loop piping. It is clear from the table that the noise levels

measured on the reactor vessel at that plant are roughly equivalent to the noise levels measured on the cold legs of Indian Point 1 (see Table 1), its noisiest locations.

19. Also given in Table 2 is an indication of the relative stability of the noise levels on the operating plant. These levels are usually fairly constant over the short term, but can vary significantly over the longer term. Data in the table were taken under operating conditions over a period of three years.

Acoustic emission testing during heatup

20. To monitor the reactor vessel for acoustic emission during the plant heatup, modifications were made to the normal heatup procedure to gain maximum signal-to-noise ratio during the crucial period of pressurization, when most acoustic emissions would be expected.

21. Heatup was begun by turning on the four primary coolant pumps of the two available coolant loops to provide about 5 MW of frictional heating power. Primary system temperature and pressure were allowed to gradually increase until conditions of 213°C (415°F) and 41 bars (600 psig) were reached. As had been expected, the general noise levels at this point were still rather large (see 500 psig column in Table 1).

22. The primary coolant pumps were turned off at that point, and the pressurization continued by means of the pressurizer alone. The reactor vessel could then be monitored for acoustic emission generation during most of the pressure escalation without any interference from flow noise. The 480 kw pressurizer heater banks were able to pressurize the system to 103 bars (1500 psig) in about two hours, with little drop in temperature. An approximate plot of the temperature-pressure heatup curve that was followed in this test is illustrated in Figure 5. The solid curves in the figure define the heatup bandwidth limits which must not be exceeded.

23. The success of the combination pump heatup and pressurizer pressurization startup sequence represented one immediate result of the Phase 1 test. Should acoustic emission monitoring of reactors become commercially viable, a similar plant startup technique should be employed, since a large change in pressure occurs under conditions which are much more favorable for acoustic emission monitoring. With the background noise from the pumps and water flow removed, the sensitivity and range of detection of an acoustic emission monitoring system are greatly enhanced.

24. Unfortunately, in this pressurization very few acoustic emissions were observed. Some mechanical signals originating in the head joint were detected, but there was essentially no acoustic emission from the major sections. Further testing in Phase 2, with enhanced sensitivity via couplant on the transducers, has verified such quietness.

Leak detection tests

25. Leak detection tests at Indian Point involved measurements on both natural plant leaks and artificially generated leaks. Excellent natural leak data were obtained near the end of the plant heatup procedure, when several leaks occurred during the final stage of pressurization (Ref. 4). One very good example of leak detection and location computation on the plant is provided by a leak that occurred at 104 bars (1500 psig) on a small drain line connected to the Loop 12 cold leg piping. The leak was caused by the slight unseating of a valve, and was later repaired by simply tightening that valve.

26. The data for this case is presented graphically in Figure 6. The figure is a semilog plot of leak signal level versus distance along the main coolant pipe. The leak signal contribution was separated from the total rms value by the formula:

$$V_{leak} = \sqrt{V_{total}^2 - V_{bkgd}^2} \ (rms) \qquad (1)$$

since background noise and leak signals are independent and uncorrelated. Points connected by the dotted curve represent the original data, while the solid curve represents the incident signal levels after correction for different sensor sensitivities. The leak can be clearly determined to be near Channel 23, and the drain line satisfied this requirement.

27. For testing under simulated operating conditions, a one-inch schedule 160 pipe section containing a needle valve had been welded onto the Loop 12 steam generator vent line, putting the valve about one meter (3 ft) from the main body of the steam generator (see Figure 2). By controlling the valve opening the leak rate from this artificial source could be varied as desired.

28. Table 3 shows the results obtained from this source with a small leak. Because the device leaked reactor water directly to atmosphere in this initial test phase, quantification of the leak rate was not possible, but based on laboratory experience the rate was estimated to be around 0.4 liters/min (0.1 gpm). The table shows that only the two closest sensors, Channels 19 and 20, detected the leak. Sensitivity to this leak was not as good as in laboratory measurements, but was not expected to be that high, since the leak here was occurring on accessory piping, not main coolant piping.

29. To allow detection by more sensors, the leak rate had to be increased. Because the source lacked a condenser, such an increased leak rate could be employed for only a very short time. Data from this case were therefore recorded on magnetic tape and evaluated by tape playback. The results are presented in Figure 7, a semilog plot of leak signal level versus distance. Note in Figure 7 that the data points fall reasonably close to a straight line, indicating a generally exponential decay. Such a one-dimensional type of attenuation agrees with earlier laboratory and plant tests (Ref. 2).

OPERATING PLANT LEAK MONITORING

30. The ability of acoustic leak monitoring to increase operating plant availability has in fact already been demonstrated, though, surprisingly,

neither by primary system monitoring nor directly through the leak detection process. Because laboratory data have shown that an acoustic monitoring system can also be quite sensitive to a low-pressure water leak, provided that the leak has some velocity to it (not a "drip"), it was possible to attach an acoustic monitoring system to a low-pressure plant piping system to allow continued plant operation. A representative sample of laboratory data for a low pressure case is shown in Figure 8, which plots acoustic signal versus pressure for an orifice-type leak in 5 cm (2 in) diameter stainless-steel pipe. The sensor was a high-frequency unit, and was located 2.5 m (8 ft) away, while the leak rate was almost constant over the pressure range at 1.0 cm^3/sec (0.015 gpm).

31. On the particular nuclear power plant that required monitoring, a low-pressure, 20 cm (8 in) diameter, stainless-steel piping system some 30 m (100 ft) long had been found to have defective welds, some of which had begun to leak. The problem was critical because the piping system was intended for safety injection of borated water into the reactor in case of a hypothesized reactor loss-of-coolant accident, and a portion of the piping passed through an almost inaccessible high-radiation area. Permanent solution to the problem required rerouting and replacement of the entire piping system, but until such time as that could be done, the known-defective welds were repaired and an acoustic leak monitoring system employing five high-frequency and five low-frequency sensors was installed. The monitoring system was in use for a six-month period, during which time no further leaks occurred, but it did enable the plant to continue in normal operation for that period until pipe replacement could be completed.

CONCLUSION

32. The data presented in this report have shown that acoustic monitoring of nuclear reactors can be quite useful for leak detection. Target specification for leak detection on an operating reactor, using high-frequency sensors, is approximately 80 cm^3/min (0.02 gpm) for a leak occurring directly on the main coolant piping on which the sensors are mounted.

33. Demonstration of the practicability of on-line acoustic emission measurements, though, is still in a process of gradual evolution. While work is still hampered by lack of accessibility to the nuclear components, a technique for greatly increasing the signal-to-noise ratio during startup has now been successfully implemented on a plant.

REFERENCES

1. R. Gopal, J. R. Smith, and G. V. Rao, "Experience in Acoustic Monitoring of Pressurized Water Reactors", Third Conference on Periodic Inspection of Pressurized Components, Institute of Mechanical Engineers, London, September 1976.

2. W. Ciaramitaro, et al, "Primary Coolant Acoustic Leak Detection System Development Final Report", Westinghouse Nuclear Energy Systems, WCAP-9239, December 1977.

3. R. Gopal, J. R. Smith, and G. V. Rao, "Acoustic Monitoring Instrumentation for Pressurized Water Reactors", Twenty-Third International Instrumentation Symposium, Instrument Society of America, May 1977.

4. J. R. Smith, M. B. Olex, and J. Craig, "Acoustic Monitoring Systems Tests at Indian Point Unit 1", Westinghouse Nuclear Energy Systems, WCAP-9324, May 1978.

TABLE 1
RMS ACOUSTIC NOISE LEVELS DURING PLANT HEATUP
(FOUR PRIMARY COOLANT PUMPS OPERATING)

Sensor Location	Channel No.	Output Levels, mv$_{rms}$		
		Cold, 17 bars (250 psig)	138°C (280°F), 34.5 bars (500 psig)	204°C (400°F), 105 bars (1520 psig)
PV top flange	6	430	120	101
RV bottom	14	4,500	500	148
Loop 12, hot leg	18	230	120	135
Loop 12, cold leg	23	20,000	2,500	520
Loop 13, hot leg	28	140	110	135
Loop 13, cold leg	29	27,000	1,600	350

TABLE 2
RMS ACOUSTIC NOISE LEVELS AT AN OPERATING PLANT
(WESTINGHOUSE TWO-LOOP DESIGN)

SENSOR LOCATION	OUTPUT LEVELS, mV$_{rms}$		
	July 1975	April 1976	March 1978
RV Bottom-1	700	610	530
RV Bottom-2	440	460	500
RV Top	600	700	550
SG, Loop A (hot leg)	840	910	480
Pump, Loop A (cold leg)	2,000	1,720	1,170
RV Inlet Nozzle A	1,800	1,730	1,860
RV Outlet Nozzle A	700	550	400

TABLE 3
RMS ACOUSTIC NOISE LEVELS FOR A SMALL
(~ 0.1 gpm) NEEDLE VALVE LEAK

Sensor Location	Channel No.	Output Levels, mv$_{rms}$	
		Pumps On, 229 C (445 F)	Pumps On, Valve Leak
Loop 12, hot leg	17	110	110
Loop 12, hot leg	18	101	100
Loop 12, hot leg	19	115	127
Loop 12, steam generator	20	120	155
Loop 12, cold leg	23	450	450

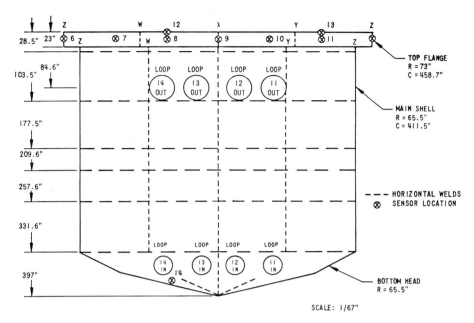

Fig. 1 Indian Point Unit 1 reactor vessel

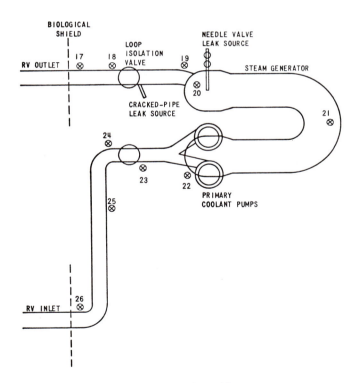

Fig. 2 Acoustic sensor locations: loop 12

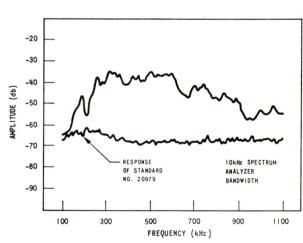

Fig. 3 Acoustic emission transducer sensitivity calibration

141

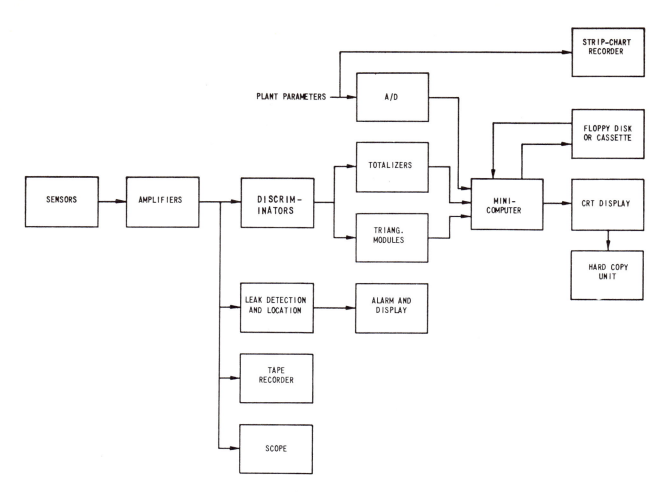

Fig. 4 Block diagram of the acoustic monitoring system

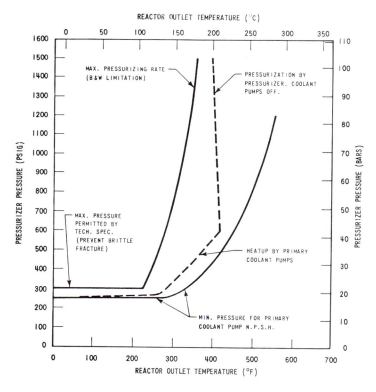

Fig. 5 Primary coolant system heatup procedure for Indian Point Unit 1

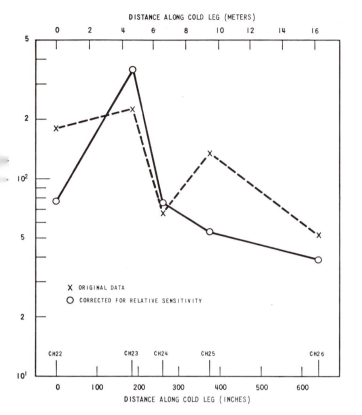

Fig. 6 Acoustic signals from a plant leak

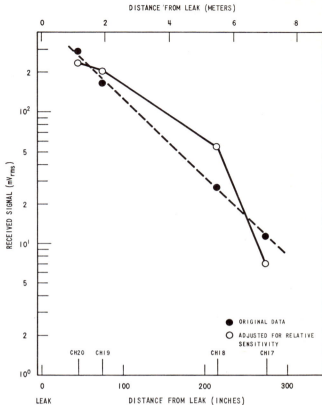

Fig. 7 Large needle valve leak data

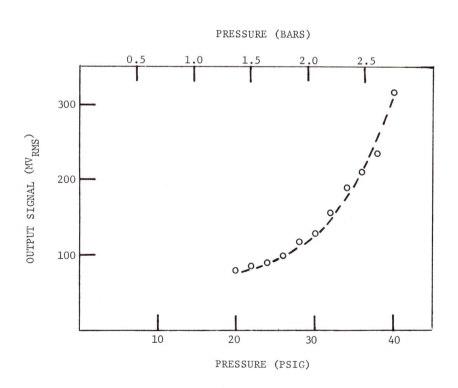

Fig. 8 Low-pressure water leak data (300-700 kHz)

143

C42/79

ACOUSTIC PROOF TESTING

D. BIRCHON, BSc, FIM, FIMarE
Admiralty Materials Technology Establishment, Holton Heath, Poole, Dorset

The MS of this paper was received at the Institution on 30 November 1978 and accepted for publication on 18 January 1979

SYNOPSIS   We are attempting to develop a method of using acoustic emmission (AE) to assess the integrity of pressure vessels during their overpressure test, and to demonstrate that no damage has been done by the test. The procedure is believed to be novel in monitoring the acoustic response at constant pressure, in a manner which can provide an internal reference response, to give confidence that the AE technique is being properly applied. The first application being studied is for the in situ proving of some high pressure air storage cylinders, but (if successful) the technique may also have merit for use during the proving of other pressure vessels, and of structures loaded other than by internal pressure. Since the technique relies upon recognition of a time and pressure-related pattern of AE response, a new design of AE monitor is also introduced, which controls the sequence of events during the proof test, and which can be readily converted to differing test requirements by replaceable logic board(s), to convert the correlation and output functions to be appropriate for other routine applications.

INTRODUCTION

1.   Acoustic emission* (AE) is a promising NDT technique (ref. 1) which has been extensively studied as a means of nondestructively inspecting pressure vessels (ref. 2) during proof tests because:

1.1 It can potentially monitor the whole of the volume of the material in the vessel from one, or a few, locations.

1.2 In the unlikely event that vessel failure during the test could occur, Harris & Dunegan (ref. 2) - among others, have pointed out that it may give sufficient warning for the pressurisation to be stopped in time to save the vessel for a local repair. (Indeed, Kellerman (ref. 3) and Peters (ref. 4) report that 75% of all serious failures of pressure vessels occur during their proof tests).

1.3 Provided that all the mechanisms responsible for crack growth can reliably be detected (see partition process, ref. 1), AE may only give warning of such defect extension processes; ie it should only tell us what we need to know, thereby reducing the amount of data to be analysed, in turn reducing both the cost, and the risk of misinterpretation of the test results.

---

*Acoustic emission is the name given to the art and science of listening to the tiny noises given out by many materials and structures when they deform or crack, and attempting to interpret the significance of the sounds heard.

2.   However, difficulties in adapting it for routine NDT use have arisen from several causes including:

2.1 The variability in timing and the wide dynamic range of the emissions recorded, leading to difficulties in obtaining quantitative reproducibility.

2.2 Inability to derive generally acceptable rules for interpreting the significance of differing levels of AE detected, plus difficulties of calibrating and standardising AE techniques.

2.3 The transient nature of AE.

3.   The last is particularly important, since unlike other NDT techniques, AE does not allow a repeat examination of any emission signal from a material and this imposes a need for high and demonstrable integrity in the AE equipment and in its method of use. Again, whilst absence of emission can hopefully be interpreted as evidence that there is nothing wrong with the structure being monitored, it should leave no room for doubt that something could have been wrong with the AE monitoring technique at the time.

4.   We are therefore attempting to improve the capability of acoustic emission for proof testing by:

4.1 Trying to improve the reproducibility of interpretation of results through seeking time related patterns of AE behaviour at constant (nominal) stress. (There have been other proposals to look at patterns of behaviour, eg Pollock (ref. 5) who suggested that the slope of the amplitude distribution (or "b" value) could sometimes provide evidence of changes in the source mechanism, and Green (ref. 6) proposed that proof tests should be redesigned, with holds, for AE.

4.2 Trying to improve confidence in the results obtained through so conducting the test that an internal standard "normal" response pattern will be observed; in this way, absence of an abnormal indication rather than an absence of emissions provides evidence of the integrity of the structure concerned.

OVERPRESSURE TESTING

5. The overpressure testing of pressure vessels and pressurised systems is an engineering tradition established at least a century ago when boiler explosions (and resulting human casualties and property damage) reached scandalous proportions. Today, no self-respecting engineer would accept or operate any pressure vessel without requiring an overpressure test to demonstrate that the system will neither leak nor be likely to fail if operated at a pressure a little below the proof pressure, (as discussed by Higginson (ref. 7) and Nichols (ref. 8)), but there is often dispute about the value of the proof pressure/working pressure ratio which should be used. Although the most common value is 1.5, values as low as 1.1 have been recommended for some cases. The balance of choice lies between:

5.1 Choosing a pressure high enough to demonstrate that any flaws present in the vessel cannot be above a given "size", (so that operating at lower pressure will give an insurance against sub-critical crack growth reaching dangerous proportions before the next inspection period) and also high enough to give the necessary mechanical stress relief (ref. 7),

5.2 yet not so high that it will in any way overload (eg reduce the fatigue strength) of the vessel.

6. Hence, no single value can be correct for all types of pressure vessel, and it is considered that there is a need for a non-destructive monitoring technique which will show that mechanical stress relief has occurred, and that no sub-critical defect growth is occurring at the test pressure, anywhere within the vessel. It is suggested that Acoustic Proof Testing (APT) may be able to provide such a capability.

THE ROLE OF AE IN OVERPRESSURE TESTING

7. Acoustic emission has often been considered for use during overpressure testing, usually to seek evidence for crack growth during an increase in pressure, or to attempt to correlate detected emission sources with those found by other NDT means (ref. 9) or for defect location during an "acceptance" overpressure test. In the familiar case of some AE sources being detected and located in a vessel during its (successfully completed) overpressure test, we have often noted the difficulty experienced in attempting to attribute severity ratings to defect locations as they are identified by AE during the test, since whatever AE parameter has been used it is quite common to observe that for any given

location, an emission assigned a relatively high "significance" may have occurred at a pressure considerably below the maximum of the pressure ramp, from a location at which no further "significant" emissions have been detected as the pressure was increased. This anomaly is considered to arise because at any one location several interdependent yet differing mechanisms of deformation (or microfracture) may be occurring as a result of co-operative response to the applied strain, and the operation of one mechanism of low emission response (eg local slip rather than movement of a Luders yield front) may trigger an event of high emission response (eg the fracture or decohesion of a non-metallic inclusion), which so reduces the local strain field that with a further increase in strain, further similar events of high emission amplitude may not occur. It follows that changing the direction of crack growth can also greatly affect the AE response of the material as shown by several workers, including Ono et al (ref. 10).

8. But perhaps the major difficulty to be overcome before AE can be considered for acceptance proof test purposes has been that in tests in which vessels (with or without defects in them) have been pressurised to failure, although the cumulative count of emissions observed during the increasing pressure ramp has often increased towards failure, the quantitative reproducibility of the relationship of the chosen emission parameter to the pressure ramp has usually eluded either duplication, or prediction from small scale tests on specimens of the same material.

9. In order to be effective and reliable for routine NDT use we must therefore demonstrate that:

9.1 The partition process (ref. 1) associated with the defect extension processes sought will always produce detectable emissions.

9.2 The background noise spectrum throughout the test period will permit the reliable detection of such signals.

9.3 The attenuation between all monitored parts of the vessel and the receiving transducer will permit detection of the signal sought.

9.4 Equipment and techniques of demonstrable reliability, adequate quantitative reproducibility, simplicity of use and acceptable cost can be produced.

ACOUSTIC PROOF TESTING

10. We have noted in our own work (and in the work of others, eg ref. 2) both a slow decay of emissions during a constant pressure "hold" at the end of a pressure ramp, and (on other vessels) emissions which may have been from sub-critical crack growth occurring at constant pressure, since these vessels finally failed. Consideration of the source of such emissions prompted the thought that the former might be due to anelastic relaxation (non-linear behaviour, as first reported by Weber (ref. 11)), whereas the latter might truly represent the resultant activity of the various deformation

and microfracture processes involved in sub-critical crack growth somewhere within the vessel (ie not necessarily from a single location or mechanism).

11. If this is true, then provided that the mechanisms involved provide detectable emissions whose magnitude is related to their rate or magnitude, one would expect the former to die away with time, but the microstructural activity of slow crack growth and the associated movement and structural disturbance of the plastic zone boundary at the crack tip will both continue and increase in rate, due to the accompanying increase in stress intensity as the crack gets longer.

12. Hence, Acoustic Proof Testing (APT), – observation of acoustic response during constant loading – may offer the opportunity to observe either or both of two time-related yet distinctly different patterns of behaviour, one normal, and one abnormal (Figure 1) whose establishment and interpretation as a matter of specifiable routine may be easier than previous methods of attempting to quantify AE behaviour in terms of emissions observed only during increasing load. In addition, it should not only indicate that the vessel concerned is fit for further service, but also that it has sufferred no significant latent damage during the overpressure test.

13. However, to accomplish this several steps are necessary:

13.1 AE techniques must be devised which can be repeatedly re-established at similar levels of sensitivity, over a considerable space of time and (if need be) in different geographical locations.

13.2 The acoustic response of the material to both anelastic relaxation and sub-critical crack growth must be demonstrable and interpreted in terms of the differing mechanisms involved. (This is the subject of work to be reported later)

13.3 The attenuation undergone by signals from all parts of the vessel to the sensor must be shown not to prevent useful detection of the signals sought.

13.4 Since the data must be time-related, the test would have to be conducted with time, not pressure, as the prime parameter.

14. Consideration of these points led to the conclusion that the most suitable pressure vessels for the first studies would be unwelded ones (since across weldments we have measured attenuation of typically 10 dB and occasionally over 20 dB, compared with 2-3 dB m^{-1} for parent steel plate), and that the AE monitor should be used to control the pressure ramp and data correlation from its own internal clock.

15. The first problem to be solved at this stage was the choice of AE parameter to be used. Possibly an energy measurement combined with some method of averaging the data over a moving time period, or Pollock's "b" value (ref. 5) may be a more suitable method than simple threshold crossing counts of the excitation of the transducer,

but our first work has shown that the customary count capability may suffice to establish whether the idea is feasible or not and whether AE techniques can be made reliable, quantitative and simple enough for normal NDT use, noting that although listening during dwell periods avoids problems due to pump noise, during use on board ship or workshops etc it would be irksome to have to impose silence and inactivity in all nearby areas during the test.

16. More confidence will also be attached to this proposed new technique when sufficient has been learnt of the correlation between differing deformation and microfracture processes in the material concerned to support the view that all significant deterioration processes can be detected within the range of microstructure concerned for any class of vessels – an aspect upon which work continues.

THE CHOSEN TEST CASE

17. The case chosen for the evaluation of the technique is an attempt to reduce the maintenance costs of high pressure air bottles of 258.6 litres (9.1 cubic feet) capacity, made from BSS970 826M31 (2.5% Ni, Cr, Mo) steel in condition T, each some 1524 mm (5 feet) long, 500 mm (20 inch) in diameter and 22 mm (0.875 inches) thick.

18. These bottles are fitted in HM Ships and Submarines in considerable numbers, and because of their shape and size are often in the confines of the vessels; hence their regular removal for inspection and overpressure test and subsequent replacement entails considerable cost and delay.

19. The service record of these vessels is so good, that it is tempting to consider using them without inspection for the twenty or thirty years in which they may be required to operate. However, caution advises that since it is not possible to eliminate the entrainment of some salt contaminated water in the air (causing corrosion and the slight risk of corrosion-fatigue or stress-corrosion cracking developing), plus the risk of some sub-critical flaw growth occurring during service, compounded by the fact that the bottles provide essential services and are required not only to remain intact but to perform their normal service despite battle damage towards the end of their lives, some demonstration of their integrity that can be conducted in situ is required.

AE DURING PNEUMATIC BURST TESTS

20. From a fracture analysis of bottles in service (generously assisted by Pellini and Loss of NRL Washington DC) it was considered necessary first to prove that if a bottle possessing the worst combination of mechanical properties (ie highest yield point/UTS ratio and lowest Izod value*) and containing a longitudinal crack was pressurised to failure, it would split

_____

*Since we are working with bottles of which some have already been in service for many years, the mechanical property data available is inevitably limited, and, in particular, fracture toughness data (or even Charpy data) was not available.

and remain essentially in one piece despite the high rate of delivery of stored energy to the advancing crack tip resulting from the pneumatic loading (see, for instance, Cowan and Nicholls ref. 12 for the effect on vessel fracture of only 10% by volume of air entrained in a pressure vessel).

21. With the co-operation of colleagues in the RN and MOD, some bottles having the worst combination of (acceptable) mechanical properties were traced and recovered from service, and tested to failure with air at 41.37 MPa (6000 lbf/in²) at PERME (Waltham Abbey). The technique used was to slit halfway through the wall, pressurise to 41.37 MPa for 15 minutes, and, if failure had not occurred by that time, to depressurise, increase the length of the slit, and repressurise to 41.37 MPa, and so on until failure occurred either during increasing pressure or during the 15 minute hold.

22. The results, tabulated in Table 1 and illustrated in Figures 2 and 3 indicated:

> 22.1 The likelihood of shatter during an in situ overpressure test was very small.

> 22.2 Acoustic emission could apparently detect sub-critical flaw growth at constant pressure.

23. If a practical AE technique were to be evolved, it must be shown to be capable of contending with background noise and other interference and it was also necessary to demonstrate that bottles could be evacuated, filled, pressurised, re-evacuated and dried again, in place. Accordingly a bank of four high pressure air bottles was monitored in a boat at Rosyth Naval Base using the AMTE/DLS (refs. 1,9) with 100 metres of superscreened signal cable between the transducers and the preamplifiers to eliminate problems due to electrical interference from other shipboard or neighbouring activity. No difficulty was experienced from pump noise (or other interference), and a useful by-product was the observation of a tiny leak in one of the bottles, an observation made from the comfort of a van on the dockside before it was observed by those conducting the test from the boat. During the test, attenuation between bottles, and between supports and bottles was also measured and found to be high enough to permit effective individual monitoring of each of a bank of simultaneously pressurised bottles to be conducted if so required, without mutual interference.

HYDRAULIC TESTS

24. The next burst test was conducted with water, at AMTE (HH) using a 305 mm long slit, 1.5 mm wide and 11 mm deep over a comparatively large number of pressure "holds" with the results summarised in Figure 4. Unfortunately, the AE counting equipment saturated during the final minutes of life of the bottle, and this, combined with the random nature of the AE events causes difficulties in discerning the idealised patterns of Figure 1b for any of the hold periods.

25. However, for the last three hold periods there is at least a suggestion that some relaxation was observed at 37.92 MPa, followed by a

larger (but reducing) amount of AE at 41.37 MPa, whereas at 42.06 MOa the rate first fell, then rose and remained high to failure, encouragingly like the manner postulated.

26. On the basis that the increase in AE count rate denoted eventual failure at the constant pressure used, it is considered that the inevitability (though not the instant) of failure could have been predicted for the last seven minutes or so of the dwell period, during which there was no other warning from the pressure gauge, or from electric resistance strain gauges fixed across the ends of the slit.

27. A further hydraulic burst test, on a similar cylinder with a similar slit, using a pressure ramp with 5 more widely spaced hold pressures gave the results detailed in Figure 5. Some decaying AE activity was observed at the beginning of the second hold period; but little during the third. At the fourth hold (intended to correspond to the working pressure of 27.58 MPa or 4000 lbf/in²) the activity continued longer, but a leak in the pressurising system (sufficiently remote from the pressure vessel that it was not detected by the AE equipment) caused the pressure to fall. Pressurisation continued, using the pump to compensate for the leak and during the fifth "hold" period there was evidence of AE activity of greater magnitude and lasting further into the hold than seen before. Finally as the pressure was raised towards the final level, 41.37 MPa a pressure of only 40.68 MPa (5900 lbf/in²) could be achieved, at which the emission rate became and remained high until the vessel failed some 5 minutes later.

PROPOSED IN SITU TESTS

28. We are therefore evaluating acoustic proof testing in situ to seek normal or abnormal response; at constant pressure. It will also be necessary for the interior of the bottles to be cleaned and inspected prior to the in situ proof test, followed by evacuation and drying of the bottle. MOD staff of DGS Bath are providing the cleaning, viewing, pressurising, evacuating and drying equipment and AMTE (HH) are contributing the AE equipment and technique.

PROTOTYPE AE MONITOR

29. Several considerations guided the design of the prototype AMTEAM*:

> 29.1 Since success demands effective time-relation of the AE and pressure data, the monitor should control the test from its own internal clock.

> 29.2 To make the use of the monitor simple, to make the use of differing correlations simple and to enable the method to be readily adapted to other pressure vessels or structures, a plug-in logic board carries out all the data correlation control and output functions, so that its replacement

---

*Admiralty Marine Technology Establishment Acoustic Monitor

by another logic board will adapt the monitor for a range of routine NDT inspections of other equipment.

29.3 It should have a wide dynamic range (100 dB) and a six decade linear counter output, be self-contained, simple to operate, provide hard copy print-out and its own regular test signals and programme to prove its operation, with an audio output since this is a very valuable adjunct to the test.

29.4 It should be small, self-contained and mains or battery operated, and highly resistant to electromagnetic interference.

29.5 Initially, it is assumed that acceptable or non-acceptable emission response can be established on count rate data alone, despite the inherent difficulties of securing reproducible count rate data close to the noise level (ref. 13), but other options are available, if required.

30. The prototype AMTEAM is shown in Figure 6; its evaluation, particularly to determine the level and duration of the pressure holds, is now under way.

CONCLUSIONS

31. An acoustic proof testing technique is proposed which may enable evidence to be obtained in real time that a vessel is either satisfactorily withstanding the overload test, or suffering some latent damage from its effects.

32. A new design of acoustic emission NDT monitor is proposed which can be readily programmed to suit differing applications in a manner considered to make it attractive for routine use by NDT engineers.

CLOSING NOTE

33. The work is still in progress, but if successful for the chosen monolithic steel vessels it may be applicable to other materials and structures, provided that the partition processes involved enable the flaw growth mechanism(s) sought to be reliably detected. It is considered that the existence of detectable anelastic relaxation as a result of being held at a sustained high load could enable a comforting internal "standard response" to be obtained from structures during their overload test, encouraging confidence in the integrity of the AE technique employed, always provided that the relaxation can be observed at pressures sufficiently below those liable to cause latent damage to the vessel - a point which must be established for each vessel concerned.

ACKNOWLEDGEMENTS

Work of this kind necessarily takes a long time, and involves the help of many colleagues both within AMTE(HH) and elsewhere within MOD, some of whom have either retired or transferred to other Establishments; it is a pleasure to acknowledge their co-operation and enthusiasm.

REFERENCES

1. BIRCHON D. The Potential of Acoustic Emission in NDT. Brit, Jnl of NDT 1978, 66-71.

2. HARRIS D.O. and DUNEGAN H. Verification of Structural Integrity of Pressure Vessels by Periodic Proof Testing. ASTM STP 515, 1972, pp 158-170.

3. KELLERMAN O.A. Present Views on Recurring Inspection of Reactor Pressure Vessels in the Federal Republic of Germany. Proc., Conf on In-Service Inspection. Pilsen, October 1966, IAEA, Vienna.

4. PETERS H. Tech Uberwach 1961, 2, 4.

5. POLLOCK A.A. Acoustic Emission Amplitudes. Non. Dest. Test. 1973 (October). 264-269.

6. GREEN A.T. Acoustic Emission and the Proposed ASME Standard: A Critique. Proc Inst of Acoustics. London 1976.

7. HIGGINSON R. Overpressure Testing of Pressure Vessels. Tech, Report 1968, 8, 29-40. British Engine Boiler & Electric Insurance Co Ltd, Manchester.

8. NICHOLS R.W. Use of the Overstressing Technique. Brit, Weld, Jnl. 1968, 15, 21 & 75.

9. BIRCHON D., DUKES R. and TAYLOR JOAN. Acoustic Emission Monitoring of a Pressure Vessel - Correlation of Results with those of Ultrasonic Inspection. Conference on Periodic Inspection of Pressure Vessels. I Mech E. London. June 1974.

10. ONO K., SHIBATA M. and HAMSTAD M. A Note on the Anisotropic Acoustic Emission Behaviour of HSLA Steels. Tech Report 78-02. Office of Naval Research. Washington DC.

11. WEBER W. Uber die Elastizitat der Seidenfaden. Poggendorff's Ann 35. 1834, 247.

12. NICHOLS R.W. (Ed). Pressure Vessel Engineering. Technology, Applied Science, London, 1971, p 194.

13. ONO K. Amplitude Distribution Analysis of Acoustic Emission Signals. Matls Evaluation. 1976, August 177-181 & 184.

TABLE 1

RESULTS OF PNEUMATIC BURST TESTS
OF 258.6 LITRE HIGH PRESSURE AIR BOTTLES

Ø Bottle	Slit Length (mm)	Maximum Pressure Reached MPa	RESULT		REMARKS
			AE Indication	COD Indication	
1	76.4	41.37	None	None	Pressure released & Slit Lengthened
	127.0	41.37	Very Slight	None	Pressure released & Slit Lengthened
	177.8	41.37 (Burst)	During "hold"	COD gauge defective	Burst after 7 mins at hold pressure
2	203.2	41.37	During last 3 minutes of pressuris- ation	An increase noted after pressure released	Pressure released and slit lengthened
	228.6	37.92 (Burst)	At increasing rate over last 5 minutes	Over last 5 minutes	Burst during increasing pressure
3	127.0	41.37	None	None	Pressure released & Slit Lengthened
	152.4	41.37	None	None	Pressure released & Slit Lengthened
	177.8	41.37	None	None	Pressure released & Slit Lengthened
	203.2	38.51 (Burst)	During last 6-7 minutes of test	During last 6-7 minutes of test	Burst during increasing pressure

Ø (1) A bottle from service representing the worst combination of mechanical properties acceptable within the specification.

(2) A new bottle.

(3) A bottle from service, rejected from further service due to internal corrosion.

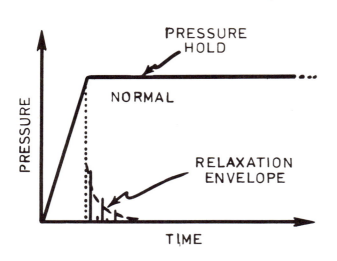
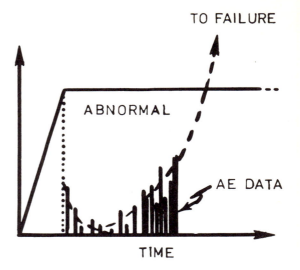

Fig. 1 Principle of acoustic proof testing

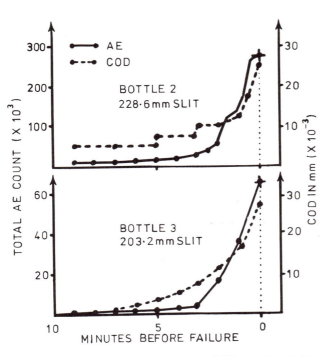

Fig. 2 Results of pneumatic burst tests (different threshold levels were used for the AE data)

Fig. 3 One of the pneumatically burst air bottles in the PERME explosion test facility (courtesy Director PERME)

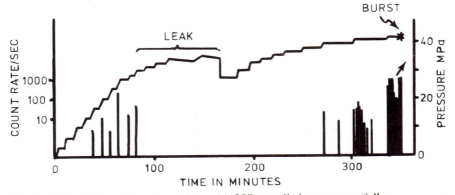

Fig. 4 Hydraulic burst test of cylinder with 305 mm slit (pressure partially released for repair of leak. AE counting equipment saturated towards end of test)

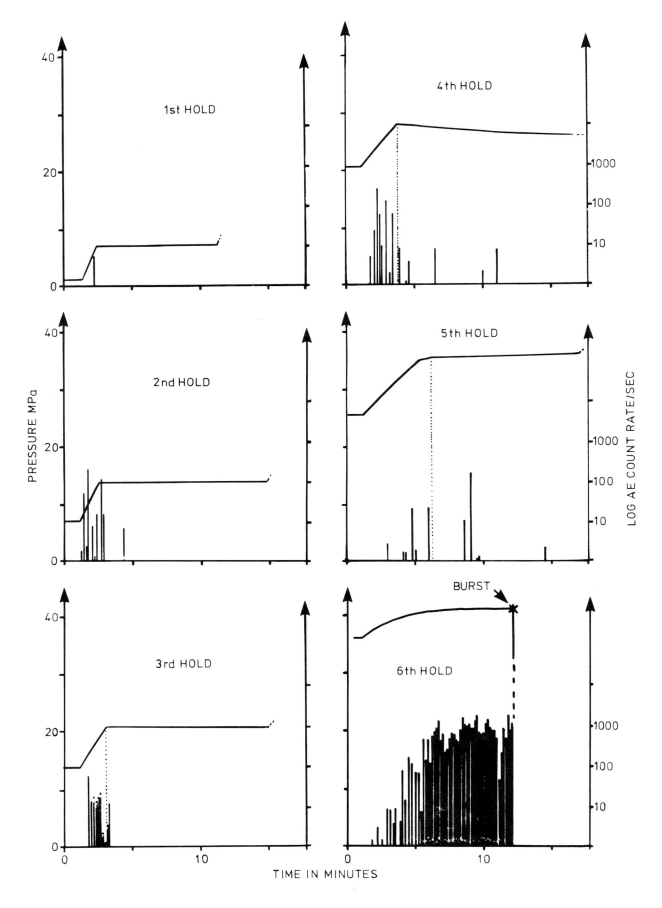

Fig. 5 Acoustic pressure test to destruction of cylinder with 305 mm slit

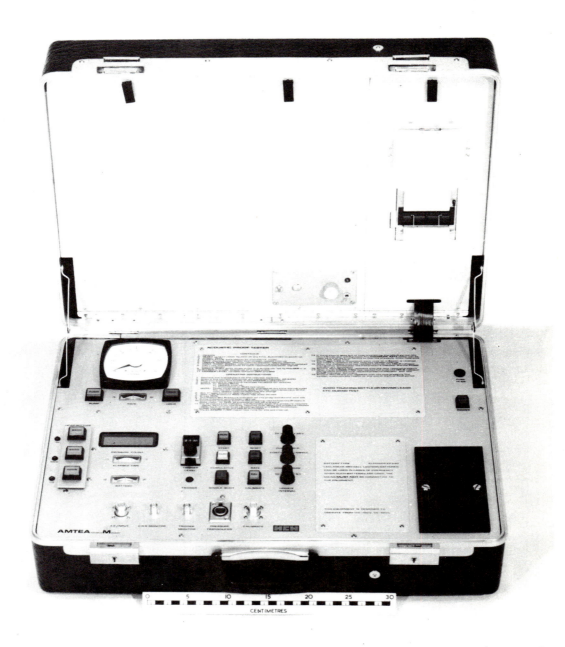

Fig. 6   The prototype acoustic monitor

C43/79

ACCURATE TECHNIQUES FOR DEFECT SIZING IN PRESSURIZED COMPONENTS

M. G. SILK, PhD, BSc
NDT Centre, AERE, Harwell, Oxon.

The MS of this paper was received at the Institution on 11 October 1978 and accepted for publication on 14 December 1978

SYNOPSIS

The general comment which could be made about all forms of nondestructive examination, until quite recently, was that techniques were available which allow defects to be located but that there was little precision in defect sizing. Thus, while the safety of components can be assured by rigorous inspection procedures, this often results in expensive repairs or replacement where pessimistic assumptions have to be made regarding defect size. On the other hand, where less rigorous inspection is carried out, the size of defects may be significantly underestimated.

This unsatisfactory situation has prompted considerable development of improved NDT techniques, particularly in the field of ultrasonics. A number of these techniques are described in detail and conclusions drawn regarding their accuracy, which in many cases is better than $\pm$ 1mm, and their potential in the wide range of material thickness from which pressurized components are constructed. It is important, of course, that we do not set too precise a target for the accuracy of these techniques, since it may be equally uneconomic to inspect to too high a standard. For this reason, some thought is also given to the level of precision the newer generation of techniques should be capable of achieving. It is shown that the required error is much smaller than is often assumed.

INTRODUCTION

1. Before discussing recent developments in NDT in any depth it is useful to attempt to define a yardstick against which their performance can be assessed. Clearly it would be useful to define a required accuracy although it is obvious that this will vary from application to application. It seems probable however that some idea of the order of magnitude of the required accuracy can be gained.

2. In any structural situation there will be a certain critical crack size (a_c) beyond which crack growth will be catastrophic unless stressing conditions change. Before this stage is reached the crack will grow by some other, probably stress linked, cause such as fatigue or stress corrosion. To take the former case as an example the common formula for crack growth is:

$$\frac{da}{dN} = C (\Delta K)^4$$

where
C is a constant
ΔK is the range of the stress intensity changes.
N is the number of cycles
a is the crack depth

For an infinite medium

$$K = \sigma \left[\frac{\pi a}{2} \right]^{1/2}$$

where σ is the applied stress

3. This gives a relation for the number of cycles for the crack to grow from an initial size a_0 to a which can be expressed as

$$N \propto \left\{ \frac{1}{a_0} - \frac{1}{a} \right\}$$

4. For this reason a crack which starts at a size of $a_c/4$ will take three times as long to reach a critical size as a crack starting at a size of $a_c/2$. These formulae are very simple and cannot be taken as an up-to-date, accurate description of fatigue crack growth. Nevertheless they represent the kind of crack growth law which is applicable and, if we take some round figure values, explain why, if a_c is taken to be 10mm a crack of 8mm depth may be regarded as serious whereas a crack of 4mm depth may be tolerable. Their relative fatigue lives are thus 1:6.

5. These values of 8mm and 4mm crack depth, representing serious and tolerable crack sizes, have been used in the following calculations since they seem to be of the right order of magnitude for many existing structures with wall thicknesses in the range from 15mm to 30mm or so. Some relative increase - but not much - may be allowable in thicker materials while even smaller sizes pertain to thin tubes, for instance.

6. The nondestructive testing problem thus becomes one of distinguishing between 'serious' cracks of 8mm depth or greater and 'tolerable' cracks of depth less than 4mm. There are two mechanisms by which this distinction may be

foiled. In the first place small cracks of depth 4mm or so may simulate larger cracks causing a spurious identification of a 'serious' crack. Secondly, large cracks of depth 8mm or more may simulate small cracks and thus not be identified. The former is a nuisance and may lead to expensive and unwarranted shut downs and repairs. The latter may lead to failure.

7. In reality, of course, the defect located may take any size but it is simpler, and will give an answer of the correct order of magnitude, to retain the two particular crack sizes mentioned above. If we assume that the natural inaccuracy of the sizing technique gives rise to a standard deviation of σmm with a normal distribution it is clear that 2.3% of defects of size 8mm will look like defects of $(8-2\sigma)$mm depth while .135% will look like defects of $(8-3\sigma)$mm depth etc. The level chosen will depend on the failure rate considered acceptable, bearing in mind that it is probable that the estimate of the critical defect size is also conservatively based. If a failure rate of 1 in 1000 is acceptable for the test the condition that 8mm defects are distinguishable from 4mm defects to this degree can be written

$$8 - 3.1\sigma > 4$$

giving $$\sigma < 1.3\text{mm}$$

8. This is a reasonable value for σ if it can be accepted that 1 in 1000 of the 4mm defects will mimic 8mm defects. Since presumably, the population of 4mm defects will considerably exceed the population of 8mm defects many indications will however be false alarms. It would thus be prudent to apply the additonal condition that at a depth of $(8-3.1\sigma)$mm the probability of simulation by the 4mm defect population has fallen by 90%. This may be written

$$8 - 3.1\sigma > 4 + 1.3\sigma$$

giving $$\sigma < 0.9\text{mm}$$

9. Thus, without taking unusually small values for defect size and without putting very onerous restrictions on the failure rate of the test, a requirement for an accuracy in testing of less than 1mm standard deviation can be demonstrated. Accuracy requirements as poor as $\pm$ 2mm (or even $\pm$ 3mm) are often assumed for achieving this separation! The calculations are not rigorous and could be improved in specific instances where the test requirements would be better known. Nevertheless it seems clear that defect sizing techniques which do not have the potential to achieve a standard deviation in the region of 1mm or better, are really relying heavily on the conservatism of the fracture mechanics calculations to prevent actual failure. For thin materials accuracy requirements are correspondingly greater and for (say) 3mm plate or tube a standard deviation of 0.05mm would not be too precise.

10. For the acceptable operation of non-destructive defect sizing techniques therefore it is necessary to look for accuracies of better than 1mm and the techniques discussed below are reviewed in this light. For

comparison the common decibel drop techniques appear to give rise to standard deviations of 3mm or more while amplitude based techniques seem even worse than this (Ref. 1).

NOVEL ULTRASONIC TECHNIQUES

11. In a paper of a reasonable length it is not possible to include a description of all potential techniques. A selection has therefore been made which, it is hoped, includes the most interesting new approaches which have the potential to meet the requirements of accuracy and to be applied in practical situations. A fuller review is given in reference 2.

ULTRASONIC HOLOGRAPHY

12. Ultrasonic holography provides a means by which defects can be visualised and a permanent record obtained. This may be presented in a form similar to a C-Scan but the technique is capable of much more than this. Thus the permanent record is in the form of a hologram which, when reconstructed, provides an essentially 3-dimensional view of the object. By focussing upon a particular defect both its lateral size and through thickness position and extent can be estimated. Moreover the technique is perhaps unique in that its lateral resolution is relatively unimpaired in large thicknesses of material. Most commonly a normal longitudinal wave beam is employed but angled longitudinal and shear wave holograms are possible and may give better estimates of depths.

13. The basic operation of the technique can be shown with reference to Fig. 1. For simplicity the optical case is described. We may then consider a ray propagating from an arbitrary point on the object at an angle α with respect to the X direction. This can be resolved along the Z axis as a wave propagating according to the equation.

$$A\cos(\omega t + k\sin\alpha z)$$

where A is an arbitrary amplitude and ω is the ultrasonic frequency. If a reference beam, of assumed amplitude unity, is also present as shown in the Fig. this can be represented by

$$\cos(\omega t - k\sin\beta z)$$

14. Suppose we now record the product of these intensities along the Z axis the recorded intensity will be given by

$$\frac{A}{2}\cos(k[\sin\alpha + \sin\beta]z) + \frac{A}{2}\cos(2\omega t + k(\sin\alpha - \sin\beta))$$

where the normal trigonometric relationships are assumed. The second term remains as a propagating wave but the first term can be seen to represent a stationary intensity pattern with respect to time which can be recorded photographically. To facilitate this an arbitrary constant intensity of illumination (B) is added so that the total is always positive.

15. For reconstruction the record or hologram is illuminated with a beam at the same angle as the original reference beam. The resultant

intensity distribution is now given by the product

$$\cos(\omega t - k\sin\beta z)\left\{B + \frac{A}{2}\cos[k(\sin\alpha + \sin\beta)z]\right\}$$

which may be rewritten

$$B\cos(\omega t - k\sin\beta z)$$
$$+ \frac{A}{4}\cos(\omega t + k\sin\alpha z)$$
$$+ \frac{A}{4}\cos[\omega t - k(\sin\alpha + 2\sin\beta)z]$$

16. Only a single ray has been considered but it is clear that the process can be generalised for all rays from the object and for each ray three terms of this type will be produced in the hologram.

17. The group of terms of the general form $B\cos(\omega t - k\sin\beta z)$ represents direct transmission through the hologram and is of no interest. The terms of the general form $\frac{A}{4}\cos(\omega t + k\sin\alpha z)$ are, however, identical with their respective input rays in all but amplitude and thus form a virtual image of the object. This, being identical to the original, is the most commonly used image but the final group of terms also respresents an image at a higher angle. This is a real image and thus reversed but has nevertheless occassionally been used.

18. Turning now to ultrasonics the equations governing the production of the hologram are equally valid. The approach taken at Harwell has been to record the hologram on sensitized paper using a probe scanning systems (Ref. 3). A refinement possible at ultrasonic frequencies is that the reference beam can be neglected and its effect replaced by suitable modulation of the received signal.

19. For the purposes of reconstruction a photograph of the hologram is taken and a slide prepared from this acts as the hologram in an optical system. In defining the scanning process, generating an artificial reference beam and changing from ultrasonic wavelengths to optical wavelengths at the hologram stage a substantial step has been taken away from the simple theory described above and the justification of this is detailed in reference 3.

20. The response of this system to a real crack-like defect is shown in Fig. 2. It is seen that as with all other techniques which rely on pulse reflection, the image produced of the crack consists of a series of highlights and a conscious decision must be made to interpret these as a single entity rather than as a number of smaller flaws. In this sense the technique offers little advantage over C-Scan.

21. Where a substantial advantage does occur is in the examination of defects deep within materials where a C-Scan approach would be seriously affected by the effects of probe beam spread. One of the consequences of holographic reconstruction is that the effect of beam spread is virtually eliminated from the image and the resolution of objects at a depth of (say) 250mm in steel remains comparable to that

achievable nearer the surface (Ref. 3).

22. A second major advantage is that the holographic process provides a realistic means of providing an ultrasonic image. The focussing of this image is thus different for different depths in the specimen. Observation of the position at which recognisable defects come into or go out of focus thus can provide a direct and quantitative estimate of the position of the defect within the specimen and its through thickness extent. This three dimensional information is moreover provided from a 2-dimensional record - the hologram. A comparison between holograms taken with different beam directions can complement and improve the precision of through thickness measurement.

23. The holographic technique thus appears to be most advantageous in the examination of thick sections in which the effects of beam spread would otherwise be troublesome. The potential accuracy of the technique is usually better than C-Scan over the whole range of thicknesses and is in the region of $\pm\lambda$ for the lateral extent of defects and $\pm 3\lambda$ for the direction along the beam

ULTRASONIC SCATTER

24. Two longitudinal wave probes, which are constructed so as to generate an ultrasonic beam at an angle, are placed on the surface of a block of material. They are placed facing one another, but their separation is such that they are not in a position to transmit a large back wall echo. An observable ultrasonic pulse can nevertheless be transmitted between the probes. This has been studied by Wustemberg and his co-workers (Refs. 4 and 5) and the transmitted energy is attributed to grain boundary scatter in the region where the ultrasonic beams are broadly focussed - Fig. 3. This explanation seems to be confirmed by other experimental work (Ref. 7). Thus when a surface opening slit or crack is present between the probes it can be shown that changing the probe separation produces a variation in signal height - Fig. 4. This can then be explained by the progressive blocking of the paths available for scattered energy as the region of focus is brought nearer the surface.

25. This approach can then be used as a means of estimating crack depth and it may be preferable to keep the probes at a fixed distance and monitor the change in signal height as the crack grows deeper into the material. The crack will, as it grows, obscure more and more of the paths taken by the scattered energy and hence a monotomic fall in signal height would be expected.

26. The use of signal amplitude in the measurement of slit or crack depth, is to some extent a retrograde step but it will be appreciated that the signal amplitude in this case is much less dependent on the less important properties of the defect than is the case for conventional ultrasonic measurements. The provision of the technique is a function of the degree of focussing at the cross-over point and the effective depth of this region is governed by the beam widths of the ultrasonic probes. Thus with better focussing greater accuracy is obtained at the expense of some corresponding increase

in data collection times. Because of the change in the beam width as a function of the distance from the probe the resolution of the technique is not constant, being a function of defect depth and probe separation.

27. The technique need not be confined to surface-opening cracks and could be envisaged as sizing internal defects, within the limits of the resolution. An accuracy of sizing of better than $\pm$ 1mm is reported. The resolution for internal defects is governed by the depth of the focussed region however and is likely to be several millimetres. The technique has been used to size defects under cladding but the details of this work do not yet appear to have been published.

ULTRASONIC DIFFRACTION

28. Over the past three years work has been carried out at the NDT Centre aimed at the improvement in the accuracy of ultrasonic crack sizing through the introduction of techniques based on transit time as a complete alternative to reflected pulse amplitude. One such technique was based on the use of diffracted bulk waves (Ref. 6) and the general testing geometry employed is shown in Fig. 5. Two probes are placed on the surface of the material one on either side of the crack. Then, if the probe separation is fixed, the transit times of the ultrasonic energy diffracted at the crack tips provides an indication of the depth of the crack. If the probes are placed symmetrically about the crack tip, and the beam entry points are separated by a distance 2S, the crack depth d is given, to a reasonable degree of accuracy by

$$d = \left\{ \frac{(c\Delta t)^2}{4} - S^2 \right\}^{1/2}$$

where Δt is the transit time in the material under examination.

This formula is valid provided the refractive index at the surface is greater than 3 and S is greater than d. If these conditions are not satisfied the effect of refraction at the specimen surface should be taken into account more rigorously.

29. The technique has the advantage that the ultrasonic wave passes through the bulk of the material and is thus not affected by surface roughness, weld beads or the attachment of other components. Additionally the parameter measured is the depth of the crack tip below the surface which is normally of most direct relevance to estimates of component safety and lifetime. It can also be shown that the error introduced by modest errors in placement or due to lateral mislocation of the crack tip is very small (Ref. 7). The locus of possible crack tip positions for a given time delay is close to an ellipse in form, the depth of which is a slow moving function in the central region between the probes.

30. With a suitable calibration the technique can be used to provide estimates of the depth of cracks and slits. Slits were used initially since their depth could be measured mechanically without breaking open the sample. However, the

technique has now been used to size a number of fatigue cracks and some typical results are shown in Figs. 6 and 7. On completion of the ultrasonic measurements the plate was broken open and the crack profile was determined optically. The agreement between the ultrasonic estimates of crack profile and the actual crack profile as subsequently determined is very good. The mean error varies from case to case depending on test geometry but accuracies better than 5% are possible. The largest errors are found to be associated with those cracks in which the crack depth changes sharply over short distances. It will be appreciated that the ultrasonic result is essentially a weighted average over a distance related to the beam spread. This will then prevent the ultrasonic data following small scale variations.

31. The diffraction technique can be applied to internal cracks and to cracks growing from the surface opposite to that on which probes can be placed. Experimental investigation confirms this and, following the analogous technique to that described earlier, realistic estimates of crack profiles have been made. Difficulties with this approach may be expected in steels with a high inclusion content from which intense reflections may be expected. However, in steel of 'engineering' quality it is found that the intensity of the diffracted signal is generally substantially greater than that from inclusion echoes.

32. It is a prerequisite for a technique suitable for practical application that stringent test requirments can be relaxed. It can be shown, for instance, that the effect of probe lift-off and of very minor changes in the probe separation can be largely discounted if one of the back-wall echoes is used to provide a standard time reference. Indeed, the requirement to hold the probes at a fixed distance apart can be relaxed almost completely (Ref. 8) if necessary, which may be relevant to the examination of T-butt welds for instance. Other work has shown that the diffraction technique is considerably more stable against the effects of changes in crack transparency than the conventional techniques.

33. Work on the use of ultrasonic diffraction to characterise volumetric defects lags behind the programme on crack sizing at the moment. Nevertheless following pioneering work by Pao and Sachse (Ref. 9) some of the potential benefits of the application of diffraction/time delay techniques to this problem have been evaluated (Ref. 10). The technique thus appears to have a continuing potential in NDT.

ULTRASONIC SPECTROSCOPY

34. It is impossible in a review of this length to do full justice to the spectrographic approach to defect characterisation. This technique can, in principle, be applied in all of the situations in which time delay is used in the estimate of defect size. The fundamental basis of the approach can be described in reasonably simple terms and from this some broad conclusions may be drawn regarding applicability etc.

35. If we take an arbitrary ultrasonic pulse it is well known that this can be represented in

terms of a distribution of frequencies as may be calculated using the Fourier transform. In effect the pulse is represented by the sum of continuous waves of varying amplitudes covering a given frequency range. The shorter the pulse in time terms the broader the range of frequencies needed to describe it and vice versa. A fuller description may be obtained from any text book on the Fourier transform. If another similar pulse is now introduced a comparably broad spectrum will be obtained from the pair of pulses but it will be seen that at some of the component frequencies, the pulse separation will be such as to cause interference. This will occur if the separation is an odd number of half wavelengths. On the other hand reinforcement will occur if the separation is equal to an integral number of wavelengths.

36. The situation is thus that we may estimate the spatial separation of the two pulses either from the time delay between their arrival times or from the separation of the interference minima or maxima in their spectrum. In fact if the latter approach is used it is often convenient to convert the spectral data into the spatial domain by another Fourier transform. The resulting data have become known as the cepstrum.

37. In the choice between a time domain or spectrographic approach to defect sizing, a number of points should be borne in mind. In the first place it must be accepted that the spectrographic approach will only provide data on the pulse separation; the knowledge regarding the order of the pulses being lost in the conversion. On the other hand the spectrogram may allow the resolution of pulses of small separation which are unresolved in the time domain. This advantage is sometimes overstated however and it must be appreciated that when pulses are poorly resolved in the time domain it will often be the case that the pulse spectrum will not be sufficiently wide to resolve more than one interference minimum accurately. Under these circumstances the use of the spectrographic technique will not provide the degree of improvement anticipated.

THE AC POTENTIAL DROP TECHNIQUE

38. This is the only recent development of sufficient potential accuracy which appears to have occurred outside the ultrasonic field. Indeed the increase in interest in this approach has been so recent that no descriptive papers are yet available. The basis of the technique is, of course, well known and the measurement is essentially one of the resistance of the specimen. The skin effect confines the electric current to the surface of the specimen and thus the presence of a crack increases the path length which the current has to take by 2d where d is the depth of the crack. Resistance changes may then directly provide an estimate of crack depth.

39. Until data are published more widely a real appreciation of the accuracy of the approach cannot be gained. Certainly an accuracy better than $\pm$ 1mm seems possible however. There is some evidence that the crack profile affects the crack depth estimates so that the relationship between crack depth and

resistance is not unique. This is to be expected since the electrical path through the material is controlled only by the distribution of resistance between the electrical inputs.

40. The appearance of this technique is of considerable importance as it provides a potential cross check on ultrasonic data. A factor of great interest is that the potential accuracy of this approach may be sufficiently great that differences in crack depth estimation between electrical and ultrasonic measurements are not simply statistical but really provide useful data on the structure of the crack tip region.

41. Having said this, it is a curious fact that this technique seems to be gaining quite such a ready acceptance in the field of practical NDT. In this type of application it must be a drawback that the result is a measure of the distance along the crack face between the opening and the tip of the crack. Only in certain cases such as when the crack orientation is normal to the surface of a flat specimen, can this measurement represent the penetration of the crack. Some 5 years ago the use of ultrasonic surface waves for sizing defects was demonstrated (Ref. 11) and despite the elapsed time this technique has found no regular use in practical NDT. This surface wave approach however appears, on the face of it, to be comparable with the AC potential drop method in every way.

CONCLUSIONS

42. A number of novel defect sizing techniques have been described in outline, all of which appear to have considerable potential not only in their precision but also for practical testing. They are not techniques which need to be bound to the laboratory.

43. It is not proposed to attempt to recommend a "best buy" since each problem in NDT will have its own particular features which may favour one approach over others. Let us instead be thankful that there are a number of alternatives. In fact this paper is not exhaustive as reference 2 makes clear.

44. Possibly there will be some surprise at the level of accuracy which needs to be achieved for realistic defect sizing. Of course the calculation cannot be regarded as fully rigorous. However the choice of defect size and the assumption of a normal distribution of errors are probably conservative.

REFERENCES

1. RUMMELL W.D. and RATHKE R.A. Detection and Measurement of Fatigue Cracks in Aluminium Alloy Sheet by NDE Techniques. Prevention of Structural Failure Am.Soc. for Metals, 1978, p. 146.

2. SILK M.G. Developments in Ultrasonics for Sizing and Flaw Characterisation. Developments in Inspection and Testing of Pressurised Components. Applied Science Publishers – To be published.

3. ALDRIDGE E.E. Ultrasonic Holography.
Research Techniques in NDT, Vol I, Ed.
R.S. Sharpe, Academic Press London, p. 133, 1970.

4. BOTTCHER B, SCHULZ E and WUSTENBURG H.
A new method for crack depth determination in
ultrasonic materials testing. Proc. 7th
World Conf. on NDT, Warsaw, 1973.

5. WUSTENBURG H and SCHULZ E. Determination
of crack depths by a scattered signal technique.
Paper presented to VDEH, 1973.

6. LIDINGTON, B.H., SILK M.G., MONTGOMERY P,
and HAMMOND G.C. Ultrasonic measurement of
the depth of fatigue cracks. Brit. J. NDT
18, p. 165, 1976.

7. SILK M.G. Sizing crack-like defects by
ultrasonic means. Research Techniques in
NDT, Vol. III, p. 58. 1978.

8. To be published.

9. PAO Y.H. and MOW C.C. Diffraction of
Elastic Waves and Dynamic Stress Concentrations.
Crane Russak, New York 1973.

10. CHARLESWORTH J.P., LIDINGTON B.H. and
SILK M.G. Defect sizing using ultrasonic flaw
diffraction. Proc. European Conf. on NDT,
Mainz 1978.

11. COOK D. Crack depth measurements with
surface waves. Brit. Ac. Soc. Spring Meeting,
Loughborough 1972.

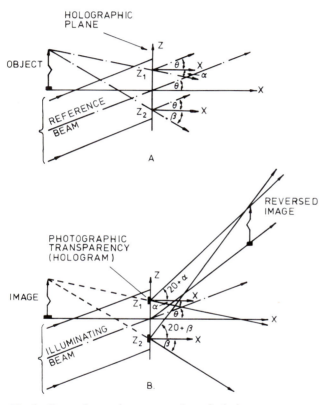

Fig. 1 Formation and reconstruction of a hologram

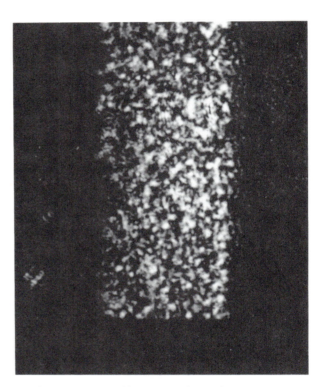

Fig. 2 Reconstructed hologram of a crack

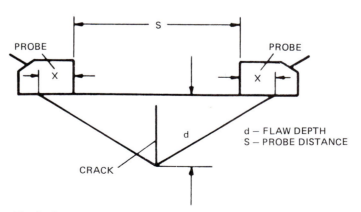

Fig. 3 Probe geometry — scatter technique

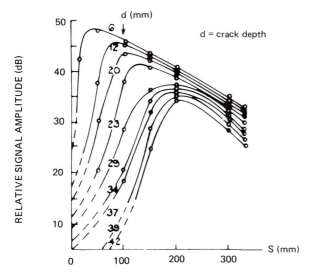

Fig. 4 Variation of amplitude as a function of probe separation 'S'

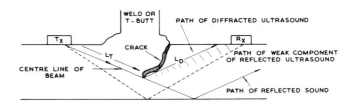

Fig. 5 Probe geometry — diffraction technique

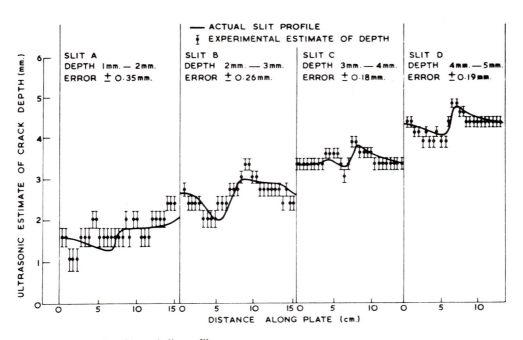

Fig. 6 Estimated and actual slit profiles

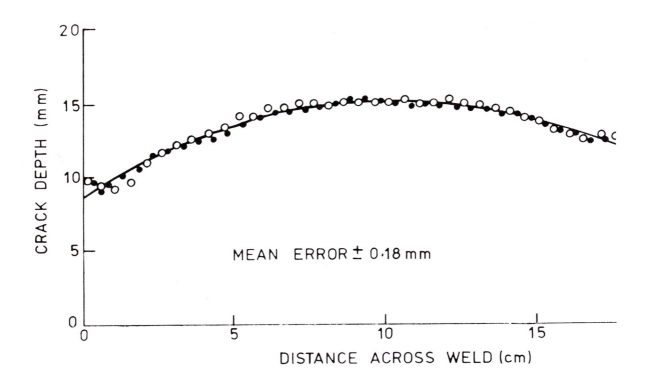

Fig. 7 Estimated and actual crack profile

C44/79

IMPROVED CHARACTERIZATION OF DISCONTINUITIES IN THICK WALLED PRESSURE VESSELS

J. R. FREDERICK, BSc, MSc, PhD, S. GANAPTHY, BSc, MSc, PhD, C. VANDEN BROEK, BSc, MSc, and M. ELZINGA, BSc, MSc

The MS of this paper was received at the Institution on 20 November 1978 and accepted for publication on 22 January 1979

SYNOPSIS   Several methods are described which are being investigated in order to obtain improved characterization of discontinuities in thick walled pressure vessels. These methods include the use of synthetic aperture focusing techniques for improved lateral resolution and narrow beam insonification for better penetration in thick sections and in weld regions. Discontinuities are imaged using grey-scale and perspective contour plotting techniques.

INTRODUCTION

1. A synthetic aperture focusing technique for ultrasonic testing, SAFT UT, has been developed which is capable of imaging discontinuities in pressure vessel steel with high lateral and longitudinal resolution, namely, about one wavelength (Refs. 1, 2, 3). With conventional focused transducers the resolution is a function of the size of the active element, assuming a constant focal length. In a synthetic aperture system, the equivalent of a very large diameter transducer is synthesized by combining the responses of a small diameter transducer which has been mechanically scanned over an area. It should be emphasized, however, that the synthetic aperture focusing process is performed after the scan has been completed. Thus the SAFT UT technique has the flexibility of focusing at any and all depths without rescanning. This is a definite advantage when operating in high radiation areas. In addition, compensation for surfaces which are not planar is straightforward with the synthetic aperture technique.

2. High signal-to-noise ratios are achieved in the SAFT UT system through the use of signal averaging. Basically, this involves averaging a large number of individual A-scans with the transducer stationary. Signal averaging has been used in the past in other types of systems with "boxcar integraters," which are analog devices rather than digital. Signal averaging in the SAFT UT system is a simple operation since each A-scan has been converted to a series of digital numbers during data acquisition. Signal-to-noise ratio improvements of 20 to 30 dB are readily obtainable. It then becomes feasible to use higher frequency transducers and to obtain greater sensitivity for detecting small discontinuities.

3. In a synthetic aperture system it is necessary that a discontinuity be scanned over a large subtended angle in order to obtain the optimum resolution. This is accomplished in the SAFT UT system by the use of a wide beamwidth transducer or by a "spot-light" scanning mode in which a narrow beam transducer having additional degrees of freedom is used. A wide scanning

angle has an added advantage in that discontinuities which have orientations that are not optimum can still be imaged by the SAFT UT system in a single scan. In essence, the ultrasonic beam is incident upon a discontinuity at many angles simultaneously, thus improving the reliability of the inspection results.

4. The use of broadband highly-damped transducers and digital deconvolution techniques makes it possible to identify discontinuities located close to another reflector, such as a front surface interface or another discontinuity.

5. SAFT UT is inherently quantitative and volumetric. The digital nature of the SAFT UT data acquisition, processing, and display makes it ideally suited to quantitative analysis. The speed and accuracy of the minicomputer assure reliability and repeatability. A two dimensional scanning apparatus makes possible a truly volumetric scan of the material. When combined with an interactive display system, the operator is able to thoroughly scrutinize suspect areas. In addition, since the system records the RF A-scans in digital form, they can be stored indefinitely on magnetic tape for comparison with subsequent inspections. Such a capability is of extreme importance in monitoring a growing crack.

6. Discontinuities are sized and located independently of the relative amplitude of the echoes from the discontinuities. Unlike the conventional UT approach in which the amplitude of the echo from a discontinuity is used to determine the size of the discontinuity, the SAFT UT system uses the image of the discontinuity's boundaries to determine its size. In this respect the amplitude of response is of secondary importance. As long as enough energy is reflected by the discontinuity so that it can be detected, then an accurate determination of its size is possible.

SYNTHETIC APERTURE IMAGING

7. Synthetic aperture imaging can be illustrated by considering the case of a transducer which is scanned in a straight line over a test specimen containing a point target. As shown in Fig. 1,

the echoes for those scan positions that are laterally displaced from the target location are delayed in time due to the greater distance they have traveled. The locus of echo phase shifts created by the scanning process is generally hyperbolic in form with its apex positioned at the location corresponding to minimum echo travel time.

8. If the scanning and surface geometries are known, it is possible to accurately predict the shape of the locus of the echoes. If the individual A-scan signals are now shifted forward in time by an amount equal to their predicted phase delays, they will come into phase with one another. Then, if the signals are coherently summed, the result will be a single large response as shown in Fig. 2. The A-scan signals included in this process constitute a synthetic aperture. The same process is repeated with a new synthetic aperture displaced by one scan position from the previous aperture as shown in Fig. 3. In this case, however, the phase shift compensation does not produce a set of in-phase echoes. Destructive interference occurs during summation which results in a significantly smaller response. The process is directly analogous to that which occurs in a conventional transducer with a focusing lens. Repeatedly shifting the new aperture by one scan position and producing a new output signal each time has the effect of producing a scanning synthetic aperture.

9. The achievable lateral resolution ρ_x will be approximately equal to $[\lambda/(4 \sin(\Delta\theta/2))]$, where λ is the ultrasonic wavelength in the medium in which the target is located, and $\Delta\theta$ is the total angle over which the target is insonified. If we assume an insonification angle of 30°, a reasonable value for a focused transducer in steel, then ρ_x is approximately equal to λ, or 2.6 mm (0.10 in) in stainless steel at 2.25 MHz.

DATA COLLECTION FOR A RASTER SCAN

10. All data collection and management is currently done digitally with an Interdata 7/32 minicomputer. A flow chart for the system is shown in Fig. 4. During data acquisition a test specimen is immersed in a tank of water and scanned under computer control by a single scanning transducer operating in pulse-echo mode. Highly damped, focused transducers are used in a raster scan. Scanning is done in a single plane with the test specimen oriented so that the focal point of the transducer is located on the specimen surface.

11. At each scan position the minicomputer causes the ultrasonic transducer to be pulsed. The resulting A-scan RF signal is sampled and converted to a digital value at equal time intervals as shown in Fig. 5. The time delay between the pulsing of the transducer and the first sample (ID), the subsequent delay increment between successive samples (DI), and the number of samples (NP) are recorded for future use. Additionally, the operator may choose to employ signal averaging to reduce the random noise introduced by wideband amplifiers in the system. In this case, at each scan position, the transducer is pulsed repetitively and the resulting sampled signals are averaged. For random noise a signal-to-noise ratio improvement equal to the square-root of the number of signals averaged can be achieved.

12. Thus far, the discussion of the synthetic aperture focusing technique has been described in terms of a linear scan. In reality, the typical data scan is two-dimensional and requires a two-dimensional aperture. The locus of echoes from a point target is therefore actually a hyperboloid and the phase shifts are calculated accordingly.

DATA COLLECTION FOR SPOTLIGHT MODE SCANNING

13. In the spotlight-mode of scanning a narrow beam-width transducer is controlled so that, as it scans a test specimen, a small region within the test volume is being continuously insonified as shown in Fig. 6. The cross marks shown on the series of concentric circles depicts the scanning positions required to obtain data over one small volume within the inspection volume. Scanning is not limited to the path shown in Fig. 6. Apertures of any desired shape may be synthesized and modifications may evolve which are particularly suited to certain flaw types and surface geometries.

14. Data collection is depicted in Fig. 7. In this figure the transducer is scanning a circle of radius RR. RR and F, the depth of the target, are under operator control. The computer calculates the required transducer position so that a waveform made up of NP digital points can be acquired and stored in the computer memory. Also, during scanning the distance the sound travels through the water is held constant. The timewise location of the NP points is calculated by the computer so that the centerpoint is at the target. The output of a scan is thus a set of A-scans, each of which is centered on the target regardless of the scanning circle diameter. The target location, of course, is the same for each circle within a set of concentric circles. Also, the number of A-scans collected around each scanning circle increases with the diameter of the circle so that the linear spacing between successive A-scans is approximately constant.

15. The synthetic aperture processing is now straightforward. For each target location (that is, for each set of concentric circles) all the A-scans from each scanning circle associated with that target are simply added together. The resultant sum is then divided by the number of A-scans that make up the sum. The result is a processed A-scan NP points long for each target location. The centerpoint of this A-scan is in true focus.

16. Given a set of processed spot-light mode A-scans covering the volume of interest, one can then apply all the display techniques, such as perspective-contour and grey scale, which have previously been developed for use with conventional SAFT UT. Preliminary results indicate that the results will be equivalent to those obtained with raster scanning which are presented below.

DATA PROCESSING AND DISPLAY

17. Figure 8 is a flow chart of the digital routines employed to generate SAFT UT flaw images. The scan program collects data as discussed above, and its output is a raw data file that is stored on a magnetic disc. As an

example of the acquisition time required a single A-scan signal made up of 250 points, such as the scan used to generate data for Fig. 10, and summed 20 times to improve the signal-to-noise ratio, takes approximately 5 seconds to collect. Synthetic aperture processing to the same signal then requires as long as 19 seconds, depending on the aperture size.

18. An envelope detection routine is used to remove the RF waveform leaving only the amplitude envelope. This is done digitally with the aid of the Fast Fourier Transform. The time required for a 250 point signal is approximately 2.5 seconds.

19. Current techniques for the display of an ultrasonic image require transforming the output of the envelope detection routine from a 4-dimensional data set (X, Y, Z, and echo amplitude) to a three-dimensional data set (amplitude plus two spatial dimensions). This transformation is done by the PLOTGEN program, and the output data is then suitable for display as a position-amplitude plot, a contour plot, or a grey-scale display. The processes of PLOTGEN and display occur with sufficient speed that the user can obtain the desired results in an interactive fashion at the computer terminal.

PERSPECTIVE-CONTOUR PLOT

20. A perspective-contour plot is a composite picture of a sequence of contour plots, each contoured at the same intensity level, rotated as user defined, and displayed after hidden line elimination. Given an inspection volume defined by three spatial coordinates, X, Y, and Z, one can construct a perspective-contour plot by using any set of either XY, XZ, or YZ planes that intersect the volume. For example, Fig. 9 shows two different perspective-contour plots of the same inspection volume. Each plot is contoured at the same intensity level and rotated by the same amount. The difference between the two plots is the number and type of planes that compose the plots. The top plot is composed of four XY planes while the bottom plot is made up of nine XZ planes.

21. The computer algorithm which produces a perspective-contour plot has been implemented as an interactive software program which uses a command language to set up the various necessary parameters. The user-defined parameters remain in force until specifically changed. This allows the user to produce a series of plots for various values of certain parameters while holding the other parameters constant. For example, a user might specify the planes he wants, the contouring level desired, and then proceed to produce a series of plots for various rotations. It is also possible to specify a region within the inspection volume to be enlarged for closer scrutiny. After enlargement, rotation can again be made to optimize the view. Other examples of the perspective-contour plot occur throughout the remainder of this report.

RESULTS

Images of Slag

22. Perspective contour plots of a test specimen made from pressure vessel grade A 533B steel with multipass, 60° included angle, vee-butt welds are shown in Fig. 10. The weld contains pieces of slag sandwiched in between the weld passes. The scan data were processed using an algorithm that results in images at all depths in the test specimen being in focus, instead of just at one depth.

23. Figure 10 shows two plots and an actual cross-section of the specimen. The photograph has had the weld and scan area artificially enhanced for clarity. The plot at the upper left of the figure is a perspective-contour plot with no rotation. The plot is made up of ten X-Y planes, each separated in the Z-direction by 0.16 mm. The plot at the bottom is identical in contact but is rotated to show the discontinuity in perspective.

24. The image shown in the plots has been verified to correspond closely (within 1λ) to the actual flaw dimensions in all but the Z-direction. The Z-direction dimensions are not accurately protrayed because the slag will not satisfactorily transmit sound.

Images of Porosity

25. Two 5 MHz scans were taken of a weld specimen which contains small voids in the weld region. The specimen was made from reactor pressure vessel grade A 533B steel with multipass, 60° included angle, vee-butt welds. The scan increment in both the X and Y directions was 1.27 mm (0.05 in), and Z-direction data points were taken at 0.30 mm (0.012 in) intervals. The data were processed using the multi-depth focusing algorithm.

26. The two scans were made adjacent to one another so that the perspective contour plots shown in Fig. 11 fit end-to-end as indicated at the top of Fig. 11, which is a perspective-contour plot with no rotation. A metallographic cross-section made after scanning was completed is shown at the bottom of Fig. 11. This metallographic section and other similar ones made at different locations confirm the ability of the SAFT UT system to accurately locate this type of small void. However, it is beyond the resolving power of SAFT UT to clearly define the individual voids within the cluster located at the intersection of the two scans (Fig. 11).

Images of Cracks

27. Figure 12 shows perspective contour plots of a weld containing cracks. The specimen was made from reactor pressure vessel grade A 533B steel with multipass, 60° included angle, vee-butt welds. The volume displayed shows cracks which were introduced by inserting small pieces of copper tubing in the weld during welding.

28. Figure 12 shows two plots and an actual cross-section of the specimen. In the actual cross-section the cracks were enhanced by the use of a liquid penetrant and hence appear to be wider than they actually are. The perspective-contour plot at the bottom includes echo signals only from the area of interest which is toward the bottom of the inspection volume.

29. With few exceptions the plots indicate only the locations of sound reflected from the crack and points. Moreover, as seen from the actual

cross-section the flaw is actually made up of
several small cracks with multiple side branches.
The result, as shown in the perspective-contour
plot, is a very complicated image. While this
image does not clearly show the cracks as they
appear in the cross-section, it may be sufficiently unique to allow identification of
similar flaws in future tests.

IMAGING OF THIN EDM SLOTS IN A HEAVY SECTION
STEEL BLOCK

30. A test of the SAFT UT system's imaging
capability was carried out with the nozzle
"drop-out" block shown in Fig. 13. The block,
made of A 533B pressure vessel steel 191 mm
(7.5 in) thick has six narrow slots machined
into the bottom surface as shown. This was
done by the electrical discharge machining (EDM)
process. The slots were 0.25 mm (0.01 in) wide
6 mm (0.25 in) deep and of various lengths.

31. A single, 2.25 MHz scan was made of the
volume enclosed by dotted lines. All scanning
was performed in the Z-Y plane so that the
cracks were viewed edge-on using longitudinal
waves. The raw data for each X-Y position was
derived from an average of 36 A-scans to reduce
electronic noise. The time increment between
Z-direction data points was 100×10^{-9} sec
which is equivalent to about 0.63 mm (0.025 in).
A 25 x 25 aperture (625 elements) was used for
processing. This results in an effective total
aperture angle at the bottom of the block of
18°.

32. Figure 14 is an X-Z position-amplitude plot
of the processed data. This is a projection
through the entire volume. It indicates the
presence of several small discontinuities
distributed throughout the volume. The large
echo due to the back surface is slightly beyond
the processing range in the Z-direction and
therefore does not appear in this plot. The
slots at the bottom of the block are likewise

not apparent. To make these slots visible a
16 level grey scale display unit was used to
search individual data planes and to increase
the contrast in the available data. Results of
this search are summarized in Fig. 15. Each of
the photographs shown here is equivalent to
position-amplitude plots at the depth indicated,
where the lightest portions are areas of
highest amplitude. The amplitude range for
these three planes is from 2 to 11 (arbitrary
units) out of a total range available in the
data acquisition system of 0 to 256.

33. A comparison of the photographs of Fig. 15
with the drawing of Fig. 13 indicates that the
images are indeed constructed from the longitudinal
waves reflected from the tips of the slots.
The quality of these images is poor, however, and
not all the slots are seen. Moreover, it would
be difficult to identify the type of flaw based
merely on the image presented.

CONCLUSIONS

34. The images presented here demonstrate the
ability of SAFT UT to quantitatively characterize
discontinuities in metal. However, they are
only interim results of an on-going effort.
Current activities include improvements in all
phases of the system. A new processing
algorithm has recently been developed so that
the output data will be focused at every plane
instead of a single plane. The display systems
are being improved with the goal of presenting
simulated 3-dimensional grey-scale images with
interactive control so that it will be possible
to zoom in on and rotate any desired portion of
the inspection volume. Concurrent with these
efforts is the on-going task of validating the
ultrasonic images by comparing them with the
sectioned test specimens. This procedure
together with a quantitative description of the
capabilities of SAFT UT is necessary for field
implementation.

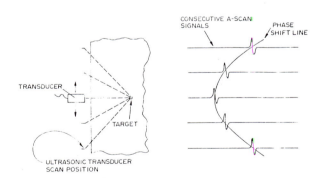

Fig. 1   Scanned transducer response

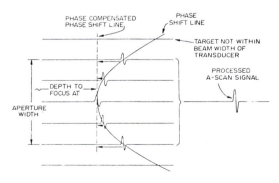

Fig. 2   Single aperture phase correction

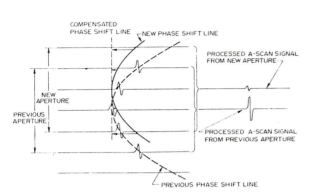

Fig. 3 Consecutive synthetic apertures showing resolution improvement

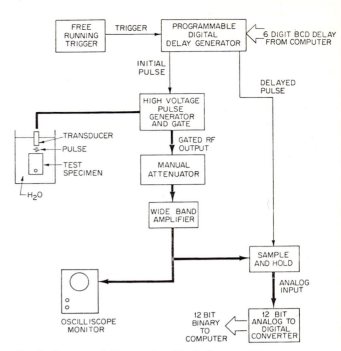

Fig. 4 Data acquisition system block diagram

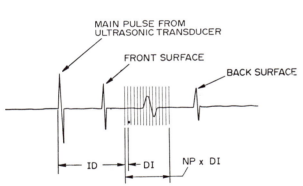

Fig. 5 Sampling of an A-scan RF signal

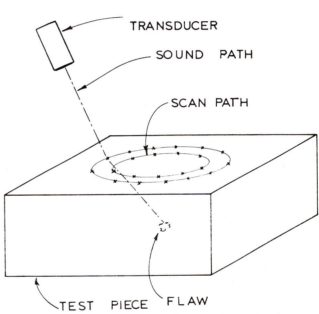

Fig. 6 Scan path followed in spotlight-mode scanning

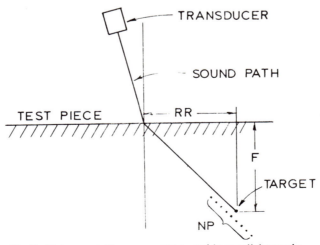

Fig. 7 Data acquisition parameters used in spotlight-mode scanning

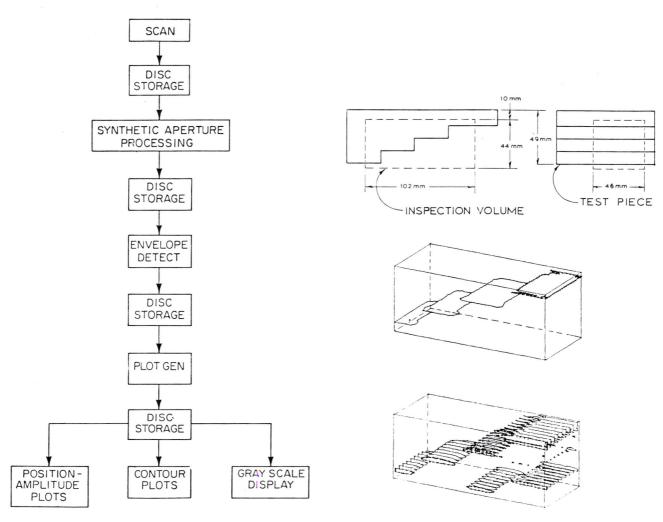

Fig. 8   Data processing block diagram

Fig. 9   Examples of the perspective-contour plot

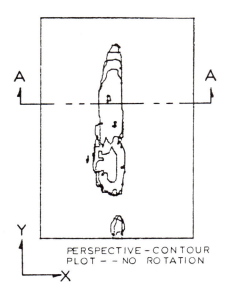

PERSPECTIVE – CONTOUR
PLOT – – NO ROTATION

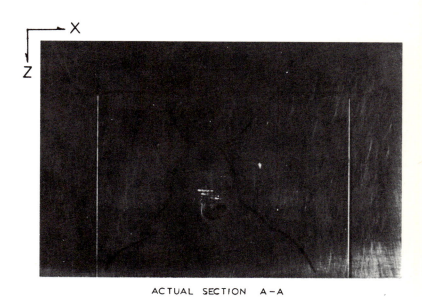

ACTUAL SECTION A – A

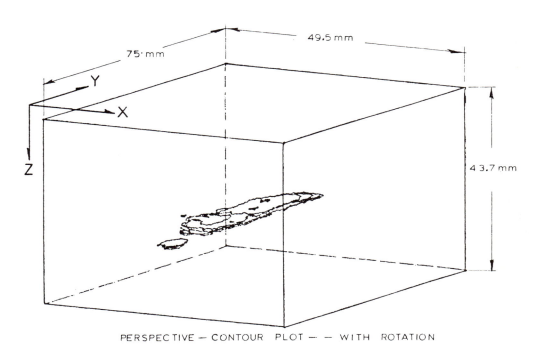

PERSPECTIVE – CONTOUR PLOT – – WITH ROTATION

Fig. 10 Perspective-contour displays of a slag inclusion

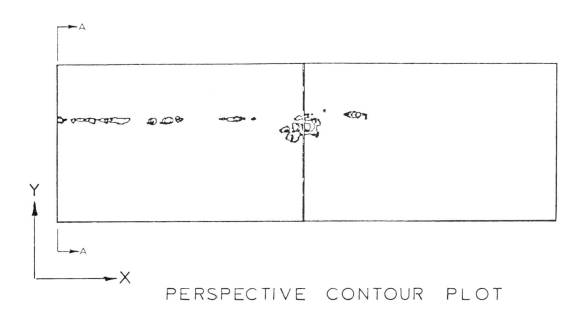

PERSPECTIVE CONTOUR PLOT

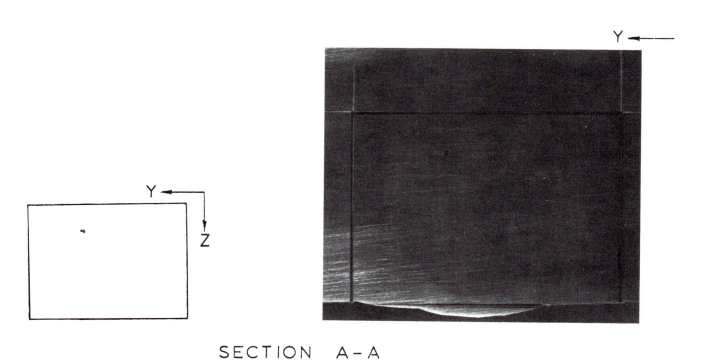

SECTION A-A

Fig. 11 Images of porosity

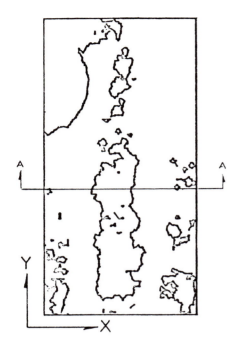

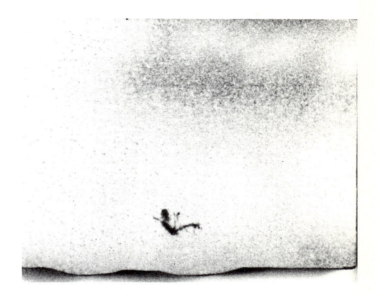

ACTUAL SECTION A-A

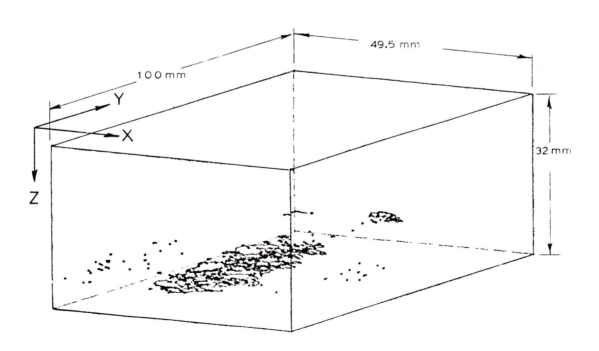

Fig. 12 Images of cracks

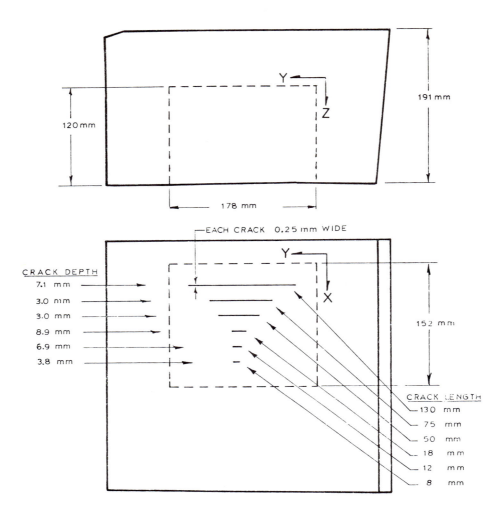

Fig. 13 Nozzle 'drop-out' block showing locations of EDM slots

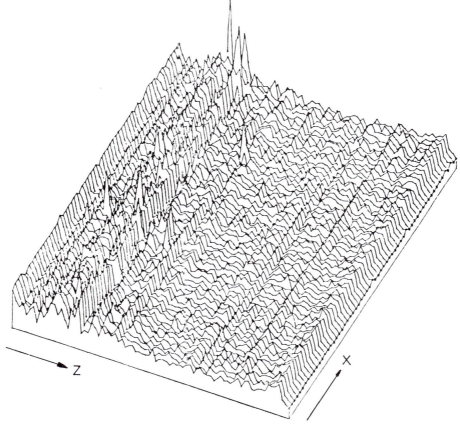

Fig. 14 X-Z position amplitude plot of SAFT UT processed data from the EDM slots in the 'drop-out' block shown in Fig. 13

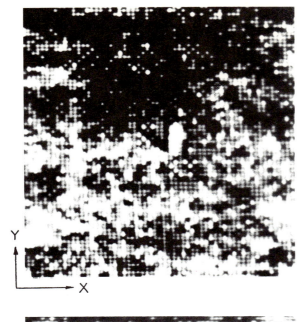

PLANE 384
183 mm DEEP

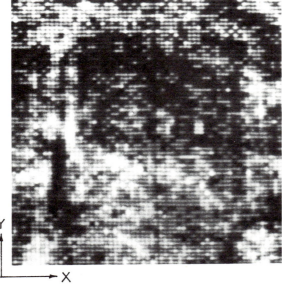

PLANE 390
185 mm DEEP

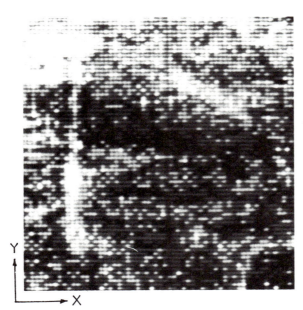

PLANE 392
186 mm DEEP

Fig. 15 Grey-scale displays of SAFT UT processed data from
the EDM slots in the 'drop-out' block shown in Fig. 13

173

DEFECT CHARACTERIZATION WITH COMPUTERIZED ULTRASONIC HOLOGRAPHY

J. H. FLORA, BS, MS, A. E. HOLT, BS, MS, and J. W. BROPHY, BS, MS
Babcock & Wilcox Company, Lynchburg Research Center

The MS of this paper was received at the Institution on 12 September 1978 and accepted for publication on 22 January 1979

SYNOPSIS Ultrasonic holography is the prime candidate method for characterizing defects in thick-walled pressure vessels. Defect images have been reconstructed from holograms by the optical method and more recently by using a minicomputer. Image reconstruction with a minicomputer provides several advantages in nondestructive examination. The advantages include: (1) a reduction in the operator skill required with present optical reconstruction systems, (2) unambiguous scaling of defect images, (3) near real-time image reconstruction, (4) enhancement of images by numerical methods, and (5) construction of composite 3-D images. This paper describes minicomputer techniques for reconstructing images from pulse-echo digital holograms. Computer images of actual defects in thick-walled steel welds are illustrated and compared to optical images.

INTRODUCTION

1. Until recently, nondestructive inspection of pressure vessels has been limited to x-ray radiography and standard ultrasonic tests. Although these conventional inspection methods can be used to locate flaws, they can provide a poor measure of defect dimensions. Acoustic holography (AH) offers potential for obtaining visual images of defects in critical areas such as welds and high stress regions.[1] If good images are formed, important defect characteristics such as size, shape, orientation, and type can be defined. The development and application of AH to pressure vessel NDE can result in substantial cost savings for the nuclear industry.

2. Useful images have been obtained in several applications with commerically available AH equipment. However, this equipment is not suited for the shop environment. A high level of skill is required to operate the system and generation of acoustic holograms and reconstruction of images from these holograms is presently a tedious and time-consuming process. Furthermore, good image quality and defect sizing is difficult with the optical image reconstruction apparatus incorporated with the commercial unit.

3. Minicomputer recording, reconstruction and display of AH images offers a major step toward improvement of pulse echo acoustic holography. Advantages include simplification of AH, more accurate defect sizing, enhancement of image display and reduction in image reconstruction time. Eventual benefits include improved image quality, automatic defect characterization, and three dimensional display. Described in the following sections are the initial development and evaluation of computerized AH at the B&W Lynchburg Research Center (LRC).

GENERAL

Historical background

4. Holography was introduced in a paper written by Dennis Gabor,[2] in 1948. Gabor proposed recording the amplitude and phase of an electron beam diffraction pattern from a reflective object. In this process, a monochromatic plane wave irradiates the object. The interference pattern between the reflected wave and a reference wave of the same frequency is used to record the diffraction pattern or hologram on a photographic plate. Reconstruction of the image is provided by illuminating the hologram with narrowband light.

5. Gabor used a colinear or on-axis reference wave of the same frequency to record and reconstruct the hologram. This produces two reconstructed object wavefronts - a duplicate image of the original object (true image), and a mirror image of the object (conjugate image). Unfortunately, these images are superimposed when the on-axis reference wave is used. The reconstruction is a distorted combination of the two images. Consequently, only simple objects could be identified.

6. Leith and Upatnieks[3] discovered that an off-axis reference wave could be used to separate the true from the conjugate image. The reference wave is directed toward the photographic plate at some angle with respect to the object wave. For example, light waves from the object can be directed normal to the plate and added to the reference wave. Since the photographic plate records the illumination intensity, the phase information of the object is retained by the hologram. The off-axis reference causes a phase shift which varies with the distance measured across the hologram. This phase shift produces a separation between the true and conjugate images when the hologram is illuminated. The use of extremely coherent illumination, e.g. laser light, provides for optimum recording and image separation.

7. The principles of optical holography are valid for other forms of coherent radiation including acoustic waves. Acoustic holography involves the use of coherent sound waves to insonify an optically opaque object. Either the through transmission or reflection of these waves are detected and recorded to form an acoustic hologram.

8. Since acoustic holography can be used to obtain images of defects, it is valuable for defect characterization in pressure vessels.[4] The reconstructed image provides a means for measuring defect size, shape, orientation, and dimensions. These measurements are necessary for fracture mechanics analysis which can assure safer economic employment of thick-walled components. Described in the following is the acoustic holography technique presently used for nondestructive testing.

Pulse echo acoustic holography

9. A schematic of the acoustic holography equipment used at LRC to interrogate thick-walled materials is illustrated in Figures 1, 2, and 3. The system is composed of three basic parts: 1) mechanical scanner, 2) electronic system, and 3) image reconstruction unit. The function of each major component is described in the following paragraphs. Attention is directed to the reconstruction unit since computerization of the reconstruction process is the primary subject of this paper.

10. Scanning system. The commercially available scanning system employs a focused ultrasonic transducer as illustrated in Figure 1. The transducer, light source, and polaroid camera are mounted to scanning main frame B. The transducer serves as both a transmitter and receiver of ultrasonic energy. The ultrasonic beam travels through the coupling media, e.g., water, and is focused at a point near the surface of the test sample as illustrated in Figure 2. The electronic receiver is time gated so that signals are detected in a narrow time interval after a pre-determined time delay. By using time gating, the scanning system interrogates only a band of material at a given depth in the test object. As the transducer, A, Figure 1, is scanned over the test object surface, the focal point sweeps out a pattern in raster fashion, covering a rectangular area in the horizontal plane.

11. During the raster scan, a pulse generator gates a continuously running oscillator signal to produce short bursts of sinusoidal output signal. These output bursts are then used to drive the transducer and produce a sinusoidal ultrasonic wave. The ultrasonic wave reflected from the defect is received by the same transducer after a time delay of a few microseconds. The time-gated, received signal is amplified and detected with a phase sensitive balanced mixer.[4] The output signal from the mixer is sustained by a sample hold circuit until the next gate pulse occurs. Each sample and hold output signal is amplified and applies to glow modulator light source. The glow modulator output is transmitted by fiber optics to an exit tip, i.e. which is a fixed extension of the transducer. The fiber optic tip sweeps out the rectangular pattern

during the raster scan. The analog signal from the scanned light source is recorded on a photographic transparency to produce the hologram.

12. Optical reconstruction. The commercial acoustic holography system uses an optical bench to reconstruct the image. Figure 3 is a schematic diagram of the optical reconstruction system. After the hologram is recorded on film transparency, the film is developed and placed in a narrow chamber formed by two parallel glass plates. An optical matching fluid fills the gap between the plates to reduce light refraction between the hologram transparency and the glass plates.

13. The hologram transparency is illuminated with coherent laser light to produce a diffraction pattern on the opposite side of the hologram.[5] A lens guides the coherent light beam and directs it toward a small aperture. The aperature is adjusted manually to screen out diffraction orders[6] associated with the raster scan lines. The aperature passes only the image of interest. A lens located behind the aperature enlarges the image and directs it to a ground glass plate. The position of the laser light source is moved to vary the distance to the hologram until image focus is obtained. The hologram is also moved laterally to a position that produces the sharpest image.

14. A closed-circuit television camera scans the ground glass plate and transmits the image to a video monitor. Intensity and contrast are adjusted for image sharpness and clarity. A polaroid camera is then used to obtain hard copy of the reconstructed image from the video monitor screen.

15. The optical reconstruction system has several drawbacks and limitations as follows: 1) limits image reconstruction and display capability, 2) requires considerable operator skill, 3) is difficult to use in plant environment, and 4) is a time-consuming process. The optical reconstruction process produces a flatened, two-dimensional image containing background noise and intensity distortions. Image enhancement techniques that can reduce these problems are not practical with optical methods. For example, images obtained with two or more ultrasonic excitations cannot be readily combined unless the images are converted to a digital form. This would require excessive time for data conversion, and additional instrumentation for signal processing. Also, digital filtering techniques are not easily applied to the optical images.

16. The optical system requires a relatively high level of skill and experience for reliable utilization. Improper procedures can result in poor image quality and false characterization of defects. Furthermore, each stage of the optical unit is a source of erroneous distortion. Adjustment and calibration requires in-depth knowledge and understanding of the optical reconstruction process. With so many factors affecting the image, correct interpretation of results requires strict reconstruction process control. For example, a change in

acoustic frequency or defect depth will cause appreciable changes in image size.[1]

17. The existing optical reconstruction is a time-consuming process. After the hologram scan is completed, it takes a minimum of eight minutes to develop and set the polaroid film. After using lookup tables for the initial laser position, adjustment of the optical bench parameters is required. Adjustments in parameters, such as laser source-to-hologram distance, hologram position, image mode rejection, and video intensity usually exceeds five minutes. Obtaining a hard copy of the video image with a polaroid camera takes at least an additional two minutes. Consequently, the total time required for optical reconstruction of images is usually twenty minutes.

18. _Digital computer reconstruction._ The basic concept of the computer reconstruction is illustrated in Figure 4. A small computer controls an analog to digital converter (ADC) and records the hologram in digital form during the transducer scan. The hologram is stored in high-density digital memory, eg., magnetic disc, eliminating the need to record the hologram on Polaroid film. The size and complexity of the mechanical scanning apparatus is reduced. Linearity, resolution, and quality of the digitally coded hologram are significantly better than that for the conventional hologram recorded on photographic film.

19. Development and evaluation of the computer reconstruction process involves the use of a laboratory system consisting of five parts:
- Scanning mechanism
- Signal excitation and detection unit
- Analog-to-digital coverter (ADC)
- Minicomputer for image reconstruction
- Output devices for image display
The Holosonics 200 unit used contains the essential components of the scanning mechanism, signal excitation and signal detection. The additional laboratory equipment including the ADC, minicomputer, and display are part of the Hewlett Packard 5451B Fourier Analyzer.

20. The HP 5451B Fourier Analyzer consists of the 2100S Minicomputer, HP5466A Analog-to-Digital Converter, 4012 Tektronix terminal and display, versatec plotter, and two HP7900 Cartridge Discs. The computer is programmed either through a special keyboard located adjacent to the ADC or by inputting instructions from the 4012 Tektronix terminal. Computer instructions for the various functions including data acquisition, image reconstruction and image display are programmed in FORTRAN and in a special keyboard language. The keyboard instructions call subroutines which are efficiently programmed in the HP 2100S computer assembly language.

21. _Reconstruction algorithm._ The basic function of the image reconstruction algorithm is to propagate an arbitrary steady-state wavefront from the hologram plane back to some specified plane. In the present case, the information is propagated from the recording plane to the image plane. Squaring of the propagated wavefront results in an image representation that is useful in flaw characterization.

22. The computational method used is referred to in the literature as the angular spectrum formulation of the back propagation problem.[8,9,10] The angular spectrum method takes advantage of the discrete Fast Fourier Transform (FFT)[7] to reduce computation time. The time required for reconstruction FFT routine is proportional to N, where N is the dimension of the hologram array of complex data points. If the array is square and if N is a power of 2, i.e., $N = 2$, 4, 8, 16, etc., then the computation speed of the FFT is directly proportional to N. By comparison, direct computation time is proportional to N^2. The FFT routine takes advantage of internal symmetry of intermediate arrays and computes iterations of a recursive formula to generate the FFT coefficients. This replaces the time-consuming operation of computing each coefficient directly.

23. The basic operations of the reconstruction algorithm are: 1) Input the complex data and perform the two-dimensional Fourier transform of the data array, 2) Calculate the back propagation function, 3) Multiply the propagator with the transformed array, 4) Perform the two-dimensional inverse Fourier transform and calculate the image, 5) Display the image and if necessary, compute a new image plane for a better image focus.

EXPERIMENTAL EVALUATION

24. The computer reconstruction algorithm was evaluated by using test blocks containing artificial defects. Holograms were recorded in digital form with the Hewlett Packard 2100S minicomputer and were also recorded on Polaroid film using the flow modulator light source and Polaroid camera.

Defects in a plane

25. Figure 5 is a diagram of an aluminum test sample containing flat-bottomed holes machined in a Y-shaped pattern. The 3/6 inch (.476 cm) diameter, holes were drilled so that the bottoms of the holes are 4 inches (10.16 cm) beneath the surface of the test sample. The left leg of the Y contains three holes, located 4 mm apart including the center hole. The holes in the top right leg are 2 mm apart and the holes in the lower right leg are 1 mm apart.

26. The Holosonics 200 system was used to scan the Y shaped hole pattern and record the ultrasonic response in digital form. The test sample was placed in a tank of water with the holes on the lower surface of the sample. The scanning apparatus was supported above the tank and adjusted so that the scan plane of the ultrasonic transducer was parallel to the test sample top surface. A broadband, focused transducer was positioned about 1 inch (2.54 cm) beneath the water surface and 4 inches (10.16 cm) above the test sample. At this distance, the focal point of the ultrasonic beam is 1 or 2 millimeters beneath the surface of the test sample.

27. The holograms were obtained by exciting the ultrasonic transducer to produce a longitudinal wave in repetitive bursts, ie, modulated pulse of 2.6 MHz sinusoidal voltages. The pulse duration was 3 microseconds and the repetition rate was 400 rps. Pulse reflection from the flat-bottomed

holes were received by the transducer and amplified. The received signal is gated and sampled
for 1.5 μsec. by the ultrasonic system. The 1.5
μsec. gate rejects ultrasonic reflections from
material which lies 0.2 inches beneath the flat-
bottomed hole plane. A narrow transmission
pulse minimizes detection of ultrasonic signal
from material above the hole bottom plane. The
gated signal is mixed with a reference signal
of the same frequency, i.e., 2.6 MHz. This provides phase sensitive detection of the received
signal. The sample hole circuit retains this
voltage until the ultrasonic pulse cycle repeats.

28. The minicomputer system was programmed to
sample the detected ultrasonic signals as the
scanner moves in the positive x direction. Scan
rates were set at 3 inches (7.62 cm/sec). Analog-
to-digital conversion rates were selected to
obtain 128 samples for each positive scan. The
data acquisition system was programmed to record
a total of 128 scans. This results in a hologram
having 128 by 128 samples of received ultrasonic
signal.

29. Figure 6 is a computer display of an acoustic hologram of the Y-shaped hole pattern. This
is a gray level plot using 3 shades of grey. The
minicomputer was programmed to print one of three
characters representing different detected signal
intensities. Characters are printed only if the
detected signal amplitude exceeds a predetermined
threshold. The gray level plot of Figure 7 represents the center portion of a hologram
obtained while scanning the transducer 3 inches
(7.62 cm) in the x direction and incrementing
the scan position by 0.024 inch (.61 mm) in the
y direction.

30. Figure 7 is a 3 gray level plot of the computer reconstructed image of the Y-shaped pattern
of holes. The image was obtained by processing
the 3-inch by 3-inch aperature hologram of the y-
shaped pattern. A threshold was set in the output code so that the plotter would print only if
the image intensity exceeds a relatively low
value. Background noise is rejected by raising
the thresholding as illustrated in Figure 8.
The hole pattern positions estimated from the
binary image of Figure 8 are within 2 percent
of the actual hole positions.

31. Comparison of the computer reconstructed
image was made with the image obtained by the
optical method. The hologram recorded on a
polaroid negative was obtained by scanning the
test block a second time after the digital
hologram was recorded. This provided a mechanical scan and transducer alignment for the optical hologram that was nearly identical to that
used for recording the digital hologram. The
electronic phase shift card on the Holosonics
200 unit was activated during the recording of
the optical hologram to provide for image separation in the optical reconstruction.

32. Figure 9 is a photograph of the image
obtained by optical reconstruction. In this
case, the hologram was recorded on a polaroid
transparency while the transducer scanned the
3-inch (7.62 cm) by 3-inch (7.62 cm) aperature.
Optical parameters such as coherent light source
distance and the lateral position of the hologram negative were adjusted for optimum image
reconstruction. Figure 10 is an enlargement of
the image displayed on the TV monitor by a factor
of four.

Defects in a volume

33. The next series of figures are computer
reconstructed images of flat-bottom holes
drilled to different depths in a 4-1/8 inch
(10.48 cm) aluminum test block. Four 1/4 inch
(.64 cm) diameter holes were drilled at the
corners of a square measuring 0.47 inch (1.19 cm)
on a side as illustrated in Figure 11. The
holes were drilled at depths of 1/2 inch (1.27cm)
1 inch (2.54 cm), 1-1/2 inches (3.81 cm) and 2
inches (5.08 cm) respectively.

34. The first hologram was obtained by scanning a 3-inch (7.62 cm) by 3-inch (7.62 cm)
aperature. The scan rate was 3 inches (7.62 cm)
per second in the x direction and the step size
in the y direction was 0.24 inch (.61 mm). The
detected ultrasonic signal was sampled at a rate
of 128 samples per second. A total of 128 linear scans were recorded in digital form. The
ultrasonic pulse width was increased so that
reflections would be received from the top of
each flat-bottomed hole when the gate is set on
the shallowest hole.

35. Figure 11a illustrates the optical reconstruction of the 4 holes. The focal distance
set to approximately 2.75 inches (6.98 cm).
This focal distance is approximately half the
distance between the deepest and shallowest
hole. Similar results were realized by computer
reconstruction of the digital hologram obtained
using same ultrasonic pulse width and gate delay.

36. Substantial improvement in the optical
image of the 4-holes was obtained by recording
four consecutive holograms in digital form.
The pulse width was set to record within 1/2
inch (1.27 cm) thickness of material. The
sample gate was then adjusted to record data at
a depth of 2.25 inch (5.72 cm). Digital holograms were also recorded by setting the time
gate for reflections from 2.75 inch (6.99 cm),
3.25 inch (8.26 cm), and 3.75 inch (9.53 cm).
A scan was performed for each depth since
multiple time gating is not provided by the
Holosonics 200 instrument.

37. Computer reconstruction was performed for
each hologram using the proper focal lengths.
The images were then averaged with equal weighting. The composite image is illustrated in
Figure 11b. Background noise is reduced by
using a threshold of 7.5 percent of the maximum
amplitude observed in the composite image.
Improvement in the images of the 1 inch (2.54cm)
deep hole and the 1/2 inch (1.27 cm) deep holes
is provided by increasing the magnitude of the
digital images of the 1/2 inch (1.27 cm) and
1 inch (2.54 cm) deep holes before averaging
the images. The amplitude compensated composite is illustrated in Figure 11c.

Shear wave images

38. The images presented in the preceeding
paragraphs were obtained by gating, detecting,
and recording longitudinal ultrasonic waves.
Shear waves have also been used to record
holograms and reconstruct defect images. The
shear hologram provides an image of a defect

when the central axis of the ultrasonic energy cone is at an oblique angle with respect to the material surface. This provides a side view of planar defects such as cracks that are essentially perpendicular to the surfaces of the pressure vessel wall.

39. Optical reconstruction of shear holograms requires tilting the hologram transparency with respect to illuminating laser beam. This provides a compensating phase shift caused by variations in transducer scan position. Compensation is provided in the computer reconstruction by multiplication of each digital value in the hologram array by a phase shift factor.

40. Figure 12 illustrates a computer reconstruction of a hole drilled in the side of an inconel test block. The side drilled hole is 1.5 inch (3.81 cm) deep and located 1.4 inch (3.56 cm) beneath the top surface. A 45° shear wave was used to interrogate the defect. The somewhat irregular and spotty image of the side drilled hole is attributed primarily to scattering of the ultrasonic signal from the relatively large gains in the inconel.

Defects in a steel weld

41. A more realistic representation of actual defects that might occur in pressure vessels is provided by a test plate used in a recent Round Robin evaluation of several nondestructive inspection techniques. Defects were induced in the weld of two 5.5 inch (14. cm) thick steel plates by inserting copper strip in the weld cavity after several initial passes had been made. It was intended that cracking would occur near the copper strip. Examination by ultrasonics and radiography have verified the presence and location defects in the weld.

42. Figure 13 is a computer reconstruction of one of the defects. The image is comperable to that obtained by the optical reconstruction. The reconstruction indicates a somewhat elongated defect having several areas where the steel is not fused. Destructive examination of the defective areas is scheduled after additional nondestructive examinations are completed.

CONCLUSIONS

43. Initial algorithm development and evaluation with the laboratory system has indicated the potential of computerized acoustic holography. In its present stage of development, the computer reconstruction algorithm forms images comparable to those obtained by the conventional optical method if single pulses are received and detected. Quality can be improved substantially by forming a composite image of several holograms each corresponding to different depths beneath the surface of the test sample. It is apparent that multiple depth holograms can be recorded simultaneously to provide enhanced composite images or can be used with holograms obtained from other directions to provide 3-dimensional representative of defects.

44. Continued development should provide even better quality images. The laboratory studies also indicate that the computer reconstruction

process can be reduced to a simple push button procedure. Once the transducer-scanning system has been properly located and adjusted, the computer can take complete control of scanning and image reconstruction. By incorporating ample high-speed memory and array processing capabilities, it is estimated that AH images can be reconstructed in less than two minutes.

45. Characterization of defects should be much more reliable when computer reconstruction is used. Defect images are plotted to exact scale after the transducer scanning aperature is defined. The size of computer reconstructed images does not change when the ultrasonic frequency is altered. Furthermore, variations in defect depth do not effect image size. By comparison, the optically reconstructed images vary with both ultrasonic frequency and defect depth.

46. Adjustment and calibration of the optical reconstruction device is tedious and requires a substantial level of operator skill. In contrast, the calibration is essentially eliminated with the computer reconstruction. Experiments with the computer reconstruction indicate that images can be accurately scaled by the computer. On the other hand, the optical reconstruction method will be in error when the position and orientation of optical components, such as lenses, mirrors, and laser source are not adjusted properly. Defect scaling and characterization is also subject to operator interpretation since calculations and lookup tables are employed with the optical method.

47. The flexibility and ease in image display offered by the computerized system opens a new dimension to AH. Experiments with the image display subroutines have indicated the possibilities of accurately adjusting the size of reconstructed images. Images have been enlarged by a factor of ten to provide accurate characterization of the defects. The image can also be plotted on a reduced scale by changing the control parameters used in the display subroutine. Reduced scaling would facilitate the combination of images taken from contiguous sections of material. Several reconstructed images could be joined to form a complete replica of the defect. For example, an image of a crack more than six inches long could be obtained by acquiring two digital holograms on overlapping adjacent sections. The reconstructed images could be spliced digitally and displayed on a reduced scale to accommodate the CRT screen. By using a similar splicing technique, an image of several feet of weld could be plotted on the line printer or other suitable output device.

48. These initial developments of computerized pulse-echo acoustic holography have just scratched the surface of a store of numerous potential improvements in acoustic imaging. It is worthwhile to mention computer techniques such as spatial digital filtering, pattern recognition and automatic dimensional analysis that promise to improve computerized AH. It is the authors opinion that substantial improvements in image quality will be provided by the ability to manipulate, combine, and display digital holograms and images with the rapidly advancing small computers.

REFERENCES

1.   A. E. Holt, Acoustical Holographic Develop-
ment, Babcock and Wilcox, Lynchburg Research
Center, Lynchburg, Virginia, Report #9055,
February 1977.

2.   D. Gabor, A New Microscopic Principle,
Nature, Vol. 161, 1948.

3.   E. N. Leith and J. Upatnieks, Reconstructed
Wavefronts and Communication Theory, Journal of
Optical Society of America, Volume 52, 1962.

4.   B. P. Hildebrand and B. B. Brenden, An
Introduction to Acoustical Holography, Plenum
Press, 1972.

5.   M. Born and E. Wolf, Principles of Optics,
Fourth Edition, Pergamon Press, 1970. P. 455-457.

6.   J. W. Goodman, Introduction to Fourier
Optics, Mc-Graw-Hill, 1968.

7.   T. W. Cooley and T. W. Tukey, An Algorithm
for the Machine Calculation of Complex Fourier
Series, Mathematics of Computation, Volume 9,
p. 297, 1965.

8.   D. E. Mueller, A Computerized Acoustic
Imaging Technique Incorporating Automatic Object
Recognition.

9.   D. L. VanRooy, Digital Ultrasonic Recon-
struction in the Near Field, I.B.M. Scientific
Center, Houston, Texas, Publication #320, 2402,
May 1971.

10.   A. L. Boyer, D. M. Hiroch, J. A. Jordan, Jr.,
L. B. Laseur, and D. L. VanRooy, Reconstruction
of Ultrasonic Images by Backward Propagation,
Acoustic Holography, Volume 3, pp. 333-348, 1977.

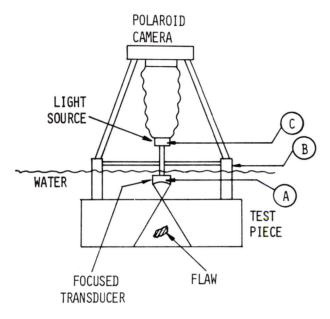

Fig. 1   Acoustic holography scanning

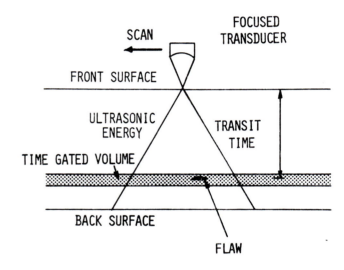

Fig. 2   Pulse echo holography in thick walled plate

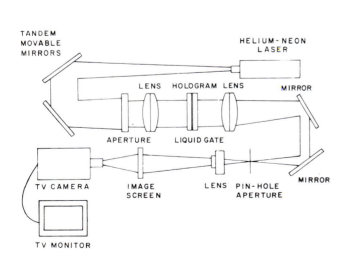

Fig. 3   Optical reconstruction unit

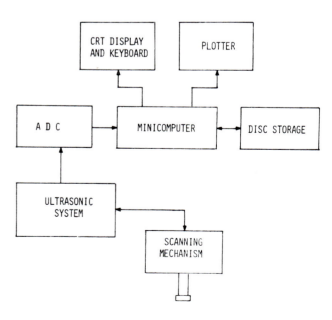

Fig. 4   Diagram of computer reconstruction system

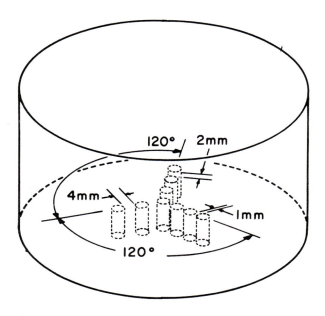

Fig. 5 Aluminum test sample containing flat-bottomed hole machined in a Y-shaped pattern

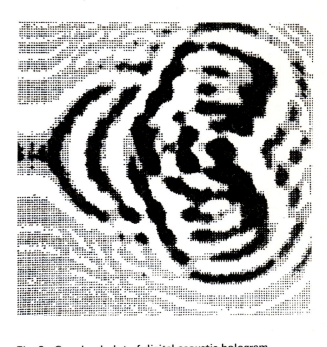

Fig. 6 Gray level plot of digital acoustic hologram

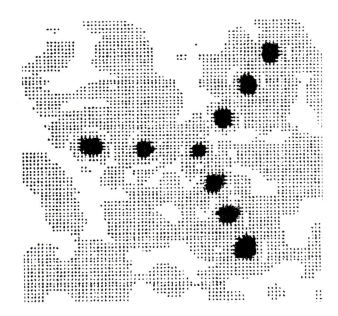

Fig. 7 Computer reconstruction of longitudinal hologram of test block containing flat-bottomed holes in Y-shaped pattern, low threshold

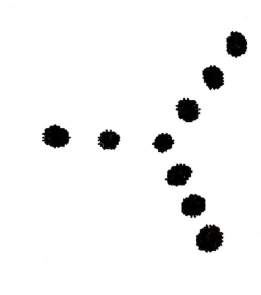

Fig. 8 Computer reconstruction of longitudinal hologram of test block containing flat-bottomed holes in Y-shaped pattern, high threshold

Fig. 9 Optical reconstruction of test block containing flat-bottomed holes in Y-shaped pattern

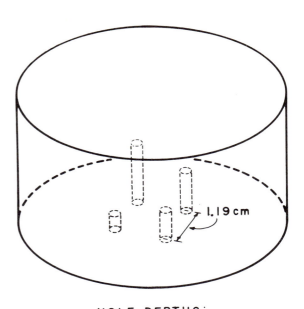

HOLE DEPTHS:
1.27 cm   3.81 cm
2.54 cm   5.08 cm

Fig. 10 Aluminum test sample containing flat-bottomed holes at different depths

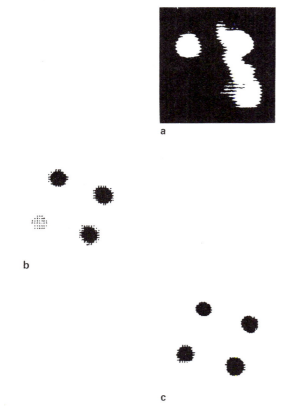

a

b

c

Fig. 11 Comparison of optical reconstruction and computer reconstruction of four holes at different depths in test block

    a Optical reconstruction
    b Computer reconstruction
    c Computer reconstruction with distance-amplitude correction

Fig. 12 Computer reconstruction of shear holograms of side drilled hole

Fig. 13 Computer reconstruction of longitudinal hologram of defect in steel weld

THE PISC PROGRAMME – A STATUS REPORT

R. O'NEIL, MIMechE, FIQA
UKAEA, Safety & Reliability Directorate, Culcheth

The MS of this paper was received at the Institution on 20 November 1978 and accepted for publication on 13 December 1978

SYNOPSIS The reliability and efficiency of ultrasonic NDE of thick steel sections is one of the remaining uncertainties in underwriting the integrity of nuclear reactor pressure vessels.

A number of thick section test plates containing implanted defects have been manufactured for test purposes. Three plates have been made available to European organisations, two with seam welds and one with a nozzle. The European inspection programme has now been completed and the plates are being destructively examined to establish the precise position of the implanted defects. The first analysis of these results should be complete this year and this paper provides a state of the art position as of December 1978.

INTRODUCTION

1.   The reliability and efficiency of ultrasonic NDE of thick steel sections is one of the remaining uncertainties in understanding the integrity of nuclear reactor pressure vessels. This paper is written to give the background to the Plate Inspection Steering Committee (PISC) programme and to indicate the lines on which the use of test plates can help in the understanding of ultrasonic examination of thick walled pressure vessels. The PISC programme is still underway and this status report therefore cannot give other than preliminary views but it is hoped to provide further information at this Conference.

2.   In 1965 in the United States, the Pressure Vessel Research Committee (PVRC) non-destructive examination programme began as part of the industry co-operative programme on heavy section steels for nuclear reactor pressure vessels (HSST programme) and in 1968 the PVRC prepared a number of welded thick section test plates for the use of ultrasonic test organisations. Various defects were implanted into the welds of these plates and these were inspected by teams from the US inspection organisations to the 1968 PVRC inspection procedure. After testing, this procedure was deemed to be unsatisfactory (inconsistent results of test teams) and a revised procedure was issued in 1970 and tested against 2 of the plates by 5 test houses. The results showed that the procedure was still not satisfactory for the same reasons and in September 1974 a revised procedure for the ultrasonic testing of the plates was issued. It was at this time that plates were offered to the Community for European Co-operation/Nuclear Energy Agency (CEC-NEA) working group on mechanical components to arrange for similar tests to be carried out.

3.   The terms of reference were to determine the capability of the US September 1974 procedure for ultrasonic examination techniques to detect flaws or discontinuities, their size, orientation and location in heavy section steel.

4.   To carry out this European programme a Plate Inspection Steering Committee (PISC) was formed initially with National representatives from eight countries, the overall programme to be managed by the Safety and Reliability Directorate of the UKAEA. Later in the programme a further two countries joined PISC. The National representatives arranged for inspection organisations from each country and a total of 34 organisations took part although only 28 carried out the examination in accordance with the formal PISC procedure.

5.   A list of the National groups and organisations taking part is given in Appendix 1. It should be noted that all participants agreed to fund the transport and NDE costs themselves with the overall management funded by the UKAEA with secretarial services from CEC and OECD. From this group sub-committees were formed from time to time to advise PISC on specific aspects.

6.   The final Destructive Examination (DE) and the associated analysis has been funded by CEC and is mainly centred at ISPRA with some work sub-contracted

to SRD (UK), Association Vincotte (Belgium), ECAN Indret (France) and MPA (Germany).

Test Procedures

7. The PISC procedure basically described the method for manual ultrasonic inspection of the welds and Heat Affected Zone of the test plates in compliance with Appendix 1 of Section XI of the ASME Boiler and Pressure Vessel code as it applies to a vessel, in service and accessible from the outer surface. The intent of this inspection was to measure the through wall dimension and length of flaws with angle beam tests.

8. By October 1976 some 13 teams from Belgium, Netherlands and the UK had inspected the plates to the PISC procedure and at a meeting of the reporting officers for each country some concern was expressed at the loss of time and confusion caused mainly by ambiguities in the procedure. It was felt that it would be unproductive to ask the remaining teams to repeat the work without some warning of, or assistance over the pitfalls in the procedure. It was decided that the most effective way of giving assistance was to rewrite the confusing clauses in the procedure with the object of clarifying the instructions and removing the ambiguities. This was done and the revised document issued in November 1976. It will be of interest to see, once the results have been processed whether there is any reduction in scatter of the results using the revised procedure.

9. The PISC procedure for calibration and examination follows closely the 1974 ASME XI procedure but with certain areas being more closely defined such as limitations on probe size and frequency. The most important change being however that for the nozzle examination scanning normal to the weld is required as well as tangential to the weld. The 1977 issue of ASME for inspection does not as yet include nozzles (it is still being written).

10. The terms of reference of the PISC project allowed all participants to use other procedures if they wished. The procedures need not necessarily be in use commercially but reporting was to be made within the format of the PISC procedures and would be assessed separately. Most teams have taken advantage of this and once again it will be of interest to compare the results of these alternative techniques with the PISC results.

11. Calibration of the test equipment is a very important part of any ultrasonic examination and at the beginning of the programme there were no known calibration blocks available meeting ASME XI requirements available for loan in the UK. Two solutions to this problem were provided, the first being

the machining of calibration features into the plates themselves and the second the provision of a high quality calibration block to ASME XI by the Japanese who although interested in the project were not able to participate directly although representatives of the Japan Steel Works have attended meetings. These test blocks have remained with the test plates throughout the project although in the event most test houses have found the machined features in the plates adequate.

The Test Plates

12. Three test plates were provided. Two nominally flat plates 50/52 and 51/53 and a plate with a nozzle welded in, 204. Deliberate flaws were introduced into all the welded sections and in addition a number of cracks appear to have been naturally formed during the fabrication process.

(a)   Plate 50/52

This plate was manufactured from 2 pieces of SA-302 Grade B, Mn-Mo by Babcock & Wilcox, the two pieces being welded together by butt weld electroslag method with the weld containing gross cracks. The welded plate measures $55\frac{1}{4}$" (140.3 cm) x $36\frac{1}{2}$" (92.71 cm) parallel to the weld x 10" (25.4 cm) thick. The plate was given a normalising heat treatment before delivery. Total weight 2.54 tons.

(b)   Plate 51/53

This plate was manufactured from 2 pieces of SA-302 Grade B, Mn-Mo by Babcock & Wilcox, the two pieces being welded together by butt weld submerged arc method with the weld containing gross cracks. The welded plate measures 41" (104.14 cm) x 36" (91.44 cm) parallel to the weld x 8" (20.32 cm) thick. This plate was not heat treated after welding and had a pronounced bend in it about the weld centre line, the difference between the edge of the plate and the centre line being raised 21.4 mm. Total weight of this plate 1.49 tons.

(c)   Nozzle Plate 204

This plate was manufactured by the Chicago Bridge and Iron Company with an 18" (45 cm) forged nozzle welded in place by manual metal arc method, the weld containing 9 discreetly implanted flaws. No information is currently presented regarding the materials used but post weld heat treatment was given. Total weight 1.91 tons.

Test Programme

13. Initially it was proposed that the test plates should reside at one location with equipment and manpower for each test house travelling to this location. However on the grounds of expense, possible damage to equipment in transit and possible Customs problems, it was decided that it would be better to move the plates to each participating country with local arrangements made for internal movement where relevant. The arrangements followed the experience of the American test house in that each test house was responsible for transporting the test plates from the previous test house. In general this worked very well, the greatest problems occurring at Customs. On average it took about 2 weeks to arrange for Customs clearance for the movement of the plates between countries.

14. The test plates arrived in the UK at the end of October 1975 and an assessment then of the overall programme timescale indicated that a preliminary report should be available about the end of January 1978.

15. On arrival the plates were cleaned and calibration features machined and became available for inspection on 5th January, 1976. The planned programme of inspection and the actual programme are compared in Table 1 below.

Table 1

| Country | Completion Date | |
	Planned	Actual
UK (Start 5.1.76)	30. 4.76	4. 5.76
Holland	30. 6.76	30. 6.76
Belgium	30. 9.76	18.11.76
Sweden & Denmark	31.11.76	20.12.76
FR Germany	28. 2.77	10. 5.77
France	30. 6.77	7.10.77
Italy	30.10.77	21. 1.78
Spain	–	19. 5.78-To Italy for DE.
Finland	–	5. 6.78-In Italy.
Preliminary Report	31. 1.78	February 1979

16. The Preliminary Report will now, as shown, be approximately 1 year later than the original proposal. The main reasons being:-

   (i)   Insufficient allowance was originally made for the problems of Customs and transport.

   (ii)   Insufficient allowance was made for Continental holiday periods (usually 1 month).

   (iii)   The addition of Spain and Finland to the participants.

   (iv)   The extent of the work necessary to carry out destructive examination for comparison purposes.

The improved facilities available at Ispra for DE allowed more detailed DE and analysis than had been originally intended. At the time of writing this paper (December 1978) destructive examination is well under way but will be explained in more detail in a later section.

Non-Destructive Examination and Results

17. In general NDE progressed by the various test houses without any major problems. The problems of confusion and ambiguity in the original procedure have already been mentioned. The original procedure allowed only for the results to be defined as raw data. A data sheet added later allowed the participating teams to give their interpretation of the raw data as defined flaws. Some confusion arose as some teams only provided interpretation and not raw data. Datum marks were provided on all plates but some teams used other references usually those preferred in their commercial practice and a great deal of work has been involved in sorting out all results into a common format for processing.

18. At the beginning of the project it was decided that confidentiality must be preserved and therefore all results sent in by the test houses have been coded so that in the final analysis only the test house concerned will know which result applies to themselves. For record purposes and as managers of the project without having participated in the examinations SRD have the complete information. All results have now been received and codified, placed in the common format and are now placed in a data bank for comparison with the results of the destructive examination of the plates. However before carrying out this examination it was necessary to have some clear idea of how this comparison was to be made and this is discussed in the next section.

Evaluation and Analysis

19. In February 1978 it became apparent that the inspection phase of the PISC would soon end and there would be a need to define the method of analysis combined with destructive testing for evaluation of the NDE results. Various discussions had taken place previously and due note had been taken of the preliminary

analysis carried out in the USA by Buchanan and the advice of the PVRC Committee during their deliberations, but the time had come for action. A subcommittee of the PISC was set up named the Evaluation Task Force, the members being restricted to the National representatives or their nominees.

20. The main questions of concern in an exercise designed to assess the reliability of a method of detecting flaws are:-

(a) Flaw detection
(b) Flaw positioning
(c) Flaw sizing

The main aim of the PISC project is, basically to estimate how well the ultrasonic inspection results to the PISC procedure can reproduce the real situation found by destructive examination. There is, inevitably, a scatter in the NDE results and it is important that an indicated defect is not ruled out because it does not exactly match the actual defect.

21. The method chosen to make a comparison of the results of all teams with the destructive examination results is as follows:-

Comparison procedure of the results of all teams with the DE results

(i) Simple statistical treatment

22. It would appear that the available data base is too small to enable a comprehensive statistical analysis to be carried out, but sufficiently large for a semi-quantitative assessment to be made. Consequently the initial analysis will be concentrated on the calculation of simple probabilities and no complicated statistical ideas will be invoked. If the results of the present exercise are encouraging it may then be in order to pursue the comparison in more depth.

(ii) The probabilities to be estimated

23. The most significant results will be:-

(a) Defect detection probability - this is evaluated by a straightforward counting method, i.e. DDP = n/N where n is the number of teams detecting the defect and N is the total number of teams. Confidence bounds can be put on this figure in the usual way by using the binomial distribution. The question arises as to how it is decided that a team has detected a particular defect and in the case of several nearby defects which one has been detected. This can only be decided by a comparison with the results of the exercise to look at the accuracy of location of the defects. It is expected that the DDP will be

a function of size, position (including orientation) and nature of the defect. It is also likely that a team will give indications of several separate defects when there is in fact only one. In this case examination of the DE results will make this clear.

(b) Probability of correct location of the defect - to be more precise, on the basis of the ultrasonic examination, a volume will be specified such that it has a certain probability of containing the position of the real defect. In this context the position of the defect is that of its centrum.

(c) Probability of correctly sizing the defect - as in most cases the projection of the defect on the Y plane (co-ordinates X-Z) is the most important, for the nozzle plate it is the R cylindrical plane (co-ordinates θZ), the ratio depth to length (B/A) could be taken as the "size" of the defect (ASME XI). Other combinations could also be taken as size representations e.g. A x B, or B/T where T is the plate thickness. These combinations can be used to estimate the relative probability of under-estimating or over-estimating the size of the defect.

Presentation of Results

24. The most important results of the PISC exercise are expected to be expressed in the following form:-

(a) The probability that a defect has been detected.

(b) The probability that a certain volume determined from the measured values contains the position (of the centre) of the real defect.

(c) The probability that, taking the size of the defect into account it has been correctly accepted or rejected.

(a) DDP - Since it is expected that this will be a function of size it is important to display this variability where possible. It is arguable whether each different type of defect should be considered separately, as they may have different responses to the probes. Results will appear typically as in Fig. 1, and any such plots which show significant variation should be presented.

(b) Positioning of the defect - This will also be a function of the size and position of the defect. The ultrasonic test results will be used to define a certain tolerance region about the measured position of the defect. The probability that this tolerance region contains the centre of the real defect will be a function of the volume of the tolerance region. Consequently we will expect to draw graphs of the type shown in Fig. 2. The error bands are evaluated in the usual way. We may deduce from such curves the size of the tolerance region needed to give say a .9 chance of including the centre of the defect.

25. The success or otherwise of this method of analysis cannot yet be assessed and will not be until some destructive examination results are available. On the basis of work carried out to date there is every reason to believe that this type of analysis will be a viable method.

26. The amount of data from this exercise is enormous and could not be handled on a manual basis. A computer program has been written and is at present being tested on the basis of the NDE results already in the data bank and certain selected results of the destructive examination carried out so far.

Destructive Examination

27. Destructive examination of the test plates is being carried out in 3 main phases with the possibility of a 4th if required. These are:-

28. Phase I - Removal of the weldment from the plates 50/52 and 51/53 and removal of the nozzle weldment from plate 204. This has been followed by intensive ultrasonic inspection and high energy x-rays to give more precise indications of the flaws before further sectioning.

29. Phase II - This phase now completed consisted of systematic transverse sectioning of the plate weldments into 50 mm slices with the nozzle plate 204 being systematically sliced into wedges giving approx. 50 mm thickness at the weld centre line. Following cutting each slice was radiographed and it is possible to detect flaws with a resolution of 0.3 mm and 0.5 mm in the 50 mm slices. Each slice was then ground, polished and etched and macro graphs produced. Fig. 3 shows a macrograph print of slice "P" from plate 51/53 and shows a major crack, 2 minor cracks and a large inclusion. The major crack does not appear to be one deliberately induced. This plate was not heat treated after manufacture and a great deal of trouble was encountered during cutting with the plate bending due to residual stresses and trapping the saw blade. It would be possible to prepare data sheets now from the information available but correct sizing and location can be improved by further sectioning.

30. Phase III - This phase has just commenced (November) and consists of further cutting of the slices to within a few millimetres of the heat affected zone giving in effect a piece of material approx. 60 mm wide x 50 mm thick. Each piece will then be subjected to intensive NDE (use of focus probes) to give more accurate sizing and location of flaws.

31. Phase IV - This phase which consists of further sectioning and reduction in size of the Phase III pieces followed by intensive NDE will probably not be necessary as sufficient information from Phase III should be available to allow accurate analysis to take place, however it is being considered to clarify certain positions where flaws diverge.

DISCUSSION

32. It is not possible at the time of writing to reach any conclusions on a programme that is not yet complete. What can be said is that international co-operation on this project has been remarkable. Every organisation taking part has recognised the importance of this kind of proving exercise and have done everything in their power to ensure the success of the programme. The destructive examination and analysis of the plates has now fallen mainly on the shoulders of the Joint Research Centre Ispra, Italy, but some portion of the work is being carried out in Germany, France, Belgium and the UK. It is planned that the work will be reported fully in a series of papers to be presented at a one day seminar following to the 1979 SMIRT Conference in West Berlin.

33. Although the PISC programme is not yet complete, the issues raised in this kind of exercise indicate a need for further work of this nature to provide a common basis for all organisations in the field of NDE and members of PISC are actively considering the possibility of continuing with a further programme of work.

ACKNOWLEDGEMENTS

34. The author wishes to acknowledge the help from his colleagues on PISC and its associated sub-committees as well as the help from the Secretariats of CEC and OECD. Finally, it should be remembered the programme would not have begun but for the generous offer of plates by the PVRC in the first place, who have also kept us up to date with information of their programme since then.

Appendix 1

Number of Participants	Country	National Representatives	Participating Organisations	Participants Town
10	United Kingdom	United Kingdom Atomic Energy Authority Safety and Reliability Directorate	Risley Nuclear Power Development Laboratories (RNPDL) (REML)	Risley
			Quality Inspection Services Ltd. (QIS)	Stockton
			Central Electricity Generating Board (CEGB)	Manchester
			Associated Offices Technical Committee (AOTC)	Manchester
			Rolls Royce and Associates Ltd. (RR&A)	Derby
			Ministry of Defence (MOD)	Keynsham
			Atomic Energy Research Establishment (AERE)	Harwell
			The Welding Institute	Cambridge
			The Unit Inspection Company	Swansea
			Babcock Inspection	Renfrew
2	Netherlands	Rontgen Technische Dienst (RTD)	Rontgen Technische Dienst BV (RTD)	Rotterdam
			Dienst voor het Stoomwegen Den Haag	The Haag
2	Belgium	Association Vincotte	Association Vincotte	Brussels
			Westinghouse Europe	Brussels
1	Denmark	The Danish Welding Institute	The Danish Welding Institute	Glostrup
1	Sweden	Studsvik Energiteknik	Tekniska Rontgen-centralen AB (TRC)	Stockholm
4	Germany	Institut fur Reaktorsicher-heit (IRS) Koln	Institut fur Zer-storungsfreie Prufverfahren ISFP	Saarbrucken
			Bundesanstalt fur Material Prufung (BAM)	Berlin
			Kraftwerk Union (KWU)	Erlangen
			Rheinisch-Westal-ischer Technischer Uberwachungs-Verein (RW TUV)	Essen

Number of Participants	Country	National Representatives	Participating Organisations	Participants Town
7	France	Commissariat a l'Energie Atomique (CEA)	Framatome	Courbevoie
			Commissariat a l'Energie Atomique (CEA)	Saclay
			Establissement des Constructions et Armes Navales (ECAN) Indret	La Montagne
			Direction Technique des Constructions Navales (DTCN)	Paris
			Electricite de France	St. Denis
			Centre d'Etudes Techniques des Industries Mecaniques	Senlis
			Institute de Soudure	Paris
2	Italy	Joint Research Centre (JRC)	JRC	Ispra
			Ente Nazionale per l'Energia Electrica (ENEL)	Milan
4	Spain	Ministerio de Industria Junta de Energia Nuclear	ARTISAE	Madrid
			Equipos Nucleares	
			Ciat Nuclear SA	
			Technatom	
1	Finland	Valtion Teknillinen Tutkimuskeskus (VTT)	VTT Technical Research Centre of Finland	Espoo

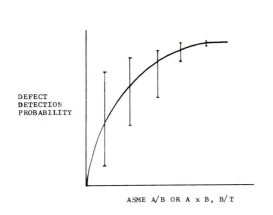

Fig. 1

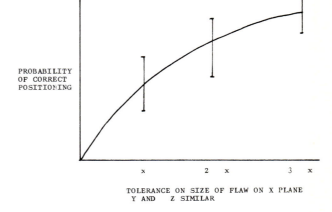

Fig. 2

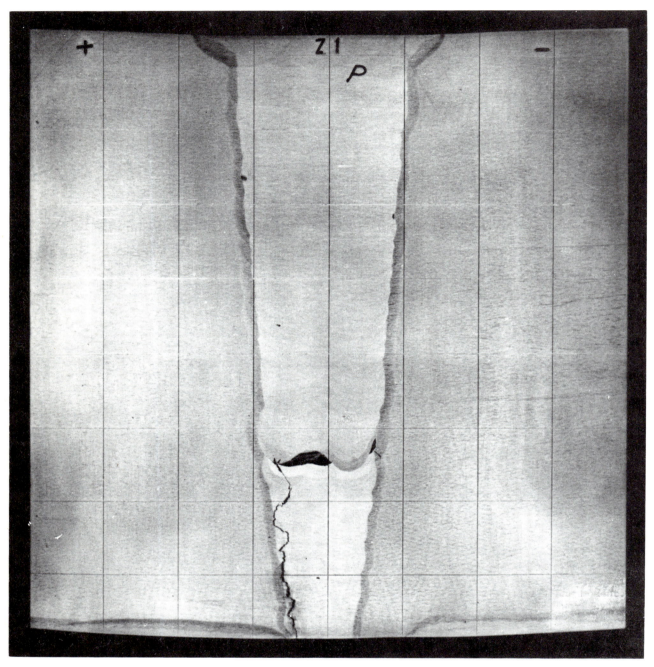

Fig. 3

CHARACTERIZATION AND SIZE MEASUREMENT OF WELD DEFECTS BY ULTRASONIC TESTING

T. J. JESSOP, BSc(Eng), ACGI, MWeldI
The Welding Institute, Abington, Cambs

The MS of this paper was received at the Institution on 30 August 1978 and accepted for publication on 16 November 1978

SYNOPSIS The application of ultrasonic testing to many different types of high integrity welded fabrication has resulted in defects being misinterpreted. The present programme of work is aimed at quantifying ultrasonic testing capability by providing detailed information on the application of both conventional and more specialized ultrasonic techniques to a large sample of known defects in ferritic steel weldments in the 38 to 95mm thickness range. This paper describes the first phase of the work dealing mainly with non-planar defects. The results emphasise the inaccuracy of conventional defect assessment techniques and recommendations are made on the use of more specialized techniques, where appropriate, to improve results.

INTRODUCTION

1. Reports from the fabrication industry on the misinterpretation of defects with conventional manual ultrasonic testing have been widespread (Ref. 1). Overestimation of defects has resulted in unnecessary, expensive and time consuming repair and underestimation may have led to significant defects being allowed into service.

2. Despite the fact that fracture mechanics and fitness-for-purpose considerations can accurately predict critical defect sizes, it is rarely possible to quantify the capability of conventional ultrasonic testing or even more sophisticated ultrasonic techniques to assess weld and associated defects. The problems stem from the large number of uncertainties involved: there is little information on how different weld defects and metallurgical features interact with ultrasonic energy; the commonly used equipment may not be capable of testing to the required standard; and the necessary degree of operator skill is difficult to define. Explicit inspection standards (particularly with regard to sensitivity level requirements) have proved difficult to formulate and testing codes have evolved using the same criteria as radiographic examination standards (Ref. 2).

3. The object of this programme is, therefore, firstly to define the problem by studying the interaction of ultrasound with weld defects using conventional equipment and assessment techniques, and secondly to establish how the more recently developed and sophisticated techniques can be used to advantage. In this first phase of the work, volumetric in-process weld defects such as porosity, slag inclusions and lack of penetration have been studied extensively in ferritic steel butt welds. Although these types of defect are generally not regarded as serious, it is important that they are subjected to equally detailed studies since they can be confused with serious defects or mask them from the ultrasonic beam. Subsequently, similar investigations will be carried out on planar defects (such as cracks and lack of fusion): metallurgical features (such as coarse grain structure): and defects in welds of more complex geometry (for example, T-butt and nozzle joints).

4. The format for Phase I was as follows:

1. Manufacture specimens with the required weld defects.
2. Ultrasonically test the specimens using conventional equipment and a range of test parameters and sizing techniques (including a test by a recognised operator to provide reference level).
3. Ultrasonically test the specimens using more specialised equipment and recently developed sizing techniques.
4. Destructively test the specimens to determine exact size, character, etc., of the defect under study.
5. Correlate the results of parts 2, 3 and 4.

Items 1, 2, 4 and 5 were undertaken by The Welding Institute. Item 3 was conducted in collaboration with The Welding Institute by the NDT Centre, Atomic Energy Research Establishment, Harwell and the NDT Applications Centre of the Central Electricity Generating Board (CEGB), North West Region, Scientific Services Division, Wythenshawe, Manchester.

EXPERIMENTAL DETAILS

Materials

5. For an experimental programme of this type, the nature and quality of test specimens is most critical. Since it is extremely difficult to simulate accurately weld defects by machining and because of the unrealistic nature of the commonly used simple machined geometries (saw cuts, flat bottomed holes etc.) it was necessary to use actual welded specimens with defects introduced deliberately but in a controlled manner, by the welder.

6. In this phase of the work, slag lines, isolated slag inclusions, porosity lines, porosity clusters and lack of penetration defects were produced for study. They were included in plain

butt welds in 38, 65 and 95mm thick O.2%C : 1.4%Mn ferritic steel plates about 500mm square. After some experimentation, ten specimens were manufactured for the test programme. Where possible, two sizes of each defect were included and in at least two positions in the weld. Both double and single sided joint preparations were used with both manual metal arc (MMA) and submerged arc (SA) welding processes. All test surfaces were machined flat.

7. The procedures used to incorporate the defects have been developed during previous work at The Welding Institute (Ref. 3). However, some procedural trials were necessary to establish adequate control over defect size and type. In general, slag defects were produced by incomplete interrun cleaning, porosity by using metal inert gas (MIG) welding with the shielding gas switched off, and lack of root penetration by alteration to the welding conditions. Slack and porosity defects were straightforward to produce, although the porosity often contained some slag.

Ultrasonic Tests

8. Conventional Ultrasonic Tests. This section was divided into two distinct parts. The first part can be conveniently labelled 'laboratory type' tests, i.e. using conventional equipment but with some degree of control over scanning variables and employing methods of recording results (see Fig. 1) rather than relying on subjective interpretation. These tests were particularly comprehensive in that test frequency (2 and 4MHz) and probe angle (45°, 60° and 70°) were systematically varied and all the commonly used defect sizing methods were attempted (dB drop, maximum amplitude and equivalent reflector). The second part, referred to as 'shop-floor' tests, employed equipment and personnel typical of those currently used in shopfloor ultrasonic inspection practice. The operator who performed this work had had several years direct experience of ultrasonic testing in the pressure vessel industry and was certificated under the CSWIP scheme. The aim of the latter tests was to provide a level of inspection with which quality control departments of fabrication companies are familiar, thus enabling direct assessment of the advantages of the more sophisticated techniques.

9. B-Scan Tests. The B-Scan display provides immediate information on the cross sectional position of a reflector and with detailed measurement and proper interpretation can provide additional information on the size and character of the defect under test. The equipment used in this investigation (see Fig. 2) is described in detail elsewhere (Ref. 4). It has been employed for site work and is portable and easy to apply. The basic ultrasonic data are derived from a conventional A-Scan flaw detector and are combined with probe position and beam angle data to produce a B-Scan display in the form of a trace on a bistable storage oscilloscope. This can be photographed to provide a permanent record. Various modes of scan using compression and shear wave probes can be used to build up a comprehensive picture of a defect. Conventional contact probes were employed.

10. C-Scan Tests (Work conducted by CEGB, Manchester). A C-Scan is a view in plan of the ultrasonic echo amplitudes received by the probe from reflections within a pre-selected volume of the specimen. The C-Scans in this work are contour maps of the echo amplitudes received by the probe as it scans over the surface of the plate. Conventional probes and flaw detection equipment were employed. This sytem is regarded as a research tool in that the scans produced are permanent records of the ultrasonic data which can be compared with the actual topography of the defects. In this way, the understanding of the interaction of ultrasonic pulses with defects can be improved. C-Scans contain good amplitude information which may be complemented by the range information available from B-Scans. An example of the C-Scans produced is shown in Fig. 3.

11. Ultrasonic Holography (Work conducted by NDT Centre, Harwell). This imaging system offers advantages in resolution which cannot be achieved with the aforementioned techniques. Where the resolution of more conventional techniques is governed by the ultrasonic pulse length, in holography the wavelength determines the limit of resolution (Ref. 5). Since this advantage is more noticeable for deeply embedded defects, holography tests have only been performed on two of the 95mm thick specimens. Again, this equipment is in the laboratory stage and is particularly time consuming to apply since it involves the manufacture of a hologram in the first instance and subsequent reconstruction on an optical bench. The system has been described in detail by Aldridge (Ref. 6).

12. Time Domain Analysis (Work conducted by NDT Centre, Harwell). In order to move away from the reliance on echo amplitude to give an indication of defect size, tests have been performed using a time-of-flight approach to defect sizing. The two probe ultrasonic time-delay method, already developed for sizing surface breaking fatigue cracks has been adapted in the present work for measuring the size and through-thickness extent of embedded volumetric defects in welds (Ref. 7). Two angle-beam longitudinal wave probes (2.5 or 5 MHz) are placed on the parent metal on opposite sides of the weld, so that their beam cones intersect in the weld region. A pulse of ultrasound transmitted by one probe produces reflected beams from the upper surface of defects and diffracted beams from their undersides which are received by the other probe (see Fig. 4). The received signal is processed and recorded to show the changes in response as a function of position along the plate. Interpretation of these records reveals information on size and location of defects in the weld, projected on a vertical plane parallel with the weld axis. This scanning and recording method sacrifices some of the accuracy of the time-delay technique in order to provide a more readily assimilated display of large amounts of data.

Destructive Tests

13. The object of the destructive tests is to obtain data on the actual size, structure, position, orientation and nature of the defects. In order to minimise the risk of losing important information, one of two methods has been adopted depending on the defect under study. For linear type defects (e.g. slag lines) a freeze-break technique was used. This involved taking a slice containing the defect(s) out of the specimen and machining a small notch directly above the defect. The slice is then cooled in liquid nitrogen and

fractured under three point loading conditions. At this temperature, brittle fracture takes place through the defect with no plastic deformation and the two halves of the defect remain intact for detailed measurement in all three dimensions, and study.

14. For cluster type defects the above approach is unsatisfactory. Therefore these defects have been dealt with by isolating each defect, completely embedded within a small block and subjecting it to radiographic testing from different angles. Sectioning can then be employed as necessary.

RESULTS AND DISCUSSION

Sizing and Positioning of Defects

15. All techniques studied gave some information on defect size and position although in the C-Scan tests the emphasis was more on characterisation aspects. In general terms positioning was within 1.5mm for all techniques and length sizing accuracy was better than 10% in most cases. Both these values are considered to be acceptable. More attention has been paid to through-thickness sizing since this is the most critical dimension when, as is usually the case, stresses are acting across the weld. Table 1 shows the accuracy of through-thickness sizing for all techniques studied.

TABLE 1. Analysis of defect through-thickness size measurements for non-planar defects. Range of sizes studied was 1.5 to 7.0mm.

		MEAN ERROR (mm)	STD DEVIATION	SAMPLE POPULATION
CONVENTIONAL TESTS	4MHz (20dB & Max Amp)	-2.2	2.2	89
	2MHz (Max Amp)	-2.6	2.4	28
	DGS	-1.8	1.6	113
	Manual Operator	-0.8	1.4	25
B-Scan		-1.3	2.2	30
C-Scan		No Through-Thickness Information		
Ultrasonic Holography		Small sample - Very Inaccurate.		
Time Domain Analysis		-0.3	1.0	24

16. It can be seen that all the conventional methods tend to undersize the through-thickness dimension by about 2mm and with a large spread of results. The B-Scan shows some improvement, but since, effectively, conventional sizing methods are employed (although without manual plotting) the similar spread of results is not surprising.

17. The inaccuracy of these results highlights the limitations of conventional ultrasonic testing. The fact that only one small facet of a non-planar defect may be sufficiently well orientated with respect to the beam to be sized by probe movement goes some way to explain the

results but they do emphasise the unreliability of probe movement methods generally even when scanning and/or display techniques are more sophisticated. Whether changes in flaw detector and/or probe characteristics would influence the accuracy of the results is open to question. It may be argued that narrower or focused sound fields would improve sizing accuracy or that short pulse/broad band transducers would be better than the rather long pulse/narrow band transducers used in the present work. Efforts will be made to assess these variables in future work.

18. Better results were achieved by the manual operator and, since conventional probe movement sizing techniques were used this, is difficult to understand. However, there are two points which should be borne in mind: firstly, unlike the laboratory tests, no negative values were recorded* and it is reasonable to assume that any negative reading found by a shop floor operator would not be recorded as such; and secondly, the manual operator is not confined to strict X-Y scanning.

19. The best results were achieved by the time domain analysis techniques which showed little tendency to undersize and a much smaller spread of results (see Table 1). The sample is small (24 readings) because the equipment was not always capable of resolving the upper and lower extremities of a small defect. Work is currently underway at Harwell to improve resolution by using shorter pulse probes and providing this is successful, the technique should be applied more widely.

20. Despite excellent lateral resolution which gave accurate estimates of defect plan position and size, the interpretation of the holography results grossly overestimated defect through-thickness (e.g. by up to 300%). This has been attributed to the generally low level of experience with shear waves (which were widely used in this exercise specifically to provide through-thickness information) as opposed to normal compression waves. Efforts are being concentrated on feeding back detailed destructive test information to assist in the interpretation procedures.

Characterisation of Defects

21. All but the time domain analysis work have contributed some information on defect character although the information from untrasonic holography was limited to predictions of defect shape. Again the latter was accurate in the plan view but distortion occurred in the through-thickness direction owing to the overestimation on size.

22. In the conventional tests records of echo envelopes have been compared qualitatively with defect structure and some correlation has been noted. However, distinction between the different defects under study can not be guaranteed because the echo envelopes are dominated by the resolution power which is in turn dependent on beam characteristics. This subject has been dealt with by Coffey (Ref. 5).

23. The B-Scan displays provided the most information on defect character because this technique has already been extensively evaluated for

_____
* Negative values for 20dB drop sizing have been reported previously by de Sterke (Ref. 8).

specific applications in the power generation industry and interpretation procedures have been defined. In the present work, reasonably accurate information on defect shape and structure has been produced for all types of non-planar defect from interpretation rules based on fundamental ultrasonic interaction considerations with some reliance on past experience. An example of B-Scan characterisation is shown in Fig. 5. Here an angled probe is scanned along a rough cylindrical defect (lateral scan) and the resulting trace shows an oblique longitudinal cross-section of the defect responses. The continuous trace A is produced by specular reflection of the sound from point 1. The region B contains many indications which correspond to reflections from facets of the defect surface (the zones marked 2 on Fig. 4). A cylindrical cluster of small reflectors may produce a similar result if they are sufficiently tightly packed but complementary information from other scans (e.g. transverse scans) and different angles might reduce the uncertainty.

24. The C-Scan equipment, being newer and laboratory based, requires more evaluation before its potential for characterisation can be fully realized. However, some judgments can be made: for example, the discrete peaks shown in Fig. 3 must be separated by at least the beam width and could either be from facets of the same defect or from smaller separate defects. Extensive records were taken of a range of linear non-planar defects (work on cluster and isolated inclusion defects will be included later) and these provide some information on defect surface structure.

CONCLUSIONS

25. Since all the work so far has been on non-planar defects the conclusions are limited to those types of defect. More comprehensive conclusions will be possible when work on planar defects is completed.

    a) A large amount of ultrasonic test data from both conventional and advanced test methods has been accumulated for approximately fifty non-planar defects of varying type and size. About half of these defects have been destructively examined and compared with the test results.

    b) Conventional ultrasonic testing is generally reliable for positioning and defect length estimation. However, through-thickness sizing which is by far the most important, is unreliable with a tendency to under-size. The effect of variations in probe angle and frequency was minimal. Some qualitative correlation between echo envelope shape and defect type was noticed.

    c) It cannot be assumed that applying simple control over scanning and recording characteristics will provide better accuracy than a good manual operator given complete freedom.

    d) The B-Scan display method was useful for providing rapid positional data and facilitated probe movement sizing. The accuracy was similar to the conventional tests. Close examination and comparison of several scans of the same defect using simple interaction models and previous experience allowed quantitative correlation with defect character. With proper education this display could be employed more widely, especially where conventional A-Scan displays prove confusing.

    e) The detailed C-Scan system requires further evaluation using a wider range of defect types and closer correlation with destructive tests. However, the system has provided accurate positional plots and some correlation with defect structure has been made.

    f) Ultrasonic holography provided accurate positional and defect length information but grossly overestimated through-thickness size. Again further evaluation using more extensive comparison with actual defect shape and size is required to define more clearly the advantages of the method.

    g) Ultrasonic time domain analysis gives the most reliable estimates of defect through-thickness although resolution limits the minimum measurable defect size. Consideration should be given to making the equipment suitable for site inspection.

ACKNOWLEDGEMENTS

26. The work at The Welding Institute and Harwell was funded by the Mechanical Engineering and Machine Tool Requirements Board of the Department of Industry. The support of the CEGB is also acknowledged.

REFERENCES

1. Anon. Major areas for research. Metal Construction 7 (11) November, 1975.
2. Young, J.G. The evolution of acceptance standards for weld defects. British Journal of NDT 18 (5) May, 1976.
3. Gregory, E.N. Making defective welds. The Welding Institute Research Bulletin 15 (7) July, 1974. pp 199-205.
4. Harper, H. Improved displays for ultrasonic testing. CEGB Digest November 1974.
5. Coffey, J.M. Quantitative assessment of the reliability of ultrasonics for detecting and measuring defects in thick section welds. I.Mech.E Conference. Tolerance of flaws in pressurized components 16-18 May, 1978. Paper C85/78.
6. Aldridge, E.E. Ultrasonic holography and non-destructive testing. Materials Research and Standards. 12 (12) December, 1972. pp 13-22.
7. Silk, M.G. and Lidington B.H. Defect sizing using an ultrasonic time delay approach. British Journal of NDT 17 (2) February, 1975.
8. De Sterke, A. Flaw tip reflection as a help in ultrasonic flaw size determination. British Institute of NDT Conference, London, 1976.

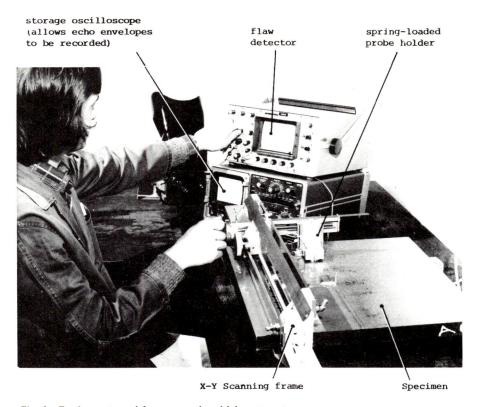

storage oscilloscope (allows echo envelopes to be recorded)

flaw detector

spring-loaded probe holder

X-Y Scanning frame

Specimen

Fig. 1   Equipment used for conventional laboratory-type tests

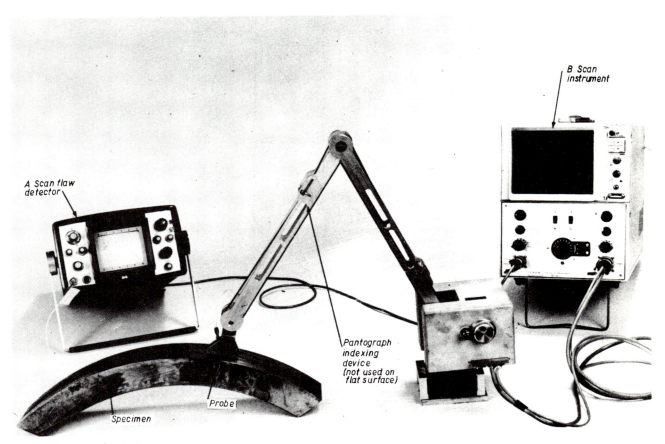

B Scan instrument

A Scan flaw detector

Pantograph indexing device (not used on flat surface)

Specimen

Probe

Fig. 2   B-scan equipment

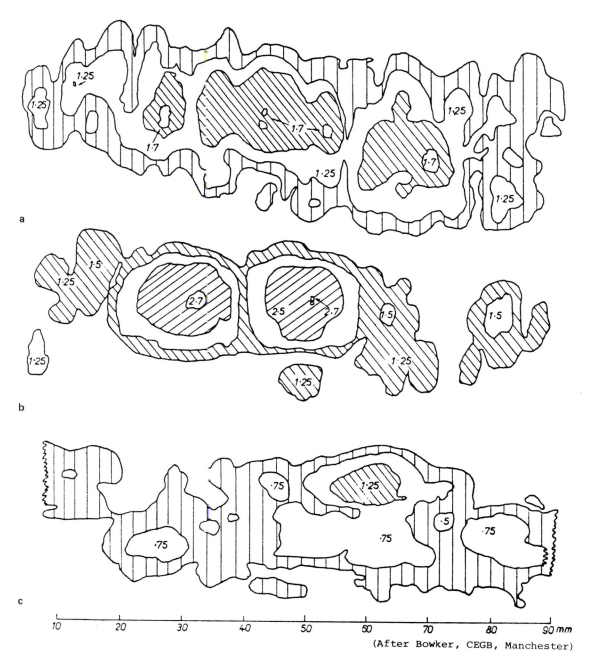

Fig. 3 C-scans with 6dB increment contours (numbers show equivalent DGS values in mm)

    a 45° probe transverse scan
    b 45° probe transverse scan from opposite side
    c 45° probe longitudinal scan

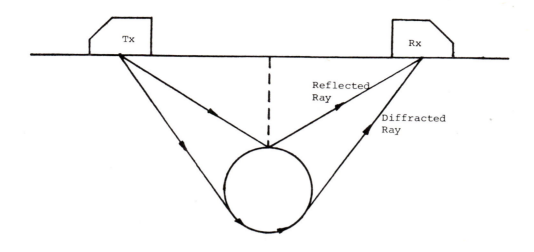

(after Charlesworth, NDT Centre, Harwell)

Fig. 4   Schematic illustration of time delay sizing applied to a volumetric (circular cross section) defect

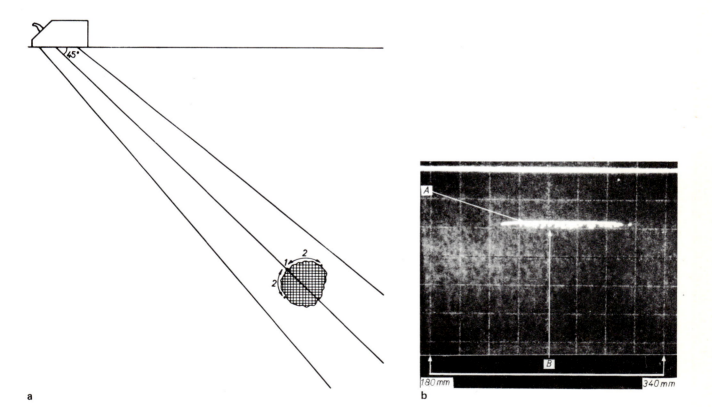

a

b

Fig. 5   Interaction of an ultrasonic beam with a rough cylindrical defect

    a   position of defect in ultrasonic beam
    b   resulting B-scan

C48/79

CRITICAL PARAMETERS INFLUENCING THE DETECTION, LOCATION, AND SIZING OF FLAWS IN STRUCTURES

S. H. BUSH, PhD
Battelle Northwest, Richland, Washington, USA

The MS of this paper was received at the Institution on 14 August 1978 and accepted for publication on 23 October 1978

SYNOPSIS The reliability of flaw detection has been assessed in terms of component geometry, metallurgical structure and non-destructive examination (NDE) variables for commercially available NDE techniques, with emphasis on ultrasonics. High detection reliability is an essential factor; however, such reliability represents a partial, not a complete solution. Fracture mechanics has been used both to optimize NDE and to assess the significance of a detected flaw with respect to the flaw size, location and orientation related to both the surface and to the principal stresses. Finally, the status of flaw detection is examined in the context of nuclear components such as piping, pumps, valves, steam generators and pressure vessels.

INTRODUCTION

1. The ideal nondestructive examination would be one that ignores all innocuous flaws and detects all flaws currently significant or with the potential to become significant in the future because of flaw size, orientation and location with respect to the system stress field. Unfortunately, reality is substantially removed from the ideal. Relatively innocuous defects such as laminations and slag inclusions are readily detected while large cracks are missed because of non-optimum flaw orientation with respect to the transducer beam angle.

2. The factors adversely influencing flaw detection by NDE have been reviewed by examining the role component geometry, metallurgical structure, flaw characteristics, NDE equipment parameters and examination procedures play in detection reliability or unreliability. Where the reliability value is considered unacceptably low, the factors leading to the low value have been examined and modifications suggested to improve the reliability. One aspect of detection that has received inadequate attention is the use of linear elastic fracture mechanics (LEFM) in conjunction with conditional detection probability as a tool to assess the safety significance of flaws in terms of flaw size, orientation and location in a given structure. This approach will be used to develop an "optimized" NDE program placing emphasis on regions of high stress intensity factor in recognition that component safety, not inspection, is the ultimate goal.

FACTORS AFFECTING FLAW DETECTION

Figure 1 presents an overview of the limitations pertinent to flaw detection. Generally these same limitations are relevant to flaw sizing and flaw location. While Figure 1 is specific to ultrasonic testing, the factors cited are pertinent to other forms of visual, surface and volumetric examinations. The location of the flaw, whether surface, near-surface, or embedded, may influence the detection reliabilities; however, the same factors cited in Figure 1 are relevant regardless of flaw location.

4. Components making up the pressure boundary of light water reactors are fabricated of ferritic or austenitic steels with some delimited regions composed of high nickel alloys. Some reactor pressure vessels may be accessible for NDE from a major portion of both inner and outer surfaces; others may be accessible from one surface only. The same is true for components such as pumps, valves and steam generators. With regard to piping, access generally is limited to the outer surface only. Both composition and accessibility influences the selection of the NDE technique(s); e.g., eddy current testing (ET), magnetic particle testing (MT), penetrant testing (PT), radiography testing (RT), ultrasonic testing (UT).

Surface Flaws

5. The work of Packman et al on surface fatigue cracks in small cylinders of steel and aluminum points out the difference in sensitivity (probability of detection) of the various NDE techniques, RT,PT,MT,UT. (ref.1,2) Figure 2 illustrates the variation in detection sensitivity for surface fatigue flaws as a function of flaw length.(ref.2) From a reliability standpoint MT and UT are preferred to PT or RT. Similar trends are observed for flaws in the aluminum cylinders which are considered as an analog for austenitic stainless steels. Several of Packman's conclusions are considered relevant to surface flaws in reactor components: (ref. 1,2):
(1) The reliability of NDE needs to be improved;
(2) For small (< 0.20 in. long) surface cracks the NDE techniques in order of preference(s) were a) UT, b) PT-Al MT-steel, c) PT-steel, d) RT;
(3) For larger surface cracks the order of preference was a) MT-steel, PT-Al, b) UT, c) PT-steel, d) RT;
(4) All NDE techniques except RT were accurate in locating surface fatigue cracks.

Embedded Flaws

6. Work by Caplan reported at the 2nd Conference in 1974 confirms the lack of sensitivity of RT and the need to expand UT examination

© I Mech E 1979

to both surfaces to improve detection reliability (ref. 3). Table 1 contains data indicating flaws in three of the five reactor pressure vessels would not have been detected if UT had been limited to the outer surface. One should not infer that everything was found; however, one can infer that RT misses more than UT and UT from both surfaces is better than UT from one surface.

7. The Doel vessel confirms the above trend. (ref. 4) During construction UT and RT examinations were from the outer surface. Later during the baseline examination UT from the inner surface detected a planar defect near a nozzle about 2.5-in. long by about 0.1-in. deep.

The Pressure Vessel Research Committee (PVRC) program

8. The PVRC sponsored extensive UT and RT on several weld specimens simulating reactor pressure vessel weldments in plates and nozzle forgings. Early work was highly variable, primarily due to variations in calibration and test procedures. An improvement in procedures led to an improvement in flaw detection. Unfortunately, the degree of improvement depended on assumptions concerning flaw size and location. If the actual flaws differed markedly in size and location from the values assumed to exist on the basis of fabrication procedure, the reliability of flaw detection would vary widely. For example, the ratio used for comparison, R_{III} is determined by dividing the number of introduced flaws found by the sum of those unfound plus the number of false indications. Values of R_{III} ranged from 0.02 to 1.52 with the primary reason for the low values being the difference in actual versus presumed flaw locations. It should be possible to reassess the UT results after the test specimens are sectioned. Until then one must consider the results as non-definitive.

Limitations in flaw detection

9. The factors influencing flaw detection are entwined with factors related to flaw location and sizing. While the following is presented in the context of detection, it generally is applicable to sizing and location. With regard to reliability of flaw detection it should be remembered that four conditions exist; namely:

Reliable
Flaw is detected that exists;
Flaw is not detected that does not exist;

Unreliable
Flaw is not detected that exists-of safety significance;
Flaw is detected that does not exist-of economic significance.

10. Figure 1 schematically touches on the several parameters influencing flaw detection. These will be examined in an attempt to assess their significance in the failure to detect flaws.

11. Theoretical considerations.

Limitations to flaw detection by UT have been developed on theoretical grounds and confirmed in several instances. (ref. 5,6) Examples of items affecting detection and sizing, either adversely or beneficially, are:

(1) Flaw roughness which reduces signal amplitude;

(2) Frequency - a reduction in frequency reduces the impact of flaw roughness by increasing signal amplitude;
(3) Number of beam angles - more than one beam angle is required or relatively large flaws may be missed;
(4) Resolution - currently resolution is limited by finite values of aperture and wave vector; however, theoretical considerations lead to prediction of higher resolution by optimization.

12. External factors.

Equipment variability has been examined and reported in conjunction with the PVRC program. The equipment testing round robin indicated that the two most important parameters were the system operating frequency spectrum observed at the instrument video detector and the beam profile or radiation pattern of the search unit in the material under consideration. (ref. 7) Another major source of variation has been the transducer. Generally the radiation pattern intensity and frequency response cannot be reproduced to within ± 20% even with two transducers from the same manufacturer. (ref. 8) In addition their characteristics decay with usage with the mean useful life being about six months.

13. An examination of data from the PVRC program illustrates the significant differences possible from one NDE team to another. Both Caustin and Herr emphasize the need to minimize this variable. (ref. 9,10) The current trend is toward automation to minimize or eliminate the human factor; however, we still lack the equipment capable of scanning and examining complex geometries. (ref. 8)

14. Calibration.

Calibration problems often are a direct result of using amplitude as a means of detecting and sizing defects. The use of flat-bottomed or side-drilled holes is misleading in that there is no good correlation with real defects. Another source of error is the difference in acoustic properties of reference blocks. (ref. 8) A factor of 10 in attenuation may occur due to the amount, distribution and size of non-metallic inclusions in ferritic blocks. The attenuation difference may be even more pronounced with blocks fabricated of austenitic stainless steel weld metal. (ref. 11)

15. Internal Factors.

Even when external variables are controlled and calibration is done with real defects the internal factors may grossly bias the results. The following factors will be considered - component surface, component geometry, material variables, and flaw characteristics (orientation, geometry, etc.).

16. The component surface is a major variable as indicated in Haines theoretical analysis, where an ideal surface was assumed; any surface degradation further reduces the signal amplitude. (ref. 5) Silk and Lidington cite the relation of surface to relative efficiency of coupling. (ref. 12) See Table 2.

Table 2
Effect of surface condition on ultrasonic coupling

Type of surface	Relative efficiency of coupling (%)
Smooth machined finish	100
Ground by hand grinder	76
Smooth but rusty	70
Adhering scale	72
Rusty and pitted	48

17. The preceding applies to contact probes. Under immersion conditions coupling should be a relatively minor problem. A condition controlled by the surface is attenuation due to temperature. If temperatures are too high or too low, conventional liquid couplants cannot be used. (ref. 13) Even where the couplants are usable, temperature is a factor since there is a 2db attenuation loss for every 25°F (14°C) increase in temperature. (ref. 14)

18. Component geometry is a critical factor. There is ample evidence that a combination of flaw location and orientation in a nozzle region may result in no detection when examined from one surface while the flaw is detected when examined from either surface. The PVRC work cited previously, Caplan's work and UT of the Doel reactor all confirm the unreliability of examination from only one surface. (ref. 3,4) Another problem with geometry is the inverse of lack of detection. A rule of thumb for systems such as piping is that geometric reflectors outnumber flaw reflectors 1000 to one, at a threshold value such as 20% DAC. (ref. 14) The severity of the geometric reflector problem is apparent from the following example; one 22-inch pipe weld required nearly 60 crew hours to plot and to analyze due to 422 separate indications essentially all of which were geometric. Another example of geometry causing problems relates to typical pipe weld joint designs. The weld root geometry may register 200% DAC while a through-wall stress corrosion crack (SCC) may register 80% DAC.

19. Material variables represent another major source of error in UT. While earlier conclusions that it is difficult or impossible to UT test the majority of austenitic materials in sections greater than 100 mm are a bit pessimistic, it is still true that such examinations are not easy. (ref. 13) The PVRC work with UT on clad and unclad test blocks clearly indicates the difficulty of detection in clad components when no compensation is made for such cladding.

20. The significance of attenuation difference unless accounted for can be substantial. For example if the attenuation difference is greater than 20 db between standard and weld, the weld will not be adequately examined. (ref. 14)

21. Another potential source of error related partially to the flaw and partially to the material has to do with regions of high stress such as may occur in the crack tip volume. Such regions may represent centers for acoustic scattering. Since local regions of high residual stress may be associated with structural discontinuities it is not necessary to associate them with flaws. There is a need for further work to establish the significance of this factor.

22. Perhaps the most significant material variable influencing UT is the discontinuity existing at the weld metal-base metal interface in austenitic stainless steel. Yoneyama et al examined UT beam behavior in austenitic shielded metal arc welds with a large dendritic grain structure. (ref. 15) The path of the wave train was determined through both base metal and weldment using 2.25 MHz and 45°C-scan. An anomaly was noted at the base metal-weld interface where a greatly enhanced transmission occurred. To date there is no satisfactory explanation for this behavior. Another anomaly occurred in the weldment. On entering the weldment the UT beam bent until it was essentially perpendicular to the surface. This behavior was attributed to the pronounced dendritic pattern where it was believed that the UT beam was channeled along the columnar grains with the grains acting as acoustic channels. As the columnar grain structure changes to one that is randomly oriented, the sound path reverts to a random walk pattern through the weld zone.

23. Flaw characteristics such as orientation and dimensions are quite sensitive to the UT conditions. The flaw characteristics per se can minimize detection when conventional UT procedures and equipment are used. Similarly, variations in equipment and technique will influence the detection probabilities. Typical UT uses only the amplitude to indicate the presence of a discontinuity. If the size of discontinuity exceeds the transducer diameter, the discontinuity cross section must be estimated because present instrumentation is not amenable to correction of beam spread and beam attenuation. (ref. 8) Factors leading to an underestimation of crack depth and, possibly, crack length include: (1) tight cracks where transmission losses differ from open cracks; (2) specular reflection from smooth surfaces, which is often the case for the lowest or latest generated section of a fatigue crack; (3) crack branching; (4) diffraction effects; (5) premature mode conversion of Rayleigh waves, possibly from irregularities down a fatigue crack face.

24. A factor that has not been definitively evaluated is the role of stress on a real or artificial flaw. A study by Corbly et al indicates that UT will under-estimate flaw depth and over-estimate flaw length for artificial flaws subjected to an applied stress. (ref. 16) The absolute accuracy increases slightly with increasing stress. With natural cracks zero stress usually permits an accurate measurement of crack depth. The effect of imposed stress is to increase the peak height and, hence, the apparent crack depth. At 20 KSI applied stress the accuracy is about 10% of that at zero applied stress. (ref. 16)

25. Flaw orientation is a critical factor. As noted in the theoretical analyses of Haines flaw tilt of about 10° is sufficient to prevent detection with a given beam angle. (ref. 5) Experimental work has confirmed that such flaws may not be detected, probably because certain flaw types and orientations do not interact sufficiently strongly with the UT wave. (ref. 6) An example is single probe analysis of lack of fusion on a square butt weld where the echo does not return to the probe. (ref. 13)

26. Several factors leading to missing a defect or under-estimating its size have been cited. An orientation related item has to do with insufficient reflection if the sound beam travels at too flat an angle in relation to the defect. Laminations may result in beam deviations and possible failure to detect flaws. Another structure leading to the missing of flaws may occur in certain weldments where "cold shuts" between each individual bead and the pass on which the bead is laid may cause a marked scattering of the sound beam because of the curvature of the cold shuts. (ref. 13)

A FRACTURE MECHANICS APPROACH

27. The conventional use of linear elastic fracture mechanics (LEFM) is to evaluate an existing flaw in a given stress field. A less conventional use is to assess the significance of various postulated flaws in different physical locations. LEFM is a powerful tool for such purposes; unfortunately it has not been used sufficiently. The following example culled from reference 17 illustrates the value of the LEFM approach in directing the NDE program.

28. The inlet and outlet reactor pressure vessel nozzles represent an excellent segment of the pressure boundary to analyze. Both fabrication defects (Hatch-1, Pilgrim-1) and operation-initiated defects (Millstone-1) have been observed in nozzles. Typical stresses during operation of a feedwater nozzle are given in Figure 3, based on a finite element analysis. (ref. 17) The thermal stresses at the inner surface are very high ($\sim$ 90 Ksi) dropping to below 0 Ksi in the first 30% of wall. The pressure stress is relatively constant, ranging from 50 Ksi at the inner surface to about 30 Ksi at the outer. Combined stresses are high at the inner surface ($\sim$ 150 Ksi) but drop below 80 Ksi in the first 10% of wall. Figure 4 illustrates the stresses in a similar nozzle during a postulated loss of coolant accident (LOCA). The pattern is similar to that of Figure 3 - very high surface stresses dropping rapidly to compressive stresses in the initial 20% of the wall since the pressure stress has been lost.

29. Figure 5 is a typical curve of stress intensity as a function of surface crack depth. As anticipated the value of K_I increases rapidly with increase in crack size. A more realistic case is for relatively small flaws located throughout a structure. An examination of either Figure 3 or Figure 4 confirms that for a given flaw size the most critical location is at or very near the inner surface. Flaws located at depths greater than 25% from the inner wall will have minimal safety significance and should exhibit little or no crack growth. In fact Figure 4 can be used to predict compressive stresses within a minute after LOCA at the 20% thickness position.

30. A selective use of fracture mechanics utilizing data from the stress report can be used to highlight regions of the pressure boundary warranting special attention because of stress levels, material properties and flaw location. In similar fashion innocuous flaws can be cataloged and the level of NDE minimized. A special case of the preceding is a near surface flaw obliquely oriented with respect to the surface. The tendency for growth or for failure initiation is reduced dramatically by such an orientation.

FLAW DETECTION PROBABILITY

31. The development of a conditional probability model for the detection of flaws is beyond the scope of this paper. Therefore, comments will be limited to a plan of attack. The Venn Diagram in Figure 6 illustrates the plan of attack covering the various cases of reliability and unreliability cited previously. The branch of major safety significance is P(E/F); the probability a flaw is not indicated, given a flaw exists. Several authors have developed the matrix of probability functions. (ref. 18-21) The major problem is the quantification of the functions. Relatively little has been published in terms of changes in acoustic signal due to the internal and external factors cited in Figure 1. A qualitative assessment of the impact of some parameters is given in Table 3 covering several NDE techniques. Silk, in an excellent paper, quantified variability due to several parameters in terms of estimates of error, (σ_n). (ref. 22) Table 4 contains estimates of error with the caveat that the values should be regarded as approximate since measurements given significantly different data would change the picture. The standard distribution (σ_n) for the sources of error in Table 4 is ±6.7dB, which is considered applicable to a clean material with a good surface finish and appropriate range-correction applied to fatigue cracks.

Table 4

Source of variation	Estimate of error (σ_n)
Defect orientation	± 3.5 dB
Defect roughness	± 3.0 dB
Interference effects	Not quantified
Transparency	± 4.0 dB
Attenuation	Not quantified
Coupling factors	± 2.0 dB
Operator variables	± 2.0 dB

NDE RELIABILITY-REACTOR COMPONENTS

32. The record of flaw detection in reactor components such as pumps, valves, steam generators, piping and pressure vessels during pre-service and in-service examinations has ranged from good to fair or poor. Several instances are known where flaws were not detected during the construction stage then were found during a pre-service baseline examination or during an early inservice examination (Hatch-1, Pilgrim-1, Doel). (ref. 3,4,7,11,14) The record has been similar with pumps and valves. Defects have been found after initial examinations. On the basis of the PVRC round robin there is reason to be skeptical concerning the detection of all flaws, particularly those with non-optimum orientations. Wüstenberg and Mundry believe it is possible not to detect fairly large flaws with UT in an non-optimum alignment, a position with which the author of this paper concurs. (ref. 23)

33. The situation is even more ambiguous in austenitic stainless steel piping containing stress corrosion cracks. Initial NDE-UT during the period of severe stress corrosion in BWR piping was relatively unsuccessful. (ref.14)

Later work with improved UT yielded better results; however, negative results were obtained during 1978 on piping known to contain through-wall cracks. Several factors may contribute to the unreliability with flaw transparency to UT being a significant one.

HOW TO IMPROVE NDE RESULTS

34. Several actions can be taken to minimize or eliminate the limitations cited with regard to NDE reliability. A very basic one would be to reduce or eliminate the dependence on signal amplitude as an indicator of defects. Recognizing that such a change is so major that early action is unlikely, other possibilities are suggested:

1. Minimize equipment variability by standardizing. Pick either tuned system or broad band, but not both;
2. Minimize the human variable through automation, special training, etc.;
3. Use as many beam angles and traverses from as many surfaces as economically feasible to minimize inherent limitations in defect detection;
4. Move toward reference blocks containing real defects not drilled holes;
5. Minimize coupling problems by suitable surface preparation;
6. Have NDE input into component design to optimize material properties, joint design, welding procedure;
7. Modify the codes to eliminate or minimize UT in the non-critical central portion of components. Based on LEFM, flaws in the region surrounding the neutral axis have minimal safety significance. This approach should eliminate a large number of geometric indicators;
8. Consider the feasibility of varying the recording and analysis thresholds as functions of stress intensity factors. This approach would drop DAC levels to 5-10% at the surface, rising rapidly as one moves away from the surface;
9. Use LEFM to establish necessary UT sensitivity flaw requirement;
10. Minimize beam attenuation, etc. through appropriate control of material quality and fabrication process, particularly welding.

REFERENCES

1. PACKMAN P.F., PEARSON H.S., OWENS, J.S. and YOUNG G. Definition of fatigue cracks through nondestructive testing. J. Mat. $\underline{4}$, 1969, pp 666-700.

2. PACKMAN P.F. Fracture toughness and NDT requirements for aircraft design. NDT 12/73, pp 314-324.

3. CAPLAN J.S. The ultrasonic shop map and its use in preservice inspection. 2nd Int. Conf Periodic Inspection, London, 1974, I Mech Eng., pp 19-28.

4. ANON. The Doel defect. Nuc. Eng'g Int., July 1974, p. 543.

5. HAINES N.F. The reliability of ultrasonic inspection. IAEA-SM-218/41, Int. Symposium on application of reliability technology to nuclear power plants, Vienna, Oct 10-13, 1977.

6. THOMPSON R.B. and EVANS A.G. Goals and objectives of quantitative ultrasonics. IEEE trans on sonics and ultrasonics SU-23, No. 5, 9/79, pp 292-299.

7. BIRKS A.S. and LAWRIE W.E. Improved repeatability in ultrasonic examination. Also ultrasonic testing system standardization requirements. Welding Research Bulletin, No. 235, Feb 1978, 16pp.

8. YEE B.G.W. and COUCHMAN J.C. Application of ultrasound to NDE of materials. IEEE trans. on sonics and ultrasonics, SU-23 No. 5, 9/76, pp 299-305.

9. CAUSTIN E.L. State of art NDE in quantitative inspection. Proc. of the interdisciplinary workshop for quantitative flaw definition. AFML-TR-74-328, June 17-20, 1974.

10. HERR J.C. Human factors in NDE, in preventing structural failure: the role of quantitative nondestructive evaluation, ASM 1975, T.D. Cooper and P.F. Packman (Editor)

11. LAUTZENHEISER C.E., WHITING A.R. and McELROY J.T. Ultrasonic variables affecting inspection. 3rd int. conf. on periodic inspection. 1976, pp 107-111.

12. SILK M.G. and LIDINGTON B.H. The potential of scattered or diffracted ultrasound in the determination of crack depth. NDT, 6/75, pp 146-151.

13. ANON. Limitations inherent in the use of ultrasonics for the examination of welds. Welding in the weld $\underline{6}$ No. 4, 968 pp, pp 31-33.

14. DAU G.J. (Editor). Proceedings of NDE experts workshop on austenitic pipe inspection. EPRI SR-30, 2/76.

15. YONEYAMA H., SHIBATA S. and KISHIGAMI M. Ultrasonic testing of austenitic stainless steel welds - false indications and the cause of their occurrence. NDT Int., 2/78, pp 3-8.

16. CORBLY D.M., PACKMAN, P.F. and PEARSON H.S. The accuracy and precision of ultrasonic shear wave flaw measurements as a function of stress in the flaw. Materials Eval. $\underline{28}$, 1970, pp 103-110.

17. ASME XI WORKING GROUP ON FLAW EVALUATION. Background and application of ASME Section XI Appendix A "Flaw evaluation procedures". EPRI-719-SR, 1978.

18. DAVIDSON J.R. Reliability after inspection. Fatigue of composite materials. ASTM-STP-569, 1973, pp 323-334.

19. JOHNSON D.P. Cost risk optimization of nondestructive inspection level. Nuclear engineering and design $\underline{45}$, 1978, pp 207-224.

20. JOHNSON D.P. Determination of nondestructive inspection reliability using field or production data. EPRI-NP-315, 10/76.

21. PACKMAN P.K., MALPANI J.W., WELLS F. and YEE B.G.W. Reliability of defect detection in welded structures. Reliability engineering in pressure vessels and piping. ASME 1975, pp 15-28 A.C. Gangadharan Editor.

22. SILK M.G. Estimates of the magnitude of some of the basic sources of error in ultrasonic defect sizing. AERE-R9023, Feb 1978.

23. WÜSTENBERG H. and MUNDRY E. Limiting influences on the reliability of ultrasonic inservice inspection methods. Paper C 112/74, 2nd conf. on periodic insp. London, Inst.Mech. Eng. 1974.

TABLE 1

DETECTION OF FLAWS IN FIVE REACTOR PRESSURE VESSELS (REF. 3)

VESSEL	UT DATA	RT DATA
RPV-1	23 indications found, eight to 400% DAC with 40° angle beam; also 60°. Lack of fusion - 12, slag inclusions - 11.	None of 23 detected with initial RT. Only four with subsequent RT.
RPV-2	14 indications in 6 of 8 nozzles to shell welds. Five greater than 100% DAC. Detected by straight beam from bore side. UT from outer surface by straight and 45° beam did not detect. Porous structure produced by solidification after liquation.	None detected by initial RT; however this had incorrect beam alignment. With correct alignment 12 of 14 were found by RT with 9 code rejectable.
RPV-3	16 indications in nozzle welds detected by straight beam (> 100% DAC) from bore side.	No indications with initial RT. More sensitive RT detected 9 of 16.
RPV-4	Low amplitude signals detected in two longitudinal welds with 45° beam from inside. 60° gave negligible response. There was a complete lack of detection from the outside. Straight beam detected nothing from inside nor did angle beam directed along the weld. Characterized as slag porosity, lack of fusion and freeze-line shrinkage or liquation cracking.	No detection in either initial or subsequent RT.
RPV-5	45° angle beam detected rejectable indications in circumferential weld between head dome and flange. Lack of fusion.	Nothing in RT's.

TABLE 3

SIGNIFICANT PARAMETERS INFLUENCING FLAW DETECTION AND SELECTION OF NDE TECHNIQUES USED

FLAW LOCATION	NDE TECHNIQUES	SURFACE CONDITONS	COMPONENT THICKNESS	SIGNIFICANT PARAMETERS		OTHER AS CITED
				COMPONENT GEOMETRY	MACRO AND MICRO STRUCTURE	
Surface	MT, PT Visual Rayleigh ET	Critical ↓ ↓	Not important ↓ ↓	Not important ↓ ↓	Not important ↓ ↓	
	UT	↓	Critical	May be important	Significant	
	RT	Not too important	↓ ↓	important	not important	
Near surface	MT	Critical ↓	Not Important	Not important	Not important	
	ET	↓ ↓	Important ↓	May be important	May be important	
	UT	↓ ↓	↓ ↓	↓ ↓	↓ ↓	Flaw orientation is critical
	RT	Not Critical	↓ ↓	↓ ↓	Not important	
Embedded	UT	Important	Significant	May be important	Significant	Flaw orientation is critical
	RT	Not important	Critical	↓	Not important	

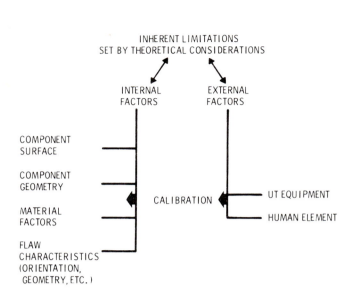

Fig. 1   Limitations to flaw detection, sizing, location

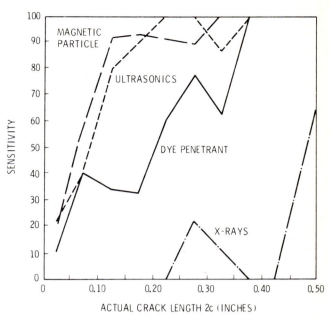

ACTUAL CRACK LENGTH 2c (INCHES)

Fig. 2   Sensitivity of crack length 2c, indication by NDE in 4330V steel (Ref 1)

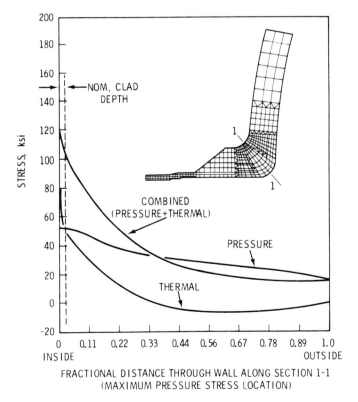

FRACTIONAL DISTANCE THROUGH WALL ALONG SECTION 1-1
(MAXIMUM PRESSURE STRESS LOCATION)

Fig. 3   Hoop stress at nozzle blend radius section

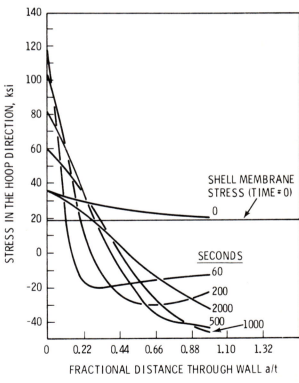

FRACTIONAL DISTANCE THROUGH WALL a/t

Fig. 4   Reactor inlet nozzle stress profiles during a postulated loss of coolant accident

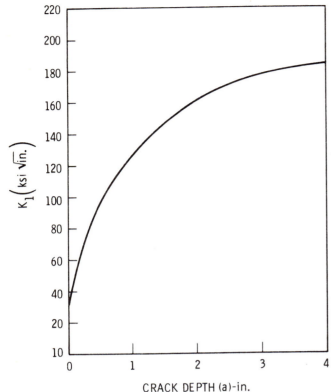

Fig. 5 Stress intensity factor vs. crack depth

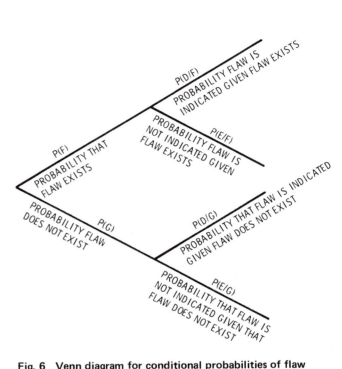

Fig. 6 Venn diagram for conditional probabilities of flaw detection

PERFORMANCES OF THE ULTRASONIC EXAMINATION OF AUSTENITIC STEEL COMPONENTS

P. CAUSSIN, Ir, and J. CERMAK, Ir
Association Vinçotte, Belgium

The MS of this paper was received at the Institution on 25 October 1978 and accepted for publication on 29 January 1979

SYNOPSIS   The performances of the ultrasonic examination of austenitic stainless steel components, 5.5 up to 92 mm in thickness, are discussed and exemplified by typical field and laboratory results. The cases of pipes, welds and castings are considered.

INTRODUCTION

1.   The ultrasonic examination of austenitic materials is regarded as of high importance, since those materials are specified for the petrochemical and nuclear plants. Ultrasonics are of general interest, because its sensitivity to planar defects, and its speed of application, made it very desirable to use. Furthermore, this technique is presently the only one which is applied for the volumetric in-service inspection of critical components in nuclear environments.

2.   It has for a long time been recognised that most of the stainless steel components can be very difficult to inspect (1 - 3). Indeed when producing stainless steels, the initial austenitic phase is preserved in grains which grow along the heat dissipation lines, up to an appreciable size (several centimeters in heavy castings). Depending upon the mechanical work done on those steels, the cast structure can remain or be refined to an extent which makes the grain quite fine. In welds a similar process is observed. Even in multi-pass welds, the grain can be very big due to epitaxial growth between passes and usually no refinement treatment is possible. Thus the first observation was that the grain size of austenitic stainless steels can lead to problems (2 - 5), since the bigger the grain, the more attenuating is the material. In practice, the grain size is such that the ultrasonic waves are more attenuated by scattering than by absorption (3, 6-8).

3.   To describe the behaviour of ultrasonic waves in austenitic steel, other microstructure characteristics must be considered. Those characteristics combine their effects in such a way that it is very difficult to draw general conclusions. The literature contains several examples of conflicting results which have not yet been explained. For instance, one of the most interesting observations of the last years is that the actual orientation of the ultrasonic beam in regard to the austenitic grain plays an important role in the total attenuation of the waves (5, 9-13). That observation was correlated with measurements indicating that the austenitic grains grow preferably along the $\langle 100 \rangle$ crystallographic axis in the direction of the heat flow and form a columnar structure. The first experiments show that the attenuation of longitudinal (compression) waves was less important when the ultrasonic beam propagated at 0 and 60° of the grain axis ($\langle 100 \rangle$ axis) and more important at 30 and 90° (9). Other reports indicate a lower attenuation at 0° (11, 13) or at 45° (5, 10). For transverse (shear) waves, it was generally agreed that the attenuation was more important than for longitudinal waves. But recently it was demonstrated (11 - 13) that transverse waves polorized parallel to the major interfaces existing in a weldment (fusion line, grain boundaries) can be less attenuated than longitudinal waves of equal wavelength.

4.   The effect of other structural parameters on attenuation is also discussed in the literature. The most-often cited parameters are ferrite content, microcracking, and carbide precipitates. In low ferrite austenitic steels, hot cracking can appear by a liquation process most often associated with thermal cycling (14). Hot cracking is thought to be responsible for an increased attenuation (3). A small amount (5 %) of ferrite prevents the initiation of hot cracks, and in that way would favour the propagation of ultrasonic waves. Besides that, it is not clear whether the presence of ferrite improves (3, 9) or has no effect (1, 6, 7,) upon the ultrasonic attenuation. Recently, it was suggested that the distribution of the ferrite could also play a role (5), but the question is still open. About carbide precipitate the situation is also quite confused. In earlier results the carbide precipitates were thought to impair the ultrasonic wave propagation when concentrating at the grain boundaries. In recent data that effect was not confirmed (1, 2, 5-7, 9, 11, 13), but it was also suggested that the effect of precipitates could be more important when the austenitic steel was heat-treated above 1100°C. Indeed in that case the precipitates can leave the grain boundaries, enter in solid solution and destroy the preferential crystallographic orientation of grains. In that process the structure would become highly attenuating in all directions.

5.   In welds and castings the grain growth along preferred crystallographic axis also influences the ultrasonic wave velocities. Indeed the actual velocities depend on local independent elastic constants of which the number inversely increases with crystal symmetry. Different sets of elastic constants must be considered according to the wave direction into the crystal (15). In Face-Centered-Cubic (FCC) crystal of a 304 type austenitic steel, the ultrasonic velocity can vary from 2120 to 4040 m/s for transverse (shear) waves and from 5230 to 6570 m/s for longitudinal (compression) waves (12). This covers the range of values which are encountered in practice.

6.   This effect of the microstructure on the ultrasonic velocities has for consequence that the velocities can vary not only from piece to piece but also in a given item. In a welded assembly, for instance, these velocities are generally different in the base material and in the weld, and can also vary from place to place in each of those weldment parts. This characteristic of the austenitic structures has two major implications :

- firstly, as the ultrasonic beams are refracted when their velocities are changing, the direction of propagation is altered and the beams can appear as bended;
- secondly, as the acoustic impedance of a medium is directly proportional to the velocities, acoustic interfaces appear when the velocities are changing. This phenomenon enhances the ultrasonic attenuation.

7.   From the hereabove considerations it appears that the metallurgical structure of austenitic steel components interact in a very complex manner with the ultrasonic waves. That interaction is not very well understood yet. Little advantage has been taken, up to now, of our present knowledge to improve the inspectability of austenitic structures. Nevertheless, that knowledge was sufficient for designing special search units and procedures which enhanced quite a bit the performances of the ultrasonic technique.

8.   Early in 1973 custom-built search units were developed and applied to the inspection of thick welds and castings. Indicative results were published in 1974 (16). Since then, a complete line of search units was developed together with special procedures for the inspection of austenitic steel items of 5 to 100 mm in thickness. More than 20 different search units were built. Their principal characteristics are as follows :

- frequency   : 1 to 4 MHz
- damping   : quality factor (- 3 dB) : 1 to 5
- transducers :
  - material   : lead-zirconium titanate (PZT) and lead metaniobate
  - number   : 1 or 2 (twin crystal probe)
  - arrangement : in twin crystal probe the transducers are mounted either side-by-side or one-behind-the-other
  - size   : 10 x 10 mm up to 20 x 20 mm
- beam spread : reduced by focusing technique using solid acoustic lenses
- shoe   : usually fixed solid shoes but in some cases a liquid column is

preferred.

The combination of characteristics used always depends on the piece to be inspected. According to our experience no general rule can be drawn. Laboratory tests should always be conducted on samples representative of the actual field conditions. On that basis the most appropriate search units (sometimes standard ones) are selected together with a precise test procedure complying with the ASME Code specifications. New search units were always and are still developed to solve peculiar field problems.

9.   In this communication, the technical description of both search units and procedures is restricted to the benefit of the discussion of the performances obtained with the present technology. The argument is supported by practical examples for which alternative non-destructive or destructive testing results are available.

THIN COMPONENTS

10.   The ultrasonic inspection of thin wall components requires special care from the operator. Indeed the flaws to be detected can easily be masked in the dead zone of the search units. Furthermore locating a flaw is generally hazardous as the beam spread is often similar to the wall thickness. The major positive point is that in those tubes the weld volume is quite small and allows to use transverse waves (TW) search units. The problem is thus very similar to the one inspecting thin ferritic steel boiler tubes. Those tubes have been ultrasonically inspected for years with success. The examination method is based upon the following principles : use miniature probes and low amplification to reduce the dead zone, do not try to locate in depth but in projection on the outer surface and make the inspection at least at full skip. In the following paragraphs, the present performances and some remaining problems are exemplified.

Ultrasonic procedure

11.   The procedure used for ultrasonic testing of pipes and welds that are equal to or less than 12.5 mm thick involves only standard probes and apparatus. The equipment includes 1 normal beam probe and 2 miniature transverse waves (TW) 45° and 70° angle probes. All probes have a frequency of 4 MHz. Depending upon the external diameter of the pipe, the angle probe shoes may or may not be shaped.

12.   A Distance-Amplitude-Correction (DAC) curve is set up from the echoes obtained on notches situated at both the outer and the inner surfaces of the tube. The exact field procedures are fully described in the written documents as required by the ASME Code.

Flaw sensitivity

13. To exemplify the sensitivity of the present technique to real flaws, results gained during field tests were selected. They are summarized in table 1. The figures 1 to 6 illustrate some of the typical A-scan representations and destructive analysis results.

14. The 5.5. mm thick tubes were suspected of intergranular stress corrosion cracking (IGSCC). The tubes were then ultrasonicaly inspected together with their welds.

Table 1

Typical results of inspecting thin components.

Tube Size			Echoes		Reflectors (destructive examination)			
Thick. mm	Dia. mm	Material	% DAC	S/N dB	Nature	Height mm	Location	Figure
5.5	165	AISI 321	> 50	> 12	IGSCC	1-5.5	in tube	1
			100	20	crack	5	HAZ	2
			100	14	crack	0.9-1.6	weld root	3
			50	12	crack	2	embedded at the weld axis	-
12	152	SS304	50	13	lack of fusion	3	weld root	4
			70	13	crack	0.3	weld root	5
			50	12	crack shaped inclusion	0.6	8 mm deep in weld	6

DAC   :   Distance - Amplitude - Correction
HAZ   :   Heat Affected Zone
IGSCC :   Intergranular Stress Corrosion Crack
S/N   :   Signal to Noise Ratio

The technique proved to be sensitive to IGSCC of a height of at least 20 % of the wall thickness. Weld defects (cracks) of the same minimum size were also detected. The echo amplitude was always higher than 50 % DAC but the cracks in the weld produced an echo 6 dB lower than equivalent cracks situated at the weld fusion line. The signal-to-noise ratio was always better than 12 dB.

15. In the 12 mm thick weld, a lack of fusion, a crack, and a crack-shaped inclusion were detected. In that case the minimum defect size was 2.5 % of wall thickness or 40 % of wavelength. For those defects, less than 25 % of the wall thickness, the echo amplitude was only a bit higher than 50 % DAC and the signal-to-noise ratio was equal to 12 or 13 dB.

16. In the hereabove described tests, which only concern fine grain base materials, the flaw sensitivity and the signal to noise ratios are similar to the ones related to the inspection of ferritic steel items for reflectors situated either in the tubes or near the weld fusion lines. For reflectors more embedded in the weld volume the performances are a bit worse but still acceptable. The real problem of inspecting thin wall austenitic steel welds comes rather from another phenomenon discussed in the next paragraphs.

Spurious echoes

17. Welds of pipes are very well known for producing a lot of spurious and geometrical indications. Very often those indications are due to features of the weld preparation, particularly those allowed to cope with pipe alignment. For example, a too short counterbore length and too sharp a taper are often the cause of geometrical indications. These can usually be recognised by an experienced operator and should not lead to erroneous results, at least with manual inspection techniques.

18. Besides the indications arising from geometrical effects, one can often find indications that may have been caused by defects. To study the situation, trials were made to gain more understanding of the nature of these reflectors and their significance, if any. Three blocks (304 type stainless steel) were investigated with ultrasonic testing, radiography, and metallographic techniques.

19. Two kinds of echoes were further investigated. The first kind consisted of a relatively high echo, the bottoms of which were surrounded by considerable background echoes ("grass") (fig. 7). Neither the radiographic examination nor the metallographic one discovered a possible reflector.

20. The second kind was a double echo that indicated the presence of two reflectors at the root of the weld. The radiographic technique did not reveal anything, but two small cracks were found at the root of the weld (fig. 8) by destructive testing. The first one, closest to the probe, was 0.2 mm high but very short. The other one was 0.06 mm high and was found existing all along the weld.

21. It is perhaps surprising that such a small defect, less than 0.1 wavelength, gave rise to a substantial echo, but it appears that this was due to its being associated with the root geometry. It was further observed that the weld transition zone had a thin layer, of which the crystallographic structure differed from the rest of the base material and of the weld. In the base material, the grains were very fine and equiaxed; in the weld, a columnar structure was developed during welding, whereas at the transition zones the grains had an intermediate size with no apparent preferential orientation.

If, as discussed previously, the ultrasonic attenuation and reflection coefficient depend on the actual orientation of the ultrasonic beam in regard to the crystallographic axis, the structure of this fine layer can constitute a semi-transparent acoustic interface. The impinging ultrasonic energy can be partly reflected and contribute to ultrasonic response of both cracks.

22. The occurrence of a double echo coming from the root of a weld was clearly correlated with actual defects but the defect reflectivity seemed to be a bit exaggerated due to the composition of the weld transition zone. The indications, surrounded by background echoes ("grass"), having a relatively small signal-to-noise ratio (6-8 dB) could not be associated with real flaws.

23. Until the present time, the aspect of echoes can be taken as an objective criterion for the evaluation of indications only with manual techniques. With those techniques, an experienced operator should make the discrimination on the spot. This practice is not compatible with "automatic" techniques, such as those used in nuclear practice. In these cases the recording criteria are only based upon an amplitude threshold level and do not allow any operator assessment.

24. This situation infers that a lot of data must be recorded and processed. A fast mechanized system can help for recording and a computer for data processing. A computer analysis based upon a threshold level is relatively simple to implement. But taking account of the indications aspects will require some kind of pattern recognition analysis. Till now such a system is not available for field tests, whereas operator assessment can be applied in situations where the requirements are more flexible, namely in non-nuclear jobs.

THICK COMPONENTS

Fine grain base materials

25. When the base material is finely grained, the ultrasonic inspection always involves both longitudinal waves (LW) and transverse waves (TW) angle probes besides a normal straight beam probe. The exact set of probes mainly depends on the thickness and the geometry of the items to be inspected. For instance, a 25 mm thick piece will be inspected using a LW 70° angle probe and TW 45° and 60° angle probes. LW 45°, 60° and 70° angle probes and a TW 45° angle probe will be used for a 100 mm thick piece. The frequency will be 1, 2 or 4 MHz depending on the particular case.

26. As usual, the straight beam probe is mainly used to detect laminar defects in the base material. When the latter is finely grained, the transverse waves can propagate without undergoing too much attenuation. So standard TW probes can be used to inspect the heat affected zones (HAZ) and the neighbourhood of the weld fu-

sion lines. In principle LW angle probes are used to inspect the weld volume. For 25 mm thick items the welds, usually V shaped, are relatively narrow in the inner half and TW angle probes provide enough sensitivity. The LW angle probe is mainly necessary for the upper half of the weld. For a 100 mm thick items, the width of the weld usually requires LW probes and the use of the TW angle probe is restricted to the inspection of the HAZ.

27. During field tests, several indications were detected but further investigations with other non-destructive techniques were never possible or well-enough documented. So the performances of the technique will be exemplified using laboratory tests.

28. Block for Commission of European Communities

In the frame of activities of the Commission of European Communities (CEC) - Working Group on Creep, an inspection of a welded sample was performed. The purpose was to detect any defects, even the smallest ones, in order to avoid their interference in the creep tests.

29. The material conformed to DIN SW 4919 standard. Two 52 mm plates were welded, using multipass procedure. Perpendicular to the main weld, there was also an auxiliary weld, which was ultrasonically and destructively examined to establish the detection level.

30. The apparatus time base was calibrated using a ferritic steel sample. The sensitivity was calibrated with auxiliary reference blocks. For transverse waves (TW) angle probes, one used a 42.5 mm deep, 1.2 mm diameter side drilled hole in ferritic steel. For longitudinal waves (LW) probes, one used a 21.5 mm deep, 4.8 mm dia. side drilled hole in austenitic cast steel (ASTM A 351 - CF8M). The hole echoes were brought up to 80 % of screen height (SH).

31. When inspecting the auxiliary weld, seven ultrasonic indications were attributed to flaws located in the base material or in the heat affected zone (HAZ). Only one indication, obtained with a LW 70° angle probe, reached 50 % SH. The other indications were all below 25 % SH. Four indications were further investigated by destructive analysis. Only the one with the highest echo amplitude was correlated with a slag inclusion situated at the weld fusine line. It was 1.0 mm high and 16 mm deep. In two other sections, for which ultrasonic indications were recorded, small slag inclusions (0.5 mm high) were found but at different locations. Those inclusions were considered as not detected by ultrasonic and the echoes were attribued to structural noise. Indeed the signal-to-noise ratio was not better than 6 dB. Four sections, for which no ultrasonic indications were recorded, did not reveal any defects.

32. During the inspection of the main weld several reflectors were detected. One was in the auxiliary weld zone. This location allows to conduct a destructive test without losing the material necessary for the creep tests.

That indication was investigated by fine cutting in passes of 0.5 mm. The shape of the related flaw and the results of the cutting examination is given at figure 9. The flaw had a length of 11 mm and a maximum height of 1.1 mm. The corresponding ultrasonic indications had an amplitude of 100 % SH with both LW, 2 MHz, 60° and 70° angle probe.

33. From these investigations, it appears that low amplitude indications cannot be correlated with actual flaws by using a simple cutting method. One should nevertheless be aware that a flaw could easily be missed in such an investigation due to either cutting beside that flaw or destroying it in the cut. High amplitude echoes were successfully correlated with an actual flaw of about 1 mm in height. This dimension seems to be about the limit of the method used.

34. AHCUISS block

In the frame of activities of the German Ad Hoc Committee for the Ultrasonic Inspection of Stainless Steel (AHCUISS) a butt-weld was ultrasonically and radiographically examined. The details will be published elsewhere. One will only give here an example of typical results. These come from the thinner part (49.2 mm) of the block.

35. The piece was multi-pass welded and its surfaces were finely machined. During welding by manual metal arc, defects were introduced.

36. Two sets of search units were used. They both involved the same normal beam probe (25.4 mm dia., 2.25 MHz) and the same longitudinal waves (LW) angle probes (20 mm dia., 2 MHz). The only difference was the transverse wave (TW) angle probes. The first set (M) contained miniature TW angle probes (8 x 9 mm, 4 MHz); the second set (N) normal TW angle probes (20 x 22 mm, 2 MHz).

37. As a standard calibration block was not available, a 61.5 mm thick austenitic weld was used. It contained three 2 mm dia. side drilled holes regularly spaced in the block thickness. In its other aspects, the inspection conformed to the ASME Code, Section XI.

38. The first observation was that the TW angle probes of the N set gave rise to more noise echoes than the M set probes. That noise was caused by reflections at the junction between the weld fusion lines and the block surface opposite to the scanning surface. This discrepancy is not well understood. Indeed less noise was obtained with 4 MHz probes than with 2 MHz probes. All other parameters (beam spread, damping) were equivalent.

39. The ultrasonic results were compared with the radiographic examinations ones and are summarized at the figure 10 which concerns the M probes set. The ultrasonic indications are represented by square or rectangular boxes, the radiographic ones by figures copied from the films. The figure 10a is the projection of all ultrasonic results on the top surface of the

block. At figure 10b, only one retained the indications which were confirmed with two different probes or with one probe but with scanning from two different directions. The figure 10c gathers the radiographic results.

40. One observes (fig. 10a) that the elongated flaws are detected, but that some porosities (circular defects on fig. 10c) are missed. When restricting the analysis to "confirmed" indications, the ultrasonic (fig. 10b) and the radiographic (fig. 10c) examinations give exactly the same results concerning elongated flaws. But the ultrasonic technique detects more reflectors when the confirmation is not asked for. Up to now it has not been possible to destructively examine that weld, and so the validity of the "unconfirmed" indications cannot be assessed. But in any case, this experiment demonstrates that when considering reflectors confirmed with two scanning methods (2 probes and/or 2 scanning directions), one achieved a reliability comparable with that obtained with radiography.

41. Statically and centrifugally made castings

To inspect a weld, when the ultrasonic beam had to pass through a cast austenitic base material, the main problem comes from the base material itself and not from the weld which is generally found less attenuating. Results were previously obtained for statically cast items with specially designed 2 MHz LW angle probes (16). New developments have recently been made for the inspection of centrifugally cast pieces. New twin crystals 1 MHz LW 45° angle probes were designed and used with success. A comparison between results gained with a standard TW 45° angle probe and the new probe is shown at the figure 11. A signal-to-noise ratio of at least 16 dB was obtained in ASME calibration blocks less than 92 mm thick.

42. No field results are available yet for publication on centrifugally cast austenitic stainless steel components. Laboratory experiments have been conducted to assess the feasibility of interpreting the indications, the results of which are reviewed in the following paragraphs.

43. The influence of the metallurgical structure on the ultrasonic beam profile has been investigated in the case of several probes, but we will only consider the one of a normal beam probe, 2.25 MHz, 25.4 mm dia. The beam profile has been measured in both ferritic and centrifugally cast austenitic steel (ASTM SA 351 - CF8A) with an electrodynamic sensor, an ultrasonic amplitude sorter, and an electrosensitive paper recorder. The normal beam probe was used as transmitter, the electrodynamic sensor as receiver. The metal path was 60 mm in both cases. The results are shown at the figure 12. One may observe that the casting highly distorted the ultrasonic beam.

44. For those measurements the frequency response of the receiving circuit has been adapted to the frequency bandwidth of the ultrasonic signals. The frequencies were found different in both cases.

For instance (fig. 13) for the ferritic steel, the spectrum of a back wall echo contained more high frequencies than for the austenitic casting.

45. The frequency spectrum of the signals delivered by the receiving electrodynamic sensor was also measured. The frequency transfer function of that sensor was unknown. No preamplifier was used, the signals were then very low and a lot of side lobe interference was present. Those measurements were not absolute, and were made for the purpose of comparison. One observed (fig. 14) that at the beam center the spectrum maximum was close to the nominal frequency of the probe. The two measurements made 2.5 mm further from that point in orthogonal directions revealed a frequency shift towards the low frequencies, but that shift was not equal in both cases. In homogeneous materials, the frequency content of an ultrasonic beam emitted by a heavily-damped transducer, as the one used in this experiment, varies from point to point in a given cross-section. The signal frequencies are the highest at the beam axis and decrease when going away from that axis. That shift only depends on the radial distance. The fact that in the hereabove experiment the frequency shift also depends on the angular position indicates an unhomogeneity of the material.

46. After having moved the transmitting probe 4 mm to the left from its original position, the receiving sensor was positioned again at the previous location of the beam center and the frequency spectrum was measured (fig. 15). One observed that the spectrum did not change. According to the hereabove consideration it should have shifted towards the low frequencies. Correlating this measurement with the beam spread distortion shown at the figure 12, it seemed that the material partitioned the ultrasonic beam in small fillets having peculiar frequency characteristics. A small displacement (16 % of the transducer diameter) of the transmitting probe would not affect the energy partition in those fillets closed to the beam axis.

47. It appears from those measurements that :
- it is possible to detect ASME calibration reflectors in centrifugally cast stainless steel blocks up to a thickness of 92 mm
- the material highly attenuates high frequency signals
- the material highly distorts the beam geometry
- the material seems to act as an acoustic fiber medium guiding the waves according to privileged paths.

CONCLUSION

48. The inspection of thin austenitic steel components was made possible by the use of both standard transverse wave angle probes and specially designed procedures. During field tests, intergranular stress corrosion cracking was detected, amongst other types of flaws in items as thin 5.5 mm. Spurious echoes were also present in those pieces. It was possible to discriminate between them when considering the shape of the echoes and the apparent location of the reflec-

tors.

49. In thicker components (above 15 mm), fewer problems were encountered with spurious echoes but special probes and procedures had to be used to inspect welds even when the base material was finely grained. In this case it was shown that the technique can detect flaws of more than 1 mm in height.

50. When inspecting centrifugally cast austenitic steel items, it appeared that the ultrasonic beams were so widely distorted that locating and characterizing reflectors with accuracy was hazardous. But in any case, the detection of existing discontinuities was possible with the sensitivity required by the ASME Code when using the specially designed probes.

REFERENCES

1. E. HOLMES, D. BEASLEY. The influence of microstructure in the ultrasonic examination of stainless steel welds. J. of the Iron and Steel Inst., 1962, pp. 283-290.

2. E. HOLMES. Ultrasonic scatter in austenitic stainless steel. Appl. Mat. Research, 1963, pp. 181-184.

3. H.V. RICHTER. Zur Ultraschallprüfung austenitischer Schweissverbindungen (Teil I-II). Die Techniek, n° 23, 1968, s. 610-619, 692-696.

4. F.A. SILBER, O. GANGLBAUER. Contribution to ultrasonic testing of austenitic steel welds. Proceedings of the 8th World Conf. on NDT, Cannes, September 6-11, 1976, n° 2B7, 8 p.

5. A. JUVA, M. HAAVISTO. On the effects of microstructure on the attenuation of ultrasonic waves in austenitic stainless steels. Proceedings of the IIW, Commission V Colloquium on Non-Destructive Determination of Type, Position, Orientation and Size of Weld Defects, Copenhagen, July 4, 1977, 12 p. Also in Br. J. of NDT, November 1977, pp 293-297.

6. E. NEUMANN, B. KUHLOW, M. ROMER, K. MATTHIES. Ultrasonic testing of austenitic components of sodium cooled fast reactor. International Atomic Energy Agency (I.A.E.A.), International Working Group on Fast Reactors (I.W.G.F.R) Specialist Meeting on "In service inspections and monitoring on Light Metal Fast Breeder Reactors (LMFBR's)". Bensberg, March 9-12, 1976, 21 p.

7. B. KUHLOW, M. ROMER, E. NEUMANN, K. MATTHIES. Ultrasonic testing of austenitic steel weld joints. Proceedings of the 8th World Conf. on NDT, Cannes, September 6-11, 1976, n° 2B6, 9 p.

8. I.N. ERMOLOV, B.P. PILIN. Ultrasonic inspection of material with coarse grain anisotropic structures. NDT Int., December 76, pp 275-280.

9.   R.M. MURRAY. Ultrasonic attenuation in austenitic steels - A review of the association's work. J. Of Research - S.C.R.A.T.A., June 1969, pp 31-43.

10.   B.L. BAIKIE, A.R. WAGG, M.J. WHITTLE, D. YAPP. Ultrasonic inspection of austenitic welds. Paper for presentation at the IAEA, IWGFR meeting on in-service inspection at Bensberg, March 1976, 6 p.

11.   D.S. KUPPERMAN, K.J. REIMANN. Effects of microstructure on ultrasonic examination of stainless steel. Paper presented at the CSNI - Specialists' meeting on the Ultrasonic Inspection of Reactor Components, Risley, September 27-29, 1976, 10 p.

12.   D.S. KUPPERMAN, K.J. REIMANN. Effect of shear-wave polarization on defect detection in stainless steel weld metal. Ultrasonics, January 1978, pp 21-27.

13.   D.S. KUPPERMAN, K.J. REIMANN, N.V. FIORE. Role of microstructure in ultrasonic inspectability of austenitic stainless steel welds. Mat. Eval., April 1978, pp 70-74, 80.

14. C.D. LUNDIN, D.F. SPOND. The nature and morphology of fissures in austenitic stainless steel welds. Welding Research Suppl., November 1976, pp 356-s, 367-s.

15.   E. DIEULESAINT, D. ROYER. Ondes élastiques dans les solides - Application au traitement du signal. Masson et Cie, Paris, 1974, pp 131-166.

16.   J.P. PELSENEER, G. LOUIS. Ultrasonic testing of austenitic steel castings and welds. Br. J. of NDT, July 1974, pp 107-113.

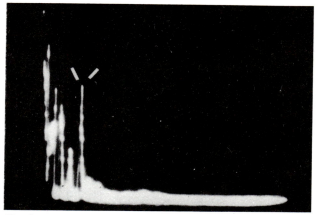

Fig. 1   Ultrasonic response of an IGSCC in a 5·5 mm thick tube obtained with a TW 70° angle probe

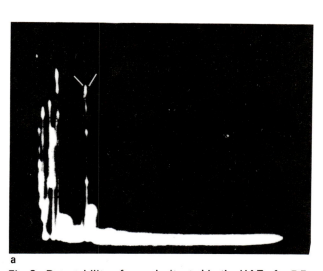

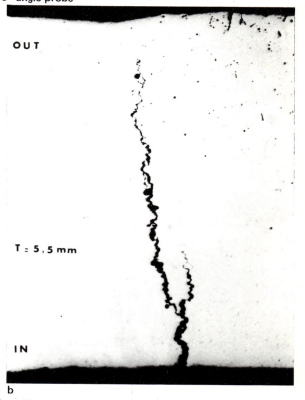

Fig. 2   Detectability of a crack situated in the HAZ of a 5·5 mm thick weld
    a Ultrasonic response obtained with a TW 70° angle probe
    b Macrography

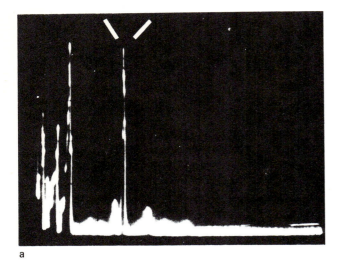
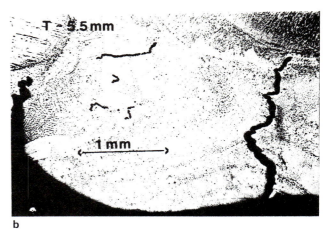

Fig. 3   Detectability of root defects in 5·5 mm thick weld
   a Ultrasonic response obtained with a TW 70° angle probe
   b Macrography

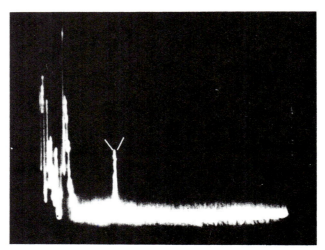

Fig. 4   Ultrasonic response of a lack of fusion detected with a TW 70° angle probe at the root of a 12 mm thick weld

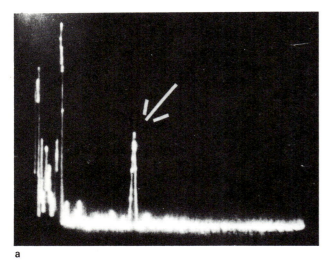
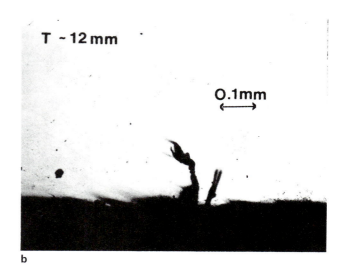

Fig. 5   Detectability of root defects in 12 mm thick welds
   a Ultrasonic response obtained with a TW 70° angle probe
   b Macrography

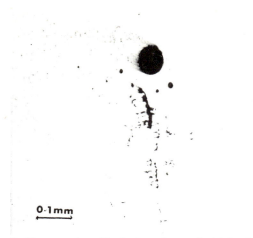

0.1mm

Fig. 6 Macrography of inclusions detected with both TW 45 and 70° angle probes

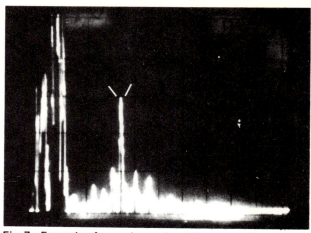

Fig. 7 Example of a spurious echo with surrounding background noise

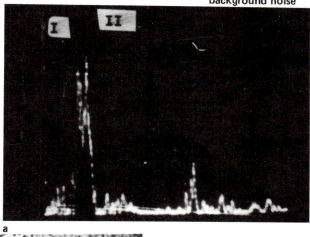

a

0.1mm

0.1mm

b

c

Fig. 8 a Ultrasonic response, obtained with a TW 70° angle probe, of small cracks situated at the root of a thin weld

b and c Micrographs of the reflectors

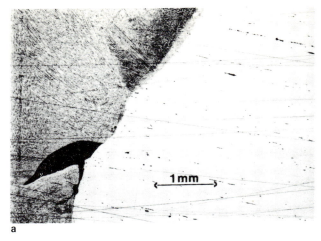

a

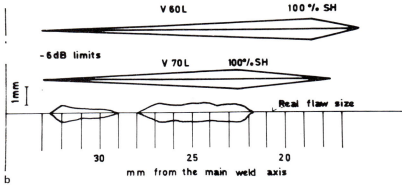

Fig. 9 Comparison of ultrasonic and destructive analysis of a flaw discovered at the fusion line of a 52 mm thick weld

    a Macrography
    b Actual flaw profile as compared with two ultrasonic indications

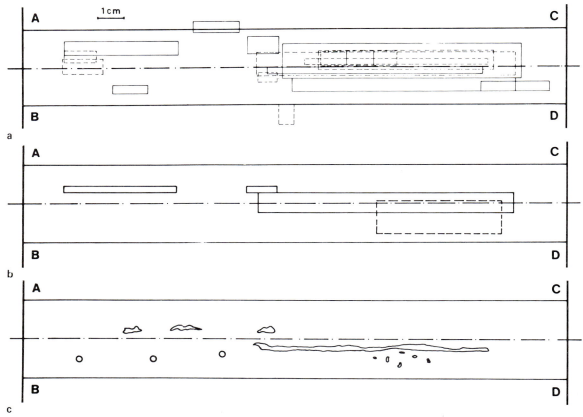

Fig. 10 Results of the ultrasonic and radiographic examinations of the thinnest part (49·2 mm) of the AHCUISS block

    a Complete ultrasonic results
    b 'Confirmed' ultrasonic indications
    c Radiographic results

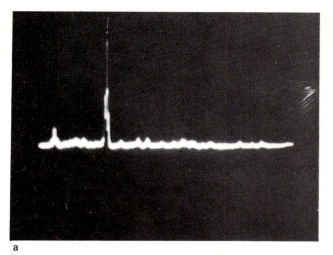
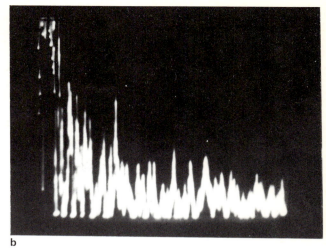

a

b

Fig. 11 Echoes from a 4·7 mm dia., 30 mm deep hole drilled in a centrifugally cast block

    a 1 MHz, LW 45° angle probe

    b 2 MHz, TW 45° angle probe (standard)

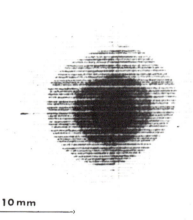
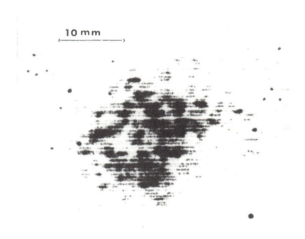

a

b

Fig. 12 Beam spread of a 25·4 mm dia., 2·25 MHz normal beam probe at 60 mm of metal path. Limits at −2, −6, and −14 dB drop

    a Fine grain ferritic steel

    b Centrifugally cast austenitic steel

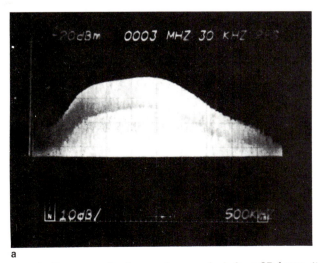
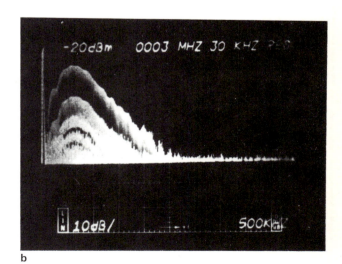

a

b

Fig. 13 Bottom wall echo spectrum analysis for a 25·4 mm dia., 2·25 MHz normal beam probe for 60 mm of metal path

    a Fine grain ferritic steel

    b Centrifugally cast austenitic steel

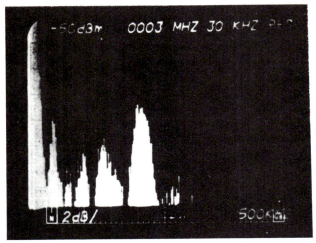

a

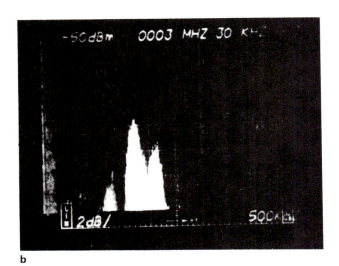

b

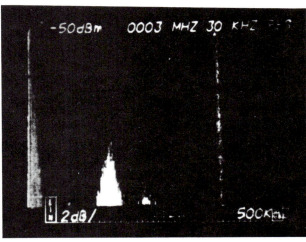

c

Fig. 14 Spectrum analysis of signals received by an electrodynamic sensor after 60 mm of travel in a centrifugally cast austenitic steel. Emitter: 25·4 mm dia., 2·25 MHz normal beam probe

    a At the beam centre
    b At 2·5 mm left of the beam centre
    c At 2·5 mm above the beam centre

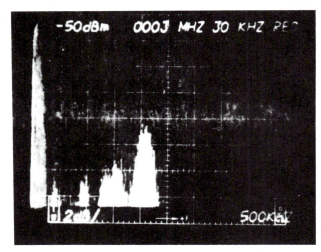

Fig. 15 Same measurement as the one of Fig. 14 a, but after
having moved the emitter 4 mm to the left

ULTRASONIC INSPECTION FOR STRESS CORROSION CRACKING IN STAINLESS STEEL

C. E. LAUTZENHEISER, BS and A. S. GREER, BS
Southwest Research Institute, San Antonio, Texas

The MS of this paper was received at the Institution on 18 October 1978 and accepted for publication on 14 December 1978

SYNOPSIS   Ultrasonic inspection of stainless steel, especially for the detection of intergranular stress corrosion cracks, has been and continues to be a problem. Commensurate with the importance of this problem, a great amount of research has been and continues to be performed. In August 1977, the Electric Power Research Institute reported results of a round robin test utilizing five different inspection companies to determine the statistical ability to detect stress corrosion cracking in stainless steel pipes. This paper reviews the results of that program and discusses significant transducer developments resulting from research conducted subsequent to that program.

INTRODUCTION

1.   In 1974, intergranular stress corrosion cracking (IGSCC) was detected in the 100-mm recirculation bypass lines of several boiling water reactors (BWR)[1]. Detection of the IGSCC by "conventional" ultrasonic techniques proved to be difficult, due both to the nature of the piping material and of the reflectors. Similar problems were encountered on BWR core spray piping in early 1975.[2,3,4] In late 1976, pressurized water reactor (PWR), low pressure, safety injection lines were confirmed to have IGSCC.[5]

2.   It is well recognized that there are acoustic problems encountered in examining austenitic materials. These problems are further complicated when the reflector in question is caused by IGSCC. Instead of presenting the relatively straight and highly reflective surface of classic fatigue cracking, stress corrosion cracks present a very diffusive face which follows the grain boundary in the material. This can cause a rather poor ultrasonic response relative to the response from typical machined calibration reflectors. Amplitudes, signal-to-noise ratios, and overall nature of the reflected signals result in difficulty in determining the size and significance of the reflectors.

ELECTRIC POWER RESEARCH INSTITUTE PROGRAM

3.   The Electric Power Research Institute (EPRI) conducted a round robin study to quantify the ability of ultrasonic inspection methods to detect the presence of IGSCCs in stainless steel piping as used in the bypass and core spray lines of BWRs.[6] This program addressed only the technique of ultrasonic nondestructive examination (NDE). It was not intended as an evaluation of the total inservice inspection and analysis process because ultimate decisions regarding the nature and consequences of reflectors are usually reached only after use of additional supplementary inspection techniques (liquid penetrant, radiography, etc.), followed by an extensive analysis of the stress, material

performance, and operating environment at the suspect area.

4.   The inspection quantification portion of the program involved five industrial teams performing examinations of both cracked and uncracked pipe samples obtained from an operating BWR. The inspections were conducted in two phases. In the first phase, the groups used their own inspection procedures, equipment, and standards. The second phase repeated the ultrasonic inspection using a reference procedure and standards defined and supplied by EPRI. (Appendix A) Both sets of procedures were within the same range of Code-acceptable inspection sensitivity. After the tests were completed, the test samples were sectioned and the actual nature, depth, and orientation of the reflectors determined. Although the data base of actual cracked and uncracked pipe samples was less than ideal from a statistical point of view, several significant results and trends can be derived from the results. These are listed as follows:

5.   First: The destructive examinations categorized the 16 test samples analyzed in the program into the following groups:

- Five test samples that had stress corrosion cracks,

- Two test samples that had lack of weld fusion and lap defects that were considered defects, and

- Nine test samples that were considered free from service-induced or major fabrication defects.

6.   In all five of the pipes containing intergranular stress corrosion cracks, the majority of inspection teams found indications and successfully identified the pipes as cracked. For each of the five test specimens with IGSCC, two inspections were performed by

each of the five test groups; one inspection was performed with anti-C clothing similar to actual inservice conditions and the other inspection was performed under laboratory conditions for a total of 50 separate inspections. The groups successfully detected and recorded flaws in 43 of these 50 pipe inspections (successfully detected is defined as an indication plotted in the proper location on the pipe with accompanying ultrasonic amplitude data). Of the 43 detections, 28 were successfully analyzed as service-induced cracks. (Successful analysis is defined as calling a plotted indication a crack. Terms such as linear indication or defect indication were not acceptable in this analysis, although this information would be useful to the plant operator in conducting further examinations).

7.   For the two specimens containing lack of fusion or lap defects, the majority of the inspection teams detected and defined these specimens as flawed, although the differentiation between IGSCC and lack of fusion was not usually made. For these two specimens, 18 separate inspections were performed by the five groups (one group did not inspect one of the specimens). All of the 18 inspections appeared to detect the defects successfully. In many cases, a group's original data were reviewed to determine the possibility that a defect was detected, but the detection was not reported on the final EPRI data form. This review found that some groups did apparently detect defects in these two specimens, but did not report them. Of the 18 inspections, 10 were successfully analyzed as lack of fusion or a crack.

8.   For each of the nine nondefective pipe samples, two inspections were performed by each of the five groups for a total of 90 inspections. Of these, 23 were called cracked. One pipe sample containing an internal fabrication reflector was involved in seven of the unsuccessful inspection calls. One sample containing severe geometric changes on the inside surface contributed to four of the miscalls. The remainder of the miscalls were randomly distributed and were attributed to other weld fabrication anomalies such as root drop-through, overlap, and irregular weld preparation geometry. The inspection teams pointed out that many of these miscalls might have been eliminated if detailed fabrication inspection data had been made available or if supplemental inspection techniques (such as radiography) had been used.

9.   The correct analysis of cracked pipe segments tends significantly to outweigh the several cases in which uncracked pipe was identified as defective. The tendency toward such overcalls indicates a conservative inspection philosophy which could result in the unnecessary repair of unflawed pipe during a reactor outage. However, this conservative approach would also allow fewer defects having the potential of growing to leak conditions to remain in the piping system. The repair of a leaking defect that causes reactor shutdown is much more costly (in terms of plant availability) than the same repair completed during normal periods of reactor shutdown.

Second: Since some of the specimens contained several defects of various depths and orientations, the individual defects were analyzed to determine what effect defect size, shape, and orientation had on defect detection and analysis. As expected, the orientation of a defect rather than its depth appears to be a significant controlling factor. The shape of defects is also important: a high depth-to-surface-length ratio contributes to lack of detection. Skewed cracks (those not parallel or perpendicular to the weld) were extremely difficult to detect in the simulated inspections using the normal field inspection procedures of scanning parallel and perpendicular to the weld. They were detected in the laboratory by rotating the transducer. However, from observations of previous field cracked pipe, skewed cracks in which no portion of the crack surface has an orientation parallel or perpendicular to the weld centerline appear to be very rare.

Third: Considering the overall results and relative performance of the groups, a conventional transducer size greater than 9.4-mm diameter (with conventional wedge) may reduce the number of stress corrosion cracks detected in some 254-mm pipe because of the physical interference between the front edge of the search unit and a raised weld crown. This observation was verified by laboratory study.

Fourth: The principal difference between these inspections and inspections for other types of defects was the way in which the data were analyzed. The groups that were most successful in correctly identifying defects plotted ultrasonic test (UT) data carefully on cross-section drawings of the weld configuration and then used the location of a suspect signal to decide upon its identity (geometry, stress corrosion crack, etc.). These groups used the criterion that a stress corrosion crack usually occurs between the fusion zone of the weld and the inside diameter counterbore (if any exists). One of these groups (Test Team A, Figures 1 and 2) used the criteria that any indication in this area is a crack and thus had the highest number of miscalls. The least successful group had the least nuclear experience which indicates that nuclear inspection is unique and requires special training.

10.   The overall results of this program indicate that pulse-echo, shear wave ultrasonics is a viable in-service volumetric inspection method for the detection of IGSCC defects in welded 300 series, core spray lines of BWR reactors (the number of bypass line samples included in the program was insufficient to draw conclusions). If the mechanism of IGSCC has started in the primary system of a BWR, and if the past historical pattern of crack growth is present (cracks at various depths and orientations and in more than one location), then there is a high probability that the presence of the IGSCCs will be detected by a well-trained inspection team using present in-service inspection methods. However, the detection and analysis of a limited number of small IGSCC defects at one or a few locations in the entire piping system could be missed

or incorrectly analyzed by this technique; this possibility is higher than desirable and needs improvement.

11. The sensitivity of the technique is not limited, as very small defects can be detected; rather, the fundamental problem in using the technique is correctly recognizing the detected signals as signals from a defect and not from other sources. The results also tend to show that the final result of the inspection process is dependent most of all upon the decision-making process of the individual inspector, and any improvement in this area would have the greatest effect on the total inspection process. Equipment, techniques, codes, and procedures influence the inspection process to a lesser degree; but there appears to be room for considerable short-term improvement in one or all of these areas.

12. The major drawback in using the ultrasonic technique is the time-consuming analysis required to distinguish defects from other signals. This creates a tendency to miscall fabrication anomalies as service-induced cracks. (Small fabrication anomalies have not been observed to grow with time in the manner of stress corrosion cracks.) This situation is aggravated by the rather low ratio of defect signal to noise and by the considerable difficulty in detecting small IGSCCs skewed to the major axis of the pipe. Previous inspection data and supplemental inspection techniques such as radiography aid in the analysis and decision-making process, but their exact contribution to improving the inspection process was not studied in the EPRI program.

SOUTHWEST RESEARCH INSTITUTE/EPRI PROJECT

13. Upon request of Southwest Research Institute (SwRI) and EPRI, sectioning of the previously discussed samples was delayed to permit SwRI engineers to perform basic measurements aimed at determining optimum search unit and instrument characteristics for the ultrasonic examination of stainless steel piping. This action resulted in a jointly sponsored SwRI/EPRI project to study the character of the reflected signal from stress corrosion cracks.

14. The results of this study indicated that the microstructure in the weld heat-affected zone (where the majority of the cracks occurred) and the nature of the crack interface displayed preferential bandpass characteristics. Frequency spectrum analysis of the RF waveform reflected from these cracks showed that, regardless of the operating frequency of the search unit, the combination of the microstructure and crack interface filtered and/or scattered the higher frequency components of the beam and returned primarily frequencies of 1.2 to 1.8 MHz.

15. In response to this information, a series of prototype search units were designed and fabricated. The design parameters were as follows:

(1) Piezoelectric element--lead metaniobate

(2) Frequency--1.5 MHz

(3) Element size--9.5 and 12-mm diameter

(4) Bandwidth--(6 dB points) 0.8 MHz (53 percent)

(5) Beam angle--45-degree shear wave

16. The performance of these prototype units was compared to search unit types typically used in stainless steel piping inspections (i.e., 2.25 and 5 MHz) with diameters of 6.3 and 12.7 millimeters. The comparison was made using the same IGSCC specimens used during the course of the EPRI round robin and SwRI/EPRI signal characteristic programs. For control, each search unit was calibrated on a Code-acceptable, basic calibration standard. A reference level distance amplitude correction (DAC) curve was established for each search unit. The prototype search units produced signal amplitudes on a series of IGSCC specimens which were 7 to 11 dB greater than that from commonly used search units. Laboratory experiments indicated nearly 100 percent more geometrical indications with the 1.5 MHz search unit than with a standard 2.25 MHz search unit.

17. In subsequent field evaluations, the prototype units continued to demonstrate improved sensitivity in detecting stress corrosion cracks. However, these field evaluations showed that the number of recordable geometrical indications increased only 34 percent over the previous inspection. Reflected signals from geometrical reflectors, such as weld crown and root, exceeding the 50 percent DAC reference level must be recorded and identified. This results in an increase in the number of hours required to conduct the examination.

18. It was postulated that a search unit, designed for improved beam directivity, could overcome the larger beam spread of the lower operating frequency of 1.5 MHz and assist in reducing the reflections from the root and crown.

19. A dual element, "pitch catch," 45-degree, shear wave configuration was designed with the transmitter and receiver elements set at a "pitched in" angle of 5 degrees (Figs. 1 and 2). The 5-degree convergence angle assigns a maximum response zone or focusing effect at a given metal path distance. Piezoelectric element sizes also determine the metal path distance for maximum response. Two designs were finalized using element sizes of 3.1 x 6.3 millimeters and 9.5 x 9.5 millimeters, optimized for wall thicknesses of 5 to 30 millimeters and 20 to 50 millimeters, respectively. Prototypes of these designs are shown in Fig. 3.

20. The design parameters were as follows:

(1) Piezoelectric elements--lead metaniobate

(2) Element sizes--3.1 x 6.3 millimeters and 9.5 x 9.5 millimeters

(3) Frequency--1.5 MHz

(4) Damping factors--5 to 6

(5) Bandwidth at 6 dB down--53 percent

(6) Angle beam shoes--45-degree refracted shear with a 5-degree "pitch angle." Material-Rexolite .

21. These units were evaluated on stainless steel piping welds at an operating nuclear power plant on welds chosen for the evaluation which had a large number of geometrical reflectors.

22. The welds were inspected in accordance with the inspection procedure being used at the plant. Three search unit types were used: single element-9.5-mm, 2.25 MHz; single element-9.5-mm, 1.5 MHz; and the prototype, dual element, 1.5 MHz.

23. The results showed the single element, 1.5 MHz units recorded 25 percent more geometrical indications than the 2.25 MHz unit, whereas the dual element, 1.5 MHz unit, recorded 5 percent more geometrical indications than the 2.25 MHz unit. Improved beam directivity also resulted in an increased sensitivity. When a comparison was made on the IGSCC specimens, the dual element, 1.5 HMz unit produced an average signal amplitude 12 dB greater than the 2.25 MHz and 4 dB greater than the single element, 1.5 MHz unit.

24. IGSCC specimens containing multiple cracks and provided by the General Electric (GE) Pipe Test Laboratory under the EPRI RP 892-1 program were used to further compare the dual element 1.5 MHz search unit to other search unit types. These pipe specimens were 300 series stainless steel, 101-mm diameter, Schedule 80, and contained nine circumferential welds. Some were half sections, and access for liquid penetrant inspection provided confirmation of cracks ranging from 2.5 mm to 152 mm in length. Fig. 4 shows the number of crack indications recorded with an amplitude exceeding the 50 percent DAC level detected by each search unit type. These data provided further evidence of the improved detection capability of the 1.5 MHz, dual element design. Continued refinements and development of this search unit design are planned during Phase 3 of the RP 892-1 program.

ONGOING RESEARCH AND DEVELOPMENT

Research and development activities continue in several different technical areas in an attempt to further improve the capability of ultrasonic inspection of stainless steel. Continued work in the analysis of frequency domain parameters and their influence on ultrasonic inspection is in progress at SwRI. Intensive research and development, primarily sponsored by EPRI, in the field of adaptive learning (ALN) or pattern recognition techniques may lead to a highly reliable method for identifying and characterizing defects as cracks or geometrical reflectors.

26. The significance of individual discriminants identified in the ALN studies is being applied to the development of new search units and instrumentation designed to enhance signal-to-noise ratios.

CONCLUSION

Ultrasonic inspection of stainless steel remains a problem worthy of continued research. However, advancements in technology over the past four years have markedly enhanced the viability of ultrasonics as an inspection method. Present research gives every indication that solutions to the remaining problems are imminent.

REFERENCES:

1. U.S. Atomic Energy Commission, Office of Inspection and Enforcement. "Through-Wall Cracks in Core Spray Piping of Dresden 2." Bulletin IE-75-01, January 30, 1975.

2. "Extra: The BWR Pipe-Crack Situation." Nucleonics Week, Special Issue, January 31, 1975.

3. H. H. Klepfer et al. Investigation of Cause of Cracking in Austenitic Stainless Steel Piping, Volume 2. Report NEDO-21000-1, 75NED35, Class 1, General Electric Co., July 1975.

4. H. H. Klepfer et al. Investigation of Cause of Cracking in Austenitic Stainless Steel Piping, Volume 1. Report NEDO-21000-1, 75NED35, Class 1, General Electric Co., July 1975.

5. Nuclear Regulatory Commission. "Stress Corrosion Cracks in Stagnant, Low Pressure Stainless Steel Piping Containing Boric Acid Solution at PWR's." Bulletin IE-76-06, November 26, 1976.

6. A Study of Inservice Ultrasonic Inspection Practice for BWR Piping Welds, EPRI NP-436-SR, Special Report, August 1977.

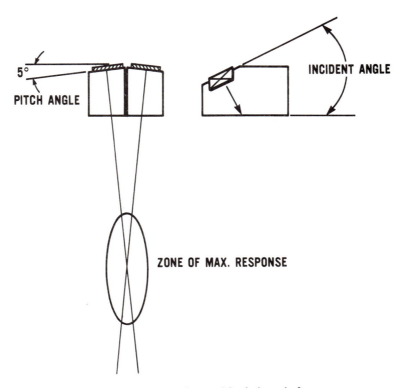

Fig. 1 45-Deg dual element transducer with pitch angle for maximum response zone

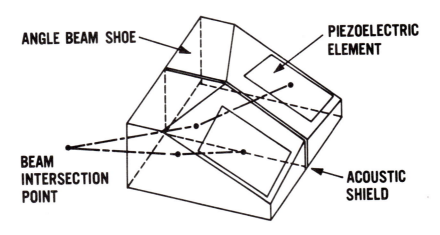

Fig. 2 Dual element transducer design for zone isolation

Fig. 3   Prototype dual element transducer

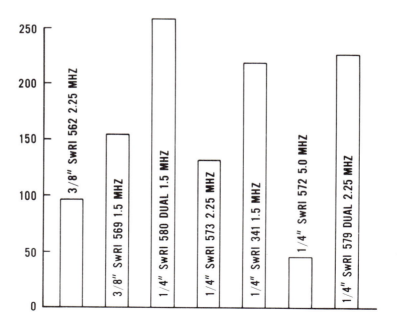

Fig. 4   Number of indications 50% DAC or greater

CONTRIBUTION TO THE IMPROVEMENT OF ULTRASONIC TESTING OF BIMETALLIC WELDS

J. P. LAUNAY, M. PETEUIL, and B. TRUMPFF

The MS of this paper received at the Institution on 22 September 1978 and accepted for publication on 14 November 1978

SYNOPSIS Ultrasonic testing of thick bimetallic welds between ferritic and austenitic steels is made difficult by the phenomenon of "scattering" which produces considerable background noise and spurious echoes. The behaviour of the ultrasonic waves (attenuation and velocity) was studied as a function of the metallurgical structure of the different parts of the weld. The results of this study enable us to determine the characteristics of special transducers which are tested on artificial reflectors (2 mm diameter holes drilled parallel to the weld axis).

1. INTRODUCTION

The circulation of heat exchange fluid between pressurised water reactor components takes place in stainless steel piping, while the components themselves are made of ferritic steel ; this choice of materials is made for both technical and economic reasons. The welding of joints between piping and components is a delicate operation due to the different characteristics of the materials concerned. The welding techniques at present in use are such that the metallurgical structures produced in the welded area, and their diversity, make conventional ultrasonic testing methods inefficient.

The present study was carried out on a flat test joint typical of a joint between ferritic and austenitic steels. The joint (Fig. 1) was made up of cast low-carbon steel on the ferritic steel side (A) ; the buttering (B) was of 24Cr-12Ni stainless steel for the first layer and 20Cr-10Ni stainless steel for the other layers, applied with a covered electrode ; the weld (C) was of 20Cr-10Ni-3Mo stainless steel, applied with a covered electrode ; the austenitic steel side (D) was of cast 20Cr-10Ni-3Mo stainless steel. Two flat blocks 50 mm wide and 60 mm thick were used ; nine holes 2 mm in diameter and of two different depths were drilled at various points in the weld parallel to the weld axis.

The logical solution adopted for the ultrasonic examination of such a weld would be to use a conventional contact tranducer, emitting shear waves with a frequency of 2MHz and an angle of incidence of 45°. This technique reveals only the holes in the ferritic steel (Fig. 2). In addition, considerable background noise and spurious echoes appear. In particular, an echo due to the corner of the ferritic steel-buttering joint is observed.

In view of these problems, a study was carried out to :

(1) link the metallurgical and ultrasonic properties of the weld,
(2) apply the results obtained to the design and choice of transducers best suited to examination of this weld.

2. ANALYTICAL STUDY OF THE METALLURGICAL AND ULTRASONIC PROPERTIES OF THE WELD

This study was inspired to a large extent by reference 1. Cylindrical test-pieces 50 mm high and 38 mm in diameter, with their axis parallel to the weld axis, were removed from the various areas of the weld (Fig. 3). In fact, the dimensions of these various areas did not allow all the samples to have the required dimensions. We had to reconstruct some parts of the joint with the same basic materials and under the same welding conditions and to transpose the test-pieces to position them correctly in the geometry of the joint (particularly in the case of the buttering). The study of ultrasonic propagation (attenuation and velocity) was carried out with the apparatus shown in Fig.4 : the cylindrical samples were placed between transmitting and receiving transducers immersed in water so that the beam passed through the diameter of the specimen. The ultrasonic propagation characteristics of the samples were studied by rotating them in the beam. We only studied longitudinal wave propagation because all the references emphasise that, for a given frequency, shear waves are attenuated much more than longitudinal waves (2), (3), (4), (5), (6).

The metallographic study was carried out on the circular faces of the cylinder in the plane of incidence of the ultrasonic beam to permit correlation with the results of the propagation study.

2.1 Metallographic study (Fig.5*)

2.1.1. Zone A : basic ferritic steel

Macrographic examination revealed a homogenous fine-grained structure (Nital etching). Microscopic examination revealed a very irregular ferrite and perlite structure in conjunction with Widmanstätten structure in places (Nital etching).

----

* In Fig. 5, "Det." means : detail

X-ray diffraction revealed anisotropy insufficiently pronounced for a significant pole figure to be plotted.

2.1.2. Zone B : manual buttering applied horizontally.

Macrographic examination revealed, on the one hand, the edges of the successive welding passes and on the other hand almost unidirectional crystallization perpendicular to the weld chamfer through-out the thickness of the buttering (boiling aqua regia etching).
Micrographic examination revealed a delta-ferrite lattice (20.7 % measured by quantometer) marking the limits of crystallization of the grains of the austenitic matrix ; these grains were columnar in shape and up to several centimetres long (boiling Murakami etching).
X-ray diffraction revealed distinct anisotropy: the dendrites are aligned in the [100] crystallographic axis of a cubic crystal.

2.1.3. Zone C : manual welding applied horizontally

Macrographic examination revealed crystallization varying about a principal axis coinciding with the axis of symmetry of the weld-chamfer, as cooling starts from the two symmetrical "V" chamfers : alignments began almost perpendicular to the edges of the chamfer and, as the distance from those edges increased, gradually rotated until parallel to the axis of symmetry of the chamfer (boiling aqua regia etching).
Micrographic examination revealed a very fine delta-ferrite lattice (15,4 % measured by quantometer) aligned in the same direction as the crystallization (boiling Murakami etching).
X-ray diffraction revealed mean anisotropy less marked than in the buttering : the grains were still aligned with the [100] crystallographic axis of a cubic crystal, parallel to the axis of symmetry of the chamfer.

2.1.4. Zone D : cast stainless steel

Macrographic examination revealed a coarse uniaxial structure with a thick ferrite lattice (16,5 % measured by quantometer) showing neither any particular alignment nor appreciable heterogeneity (boiling Murakami etching).
It was not possible to obtain useful pole figures by X-ray diffraction because of the presence of too large grains.

2.2. Study of behaviour of ultrasonic waves

For each cylindrical sample representing each area of the weld, behaviour of the ultrasonic waves was characterized by :
- overall attenuation in the case of the frequencies deemed most suitable (1, 2 and 3 MHz), as a function of the relative alignment of the ultrasonic beam axis and the structure. The transmitting and receiving transducers are conventional upright transducers (piezo, electric plate : 20 mm in diameter). For angles increasing in 2° steps, we plotted the amplification needed to produce an amplitude signal 6/7 the height of the screen, the reference amplification (0dB) being obtained with the isotropic ferrite steel sample (see para. 2.2.1).

- velocity as a function of the relative alignment of the ultrasonic beam axis and of the structure but only at the 2 MHz frequency. The measurements are accurate to within 1 %. The velocity and attenuation measurements were made from the same oscillogram.

2.2.1. Zone A : basic ferritic steel

Rotation of the ferritic sample produced negligable variations in attenuation and velocity, the latter being 5900m/s.

2.2.2. Zone B : buttering (Fig. 6a)

The variation of attenuation observed was cyclic to a very marked degree, having an approximate period of $\pi/2$.
Velocity measurements made through 180° exhibited four extremes $\pi/4$ apart ; the variation amounted to 800 m/s with a peak at 6090 m/s. The maximum velocity corresponds to the minimum attenuation.

2.2.3. Zone C : the weld (Fig. 6a)

The variation of attenuation was similarly cyclic with a period of $\pi/2$; it was, however, much less marked.
The velocity range also exhibited extremes $\pi/4$ apart. The variation amounted to 630 m/s with a peak at 5960 m/s.
The maximum velocity corresponds to the minimum attenuation.

2.2.4. Zone D : cast stainless steel (Fig. 6 a)

In the case of this material, which is the most attenuating, the results cover a considerable range without significant variations. The variations in velocity amounted to 500 m/s with a peak at 6130 m/s.

2.2.5. Behaviour of the ultrasonic waves at the interfaces of the various zones (Fig. 6 b)

Three cylindrical samples of the same dimensions as the previous ones were removed in such a way that the three interfaces - ferritic steel/buttering, buttering/weld and weld/cast stainless steel - formed diametrical planes of these samples. By studying the attenuation, the axis of least attenuation could be ascertained. All the axis of least attenuation in each zone and at each interface of the weld are summarized in Fig. 7.

2.3. Conclusions from the analytical study

-1- the axis of least attenuation are regularly observed at 45° to the grain alignment axis [100] , which agrees with references (1) and (7).
-2- frequency decrease is accompanied by attenuation decrease
-3- the more closely aligned the direction of the ultrasonic beam and the axis of least attenuation, the easier the crossing of the interface will be.
-4- a point-by-point study of the variation in attenuation as a function of the velocity of the ultrasonic waves did not reveal any simple relationship, except that the maximum velocity corresponded to the minimum attenuation.

3. OPTIMIZATION OF DETECTION OF ARTIFICAL FLAWS

The analytical study of the metallurgical and ultrasonic properties made it possible for us to determine the incidence and frequency of areas of less attenuation in each zone of the weld.

The results were applied to optimization of the detection of 2 mm diameter holes parallel to the weld-axis.

Detection of such a 2 mm diameter reflector is designed to take several factors into consideration : the crossing of the water/steel interface, the attenuation particular to each material for a given angle of incidence, the crossing of interfaces between steels, if it occurs, and variation in the path of the ultrasounds according to the angle of incidence (2), (8), (9).

The tests were carried out under water, varying the following parameters in turn : the angle of incidence into the steel, the frequency, the length of the ultrasonic pulse transmitted and the ultrasonic beam configuration of the transducer.

3.1. The effect of the angle of incidence into the steel

Using a conventional transducer (20 mm diameter) and varying its position relative to the normal to the surface of the blocks under examination, optimum detection of the five reflectors at a depth of 30 mm requires an angle of incidence in the steel between 42° and 45° if the examination is conducted from the ferritic steel side, or between 30° and 37° if the examination is conducted from the austenitic stainless steel side.

In the case of the reflectors at a depth of 57 mm, optimum detection requires an angle of incidence into the steel between 40° and 50° if the examination is conducted on the ferritic steel side (but hole no 9 in the basic austenitic stainless steel is undetected and the background noise is very considerable), whereas only hole no 6 in the ferritic steel is detected if the examination is conducted on the austenitic stainless steel side.

These results are in keeping with the axis of least attenuation revealed in the course of the analytical study of the metallurgical and ultrasonic properties of the weld.

3.2. The effect of the frequency

The best results were obtained at a test frequency of 1MHz.

3.3. The effect of the length of the ultrasonic pulse transmitted

This parameter is a function of the damping of the transducer, and of the electric pulse generated by the ultrasonic equipment. In common with reference (2) we ascertained that a well damped transducer gave better resolution and that there was an optimum electrical pulse length for each transducer used.

3.4. The effect of the ultrasonic beam configuration of the transducer

To reduce the level of background noise due to diffraction at grain interfaces the number of grains scanned must be reduced, and the ultrasonic energy concentrated in a limited volume of material (10). To this end we designed a probe focused by immersion and tested a twin probe for contact scanning.

4. DESIGN AND CHOICE OF SPECIAL PROBES

The above results make it possible to define the characteristics of suitable probes : they must emit longitudinal waves at 1 MHz at an angle of incidence of 45° into steel, they will be damped and will have a focal range which allows depths of 30 mm and 57 mm to be scanned.

4.1. Design of an immersion probe with a focusing lens

The piezo-electric element is of barium titanate (thickness : 3 mm, diameter 60 mm) and the damper is of resin highly impregnated with tungsten powder. To obtain a 45° angle of incidence into the steel, a 10° angle of incidence into the water is required. The distance from probe to test piece is 150 mm. The focus chosen is at a depth of 45 mm in the test piece and the length of the focus zone is 80 mm taking into account the incidence into the test piece. The acoustic lens is made of resin lightly impregnated with tungsten powder ; this lens is toric to correct the various aberrations to which the beam is subject, especially in crossing the dioptric plane constitued by the water/steel interface ; reference (11) enables us to calculate the two radii of curvature : $R_1 = 260$ mm and $R_2 = 160$ mm.

4.2. Choice of a twin probe for contact scanning

Out of all the probes made by Association Vinçotte (12) we selected probe no. V45 L SE 85 F-S PM5 1D2 - 20 x 15. The characteristics of which are the best suited to our study : a twin probe emitting longitudinal waves at 1 MHz at 45° angle of incidence into the steel ; the rectangular piezo-electric elements (each 20 mm x 15 mm) are side by side ; the transmitter is made of PZT and the receiver of 5 = lead metaniobate, their beams intersect at a depth of 85 mm in the steel ; damping is average.

4.3. Results (Fig. 8 and Fig. 9)

Although they need greater amplification, the echograms obtained with the holes at a depth of 30 mm are more legible if the examination is conducted from the austenitic steel side because the reflectors can be detected without interference from shear waves which are heavily attenuated. In contrast, in the case of the holes at a depth of 57 mm the best results were obtained from the ferritic steel side with the twin contact-scanning probe and from the austenitic side by immersion scanning with the focused probe.

These two special probes considerably improve ultrasonic testing of the weld by suppressing spurious echoes and reducing background noise. The focused probe for immersion scanning gives wide echoes and a spurious echo at the beginning of the echogram due to the water/steel interface, but on the other hand it makes detection of all the reflectors possible.

The resolution of the twin-contact probe is better than that of the focused probe.
However the twin-probe cannot detect hole no 9 at a depth of 57 mm in the basic austenitic steel.

5. CONCLUSIONS

-1- In this thick (60 mm) bimetallic weld, the least ultrasonic attenuation was observed at 45° to the axis of the grains wich were perpendicular to the preparation. The crystalline texture revealed the presence of a main crystallographic axis - the [100] axis - parallel to the axis of the grains alignments.
-2- The use of low frequencies (1 MHz) helps to reduce attenuation and the anisotropy of that attenuation.
-3- The use of special probes (a probe with a toric focusing lens for immersion scanning and a twin probe for contact scanning) considerably improves detection of artificial reflectors (2 mm diameter holes parallel to the weld axis) in the welded area by increasing the signal to noise ratio on the one hand and by suppressing spurious echoes on the other.

References

(1) Baikie B.L., Wagg A.R., Whittle M.J., Yapp D. "ultrasonic inspection of austenitic welds" ; J. Br. Nucl. Energy Soc., 1976, 15, July, N° 3, pp. 257-261.

(2) Pelseneer J.P., Louis G. "ultrasonic Testing of Austenitic Steel Casting and Welds" ; Brit. J. of N.D.T., 16, July 1974, pp. 107-113.

(3) Richter E.V. "Zur ultraschallprüfung austenitischer Schweissverbindungen (Teil I-II)" ; Die Technik 23. Jg. , Heft 10, Oktober 1968, S. 610-619 und 692-696.

(4) Hughes E.T. "The ultrasonic inspection of cast austenitic stainless steel for primary circuit components in Westinghouse PWR s' ; International Nuclear Industries Fair, Basel, October 7-11, 1975, colloquium A 2/7.

(5) Juva A. dip. eng., Technical Research Centre of Finland Reactor Materials Research, SF-02150 Espoo 15, Finland ; "Ultrasonic inspection of austenitic welds in the primary circuit of the Lovusa I nuclear power plant".

(6) Caussin P. Association Vinçotte, Département études, B-1640 Rhode-Saint-Genèse, Belgique ; "Ultrasonic testing of austenitic stainless steel structures, state of the art and progress report ; September 13, 1976.

(7) Juva A. and Haarvisto M. "On the effects of microstructure on the attenuation of ultrasonic waves in austenitic stainless steels" ; Brit. J. of N.D.T., november 1977, pp. 293-297.

(8) Yaneyama H., Shibata S., Kishigami M. "Ultrasonic testing of austenitic stainless welds. False indications and the cause of their occurence.";) N.D.T. International, February 1978, pp.3-8.

(9) Lotchev B. and Pawloski E. "Effect of material acoustic anisotropy on the shape of ultrasonic wave beam" ; 8th World conference of N.D.T., September 1976, 3F 10.

(10) Ermolov I.N. and Pilin B.P. "Ultrasonic inspection of materials with coarse grain anisotropic structure" ; N.D.T. International, December 1976, pp. 275-280.

(11) Flambard C. et Lambert A. "1'amélioration du contrôle des défauts par ultrasons, par la maîtrise et l'utilisation de la focalisation des faisceaux acoustiques." 8th World conference of N.D.T., September 1976, 3 J 12.

(12) Association Vinçotte, B-1640 Rhode-Saint-Genèse, Belgique.

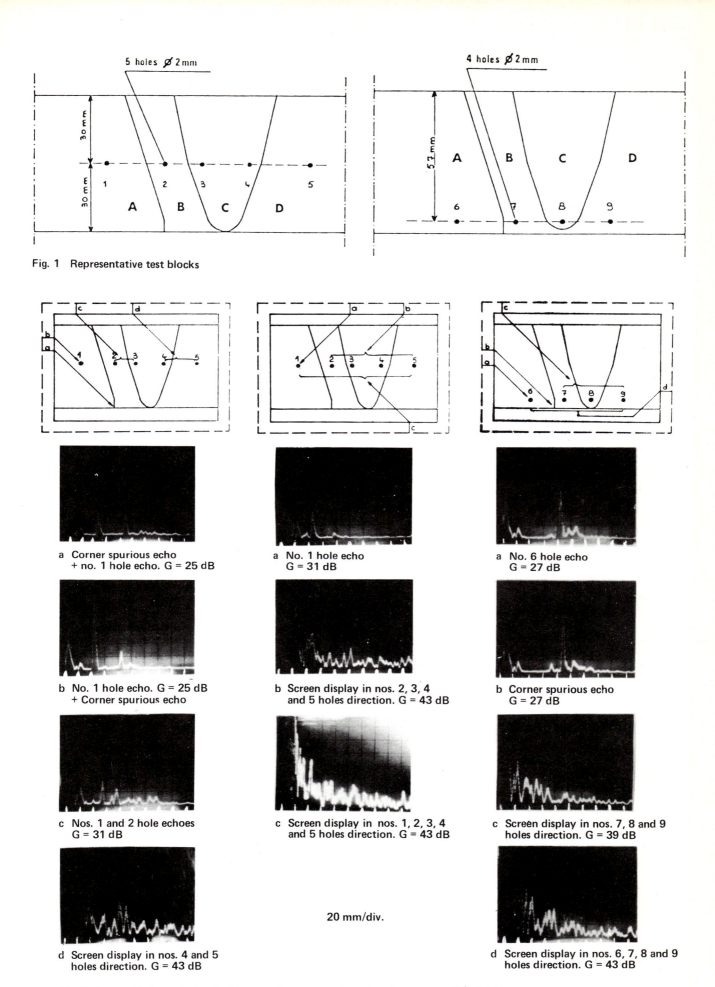

Fig. 1   Representative test blocks

a   Corner spurious echo
    + no. 1 hole echo. G = 25 dB

a   No. 1 hole echo
    G = 31 dB

a   No. 6 hole echo
    G = 27 dB

b   No. 1 hole echo. G = 25 dB
    + Corner spurious echo

b   Screen display in nos. 2, 3, 4
    and 5 holes direction. G = 43 dB

b   Corner spurious echo
    G = 27 dB

c   Nos. 1 and 2 hole echoes
    G = 31 dB

c   Screen display in nos. 1, 2, 3, 4
    and 5 holes direction. G = 43 dB

c   Screen display in nos. 7, 8 and 9
    holes direction. G = 39 dB

20 mm/div.

d   Screen display in nos. 4 and 5
    holes direction. G = 43 dB

d   Screen display in nos. 6, 7, 8 and 9
    holes direction. G = 43 dB

Fig. 2   Screen displays obtained with a usual contact angle probe, shear waves, 45°, 2 MHz

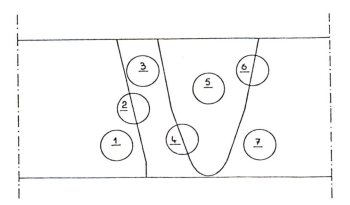

Fig. 3 Cylindrical test-pieces sampling

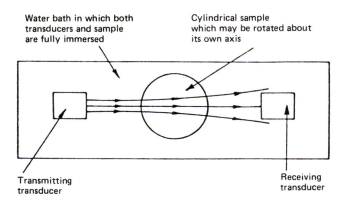

Fig. 4 Apparatus used for the study of the dependence of ultrasonic attenuation on grain orientation (Ref. 1)

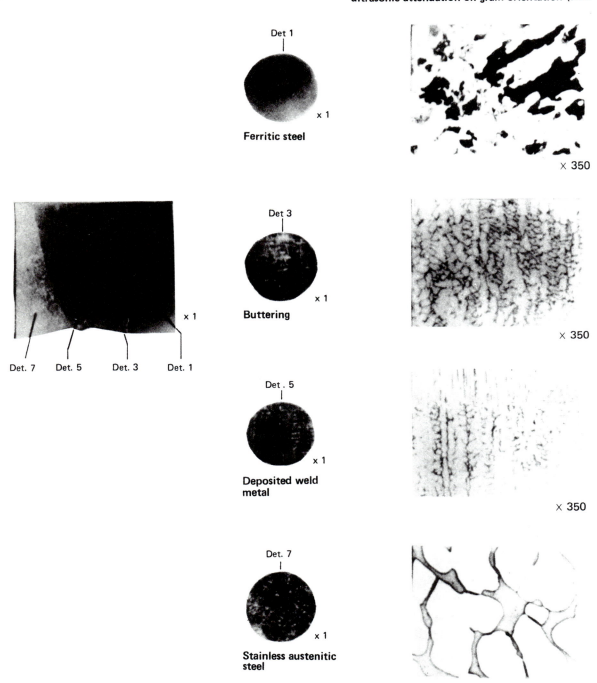

Fig. 5 Photomicrographs

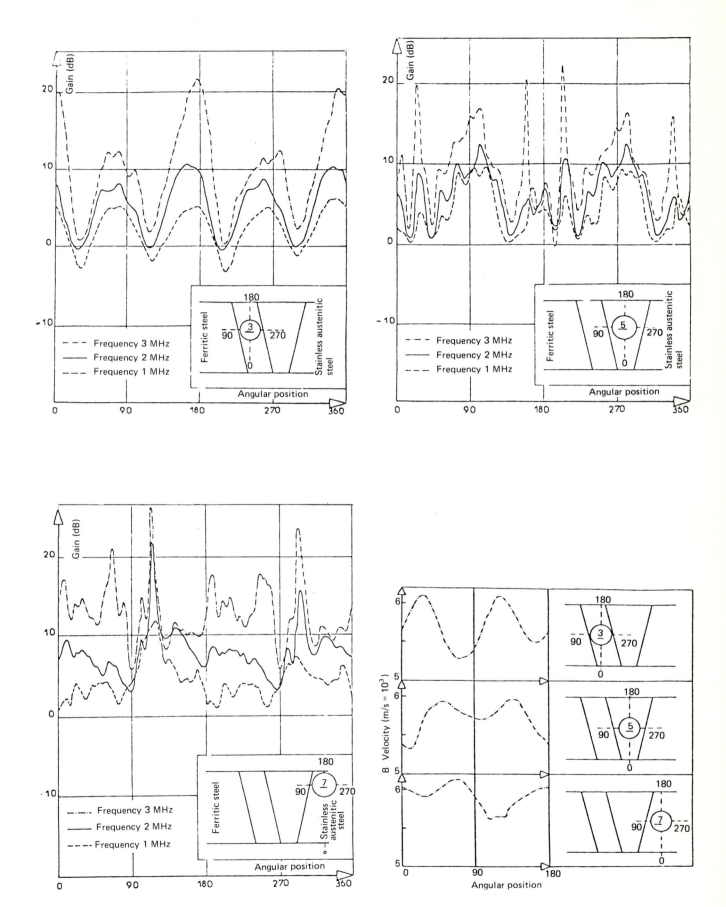

Fig. 6a Ultrasonic propagation in the different zones of the weld joint

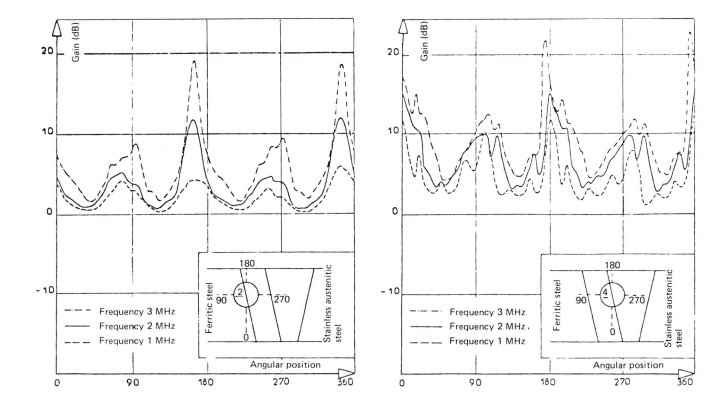

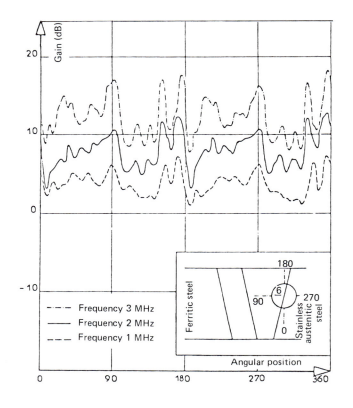

Fig. 6b Ultrasonic propagation at the different interfaces

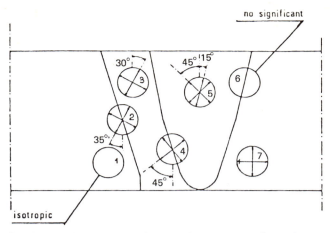

Fig. 7 Angular positions of the minimum attenuation axis

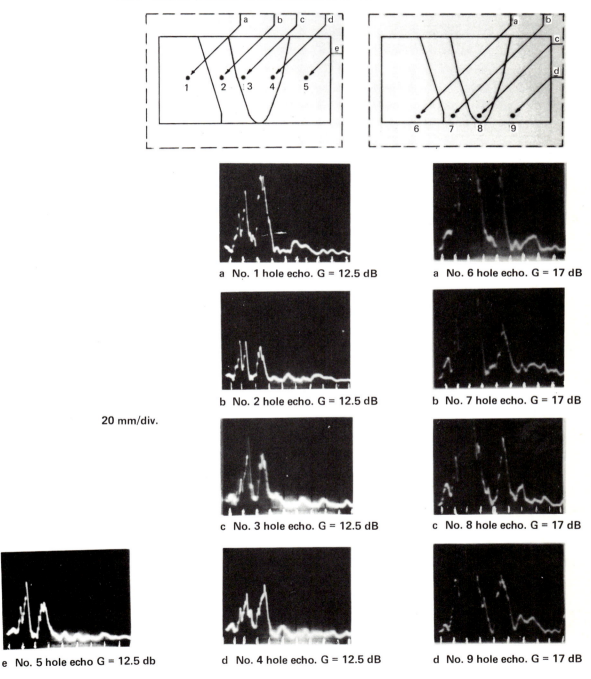

20 mm/div.

a No. 1 hole echo. G = 12.5 dB

b No. 2 hole echo. G = 12.5 dB

c No. 3 hole echo. G = 12.5 dB

e No. 5 hole echo G = 12.5 db

d No. 4 hole echo. G = 12.5 dB

a No. 6 hole echo. G = 17 dB

b No. 7 hole echo. G = 17 dB

c No. 8 hole echo. G = 17 dB

d No. 9 hole echo. G = 17 dB

Fig. 8 Screen displays obtained with the focused probe in immersion technique

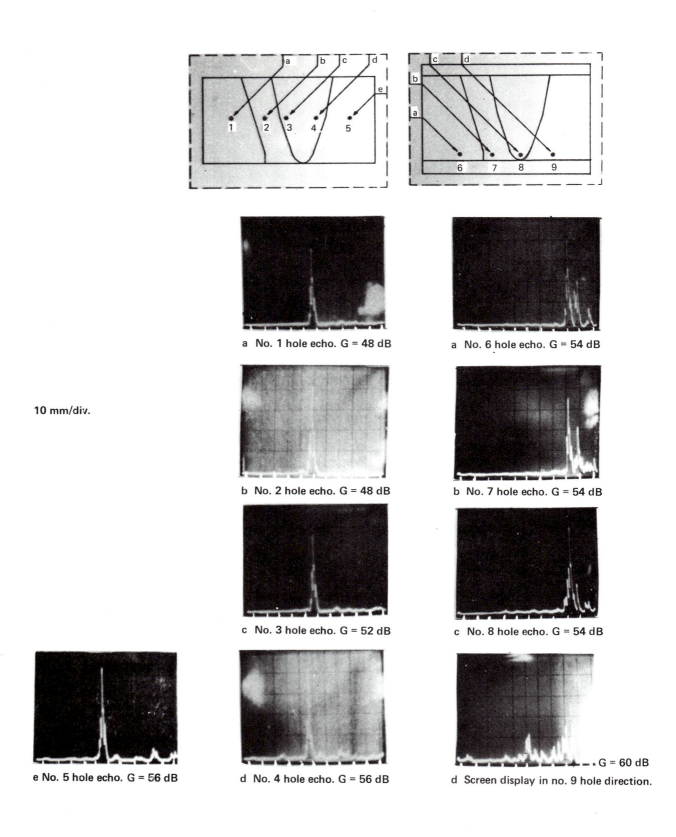

10 mm/div.

a No. 1 hole echo. G = 48 dB

a No. 6 hole echo. G = 54 dB

b No. 2 hole echo. G = 48 dB

b No. 7 hole echo. G = 54 dB

c No. 3 hole echo. G = 52 dB

c No. 8 hole echo. G = 54 dB

e No. 5 hole echo. G = 56 dB

d No. 4 hole echo. G = 56 dB

G = 60 dB

d Screen display in no. 9 hole direction.

Fig. 9 Screen displays obtained with the twin probe in contact scanning

C52/79

DEVELOPMENT OF AUTOMATIC ISI TOOL FOR DISSIMILAR METAL WELDS IN NOZZLES OF STEAM DRUMS

S. HONDA
Power Reactor and Nuclear Fuel Development Corporation
and
T. ANDO, A. KOGA, and T. UCHIHARA
Mitsubishi Heavy Industries Ltd.

The MS of this paper was received at the Institution on 31 October 1978 and accepted for publication on 20 December 1978

SYNOPSIS   This paper describes developmental works on an automatic ultrasonic flaw detection systems for dissimilar metal welds in nozzles of steam drums. In order to evaluate the system's effectiveness of flaw detectability under on-site conditions function tests were conducted using a mock-up into which were drilled several holes and notches. The test showed that this tool reduced assembly and removal time to only 5 minutes which reduces the radiation exposure served for personnel. Use of incidence ultrasonic beams from the tapered portion of the nozzle made weld metal flaw detection possible, and all flaws were detected in the nozzle mock-up. An off-line data processing program was developed which enables us to obtain the advanced data sheets using a printer and a plotter, all data being gathered from the ultrasonic multiple probes.

INTRODUCTION

1.   This work describes ultrasonic flaw detection equipment for use during Inservice Inspection (ISI), as developed by co-operation between The Power Reactor and Nuclear Fuel Development Corporation and Mitsubishi Heavy Industries, Ltd. This equipment was developed for inspecting dissimilar metal welds in the main steam nozzles (6B nozzles) of the steam drum of the "FUGEN", a Japanese heavy water proto-type reactor.

2.   The inspection of dissimilar metal welds as specified in ASME Section XI has been recognized as one of the most important ISI items. But, in the case of the dissimilar metal welds formed in the 6B nozzles, and owing to the complexity of the weld structure as shown in Fig.1 plus the difficulty of complete coverage of the ultrasonic probes, the small nozzle radius and insufficiently finished weld crown and due to the difficult working situation, ultrasonic inspection of these parts has been considered almost impossible to effectively carry out.

3.   This equipment was developed primarily to achieve the followings:

1)   To make the detection of flaws in the dissimilar metal weld in 6B nozzles possible and adequate.

2)   To reduce assembly and removal time of the equipment to the lesser figure of several minutes so as to minimize radiation hazards for the personnel involved in the examination.

3)   To reduce inspection and flaw analysis time by compiling information from the multiple ultrasonic probes and making a single record such as that of the c-scanning record.

PRELIMINARY FLAW DETECTION TEST

4.   Because of the state of the weld surface of the dissimilar metal weld crown of the 6B nozzle, which is as shown in Fig.2, it was considered insufficient to perform ultrasonic flaw detection from both sides of the top of the weld only. There-fore, in order to investigate the possibility of detecting allowable planar indications being as specified in ASME Section XI and basic calibration reflectors, 2.4mmϕ drilled holes in the weld metal, manual, preliminary flaw detection was performed by the following means:

1)   By siting the ultrasonic probes on the tapered section of the safe end where the surface state is more ideal for carrying out the test.

2)   Using a contact method where the probe is submerged in a water tank which holds the water couplant.

3)   Concentrating the effective ultrasonic energy by making use of twin probes.

4)   Multiple probes that make it possible to examine specific parts by specific probes (Fig.3) granting complete cover-age of the entire weld.

5.   The preliminary tests were performed according to the test parameters shown in Table 1, using various test specimens into which artificial flaws had been introduced.

6.   The result of the preliminary tests are summarized in Table 2. And, this summary shows successful detection of all flaws. For this successful result, information was utilized from all the probes.

As for notches, generally speaking, in the case of longitudinal waves, the reflected ultrasonic echo from notch-like defects is considerably weaker, as compared to shear waves. Nevertheless this result shows us that all notches (two in the weld metal, one in the heat affected zone) were detected, even though we used refracted longitudinal waves. This phenomenon has not been yet adequately explained.

7. From the fact that the test specimens were made by using the same fabrication procedures as that for the actual 6B nozzles, that is, same nozzle radius, same surface finishing (by hand grinding) and yet could detect all flaws in the test specimens, it was concluded that flaw detection for the 6B nozzle is possible and adequate and we started fabrication of the equipment.

PERFORMANCE OF REMOTE CONTROLLED MANIPULATOR

8. In order to examine the performance of this recently developed remote controlled manipulator, shown in Fig.4, and specifications in Table 3, the following tests were carried out.

Assembly and removal time for the manipulator

9. In order to check assembly and removal of the manipulator to and from the 6B nozzle, the assembly and removal time of the manipulator was measured in 4 operations using the 6B nozzle mock up. The procedure for the assembly of the manipulator and average values of assembly and removal time are shown in Fig.5 and in Fig.6 respectively. Test results showed that the most time consuming work is the fitting and taking off of the spacer from the 6B nozzle. Accordingly, if the spacers are fixed to the nozzles permanently, then the total time for assembly and removal will be remarkably reduced down to about 5 minutes.

Positioning accuracy of the manipulator

10. In ISI of the nuclear power plant, high positioning accuracy of the ultrasonic probes is essential. Therefore, as the first step, we performed a test on the positioning accuracy for the manipulator itself. The test was done without mounting ultrasonic probes. From this test, the positioning accuracy of the manipulator was determined to be within ±0.5mm in all cases provided a spacer was not removed. The positioning accuracy was determined by taking into account both the circumferential and axial direction in relation to the nozzle. This accuracy can be considered acceptably good.

Coupling stability of ultrasonic probes

11. In automatic ultrasonic flaw detection, technically the most important factor is the coupling between ultrasonic probes and the test surface. Therefore, by making the ultrasonic beam penetrate from the tapered portion of the nozzle safe-end, echo height as reflected from the inner surface of the safe-end during rotation of the manipulator around 6B nozzle mock-up was measured, and then coupling condition was evaluated. Also, the rotational speed effect on the coupling condition of the ultrasonic probes was checked using the same procedure. As shown in Table 4, fluctuation of the monitored echo height is within approximately 4 dB irrespective of rotation direction or rotation speed. The stability of the coupling in this case is considered acceptably good.

Because of the rough surface and narrow area available for setting the ultrasonic probes and the overlay claddidng on the inner surface, the test could not be performed from the top portion of the nozzle. Owing to the roughness of the top portion of the nozzle, the coupling condition of the probe at the top portion would be much worse than that at the tapered portion, and the result would be inadequate.

Reproducibility of flaw detection data

12. Another very important factor for carrying out ISI, which will be performed regularly if infrequently to the end of the plant life (30 – 40 years), is the reproducibility of flaw detection data. Therefore, we performed data reproducibility tests for the manipulator from tapered portion using an angle beam probe and a 6B nozzle mock-up into which artificial flaws were introduced.

The test steps are as follows:

1st step ; original scanning

2nd step ; rescanning under the same conditions

3rd step ; rescanning after removal of the manipulator (the spacer not being removed from the mock-up) and reassembly.

4th step ; rescanning after complete removal of the manipulator parts including the spacer and reassembly.

Fig.7 shows the test result. This test result indicates excellent reproducibility of echo height and echo position irrespective of removal of the manipulator from the 6B nozzle mock-up.

13. From the results of the various tests above, ease of handling of the manipulator and good reproducibility of the data were seen, and the performance of this manipulator can be stated as good. Therefore, even in a radioactive environment such as the area near the steam drum, this manipulator would be safe in application.

DATA PROCESSING

14. As the present flaw detection equipment uses multiple ultrasonic probes, if an information record of flaw detection was taken from each probe, the final data would be enormous. Therefore, a program was developed which enabled the compilation of information from all of the probes and this was represented on a recording form such as the c-scanning method on an X–Y plotter. Fig.8 shows the flow chart of the program. Fig.9 shows the presently used data processing system for this equipment.

GENERAL FUNCTION TEST

15. To confirm the effectiveness of this flaw detection equipment which is to be used for dissimilar metal welds in 6B nozzle, general function tests were performed using a 6B nozzle mock-up and the presently developed test method and equipment (Fig.10). EDM notches were made in the test weld and 4 machined drilled holes were made as shown in Fig.11.

16. For data processing, a newly developed program was adopted, but as a reference, the previously developed processing method[1] was also used. Test parameters and the probe types used are the same as shown in Table 1 but because of

attenuation, basic calibration reflectors in the base metal were not adequate, we used standard reflectors which were machined in the weld test specimen. An example of the output from the general function test on the dot printer is shown in Fig.12, and an example of the c-scan representation on an X–Y plotter is shown in Fig.13. Also Fig.14 shows a c-scan representation of the previously developed single probe scan method.

17. By conducting the general function tests, the following results were obtained:

1) All flaws in the mock-up were detected, this includes the all important EDM notch where the depth is equal to the allowable flaw size for austenitic welds as specified in ASME Section XI.

2) The flaw detection method, together with the newly developed and light weight manipulator and the data processing program all gave a good performance.

PROBLEMS REMAINING

18.

1) It proved most difficult to get good performance ultrasonic twin probes that met to a sufficient degree the specification requirements. A high performance probe is a key factor in this flaw detection and it remains a desirable product of future research.

2) The ASME specified allowable planar flaws (in this case, EDM notch) were detected with the stated accuracy but the behavior of the ultrasonic beams under conditions of stress corrosion cracks and fine fatigue cracks could be totally different from that of the EDM notch. Therefore, before putting into use this flaw detection equipment, it will be necessary to clarify flaw detectability in relation to natural cracks in the dissimilar metal weld.

3) In order to extend this equipment to cover the other types of nozzle, especially the downward facing nozzle, a means of retaining the couplant must be devised such as an enclosed type and bubble formation in the couplant must be prevented also.

CONCLUSION

19.

1) Through the preliminary tests, and the general function test, it was determined that the flaw sizes specified in ASME Section XI can be detected.

2) Assembly and removal time for the remote controlled manipulator from the 6B nozzle could be greatly reduced if a spacer is permanently fitted. The fitting of a spacer will also contribute to the reduction of radiation dose to the personnel involved in the examination.

3) The data processisng program was developed which enabled the compilation of all the information from the multiple probes and which gave a record on an X–Y plotter and produced an output on a dot printer. The effectiveness of the program was confirmed through the general function test.

4) As the result of the general function test, showing the equipment's worth and effectiveness, it seems assured that the equipment will be used for ISI work of the Japanese heavy water prototype reactor in the near future.

REFERENCE

1. M. AKEBI et al. "Development of automatic tool for the "FUGEN prototype advanced thermal reactor" THE IME, Periodic Inspection of Pressurized Components. 1976.

Table 1   Test parameters for preliminary test

Flaw detector	UM 731 (Tokyo Keiki)
Twin type Probes	Various kind (R.T.D)
Refracted angle	$45° \sim 70°$ and $0°$ (straight beam)
Frequency	4 MHz
Coaxial cable	SEKL-2 (Krautkrämer)
Couplant	Water
Standard test block	Basic calibrator reflectors specified in ASME Section XI (Base metal)

Table 2 Detected flaws in the preliminary test

Probes \ Defects*	Tapered portion									Top portion — From safe end side									Top portion — From nozzle side**								
	[A]	[B]	[C]	[D]	[E]	[F]	[G]	[H]	[I]	[A]	[B]	[C]	[D]	[E]	[F]	[G]	[H]	[I]	[A]	[B]	[C]	[D]	[E]	[F]	[G]	[H]	[I]
TR1 (straight beam)										○	○	○		○	○												
TR2 (45°)				○		○				○	○				○									○			○
TR3 (50°)				○		○																		○			○
TR4 (57°)																											
TR5 (59°)							○			○	○				○				○								
TR6 (60°)																											
TR7 (69°)																											
TR8 (63°)	○				○			○																			
TR9 (63°)																											
TR10 (70°)																											

Note; * These are the position of defects shown in Fig. 3. ([A] ~ [F]; 2.4mmΦ drilled holes, [G]~[I]; 2.1mm depth notches)
** Flaw detection from this side is almost impossible under on-site conditions.

Table 3 Equipment specifications

Items	Specifications
(1) Manipulator components	① Spacer ② Rail ③ Driver mechanism ④ Slider " ⑤ Detector " ⑥ Probe holder (max. 10) ⑦ Probe holder assembly ⑧ Control panel
(2) Operational parameters	① Operational range ; 400° ② Speed ; 4 step change
(a) rotational axis theta	o Standard incremental rate o Reduced " " No.1 o Reduced " " No.2 o Reduced " " No.3 ③ Position detection o Digital (LED) display of positions from encoder o Resolution to 1°
(b) slide axis X	① Operational range ; 100 mm coverage ② Speed ; 4 step change

Table 4 Coupling stability test

	Standard incremental rate		Reduced incremental rate (No.1)		Reduced incremental rate (No.2)		Reduced incremental rate (No.3)	
Tangential speed (mm/sec)	19		9.6		3.8		0.9	
Direction of scanning	CCW (θ)	CW ($-\theta$)	CCW (θ)	CW ($-\theta$)	CCW (θ)	CW ($-\theta$)	CCW (θ)	CW ($-\theta$)
Echo height– fluctuation range (dB)	4.2	4.0	4.3	4.5	4.2	3.8	4.2	4.3

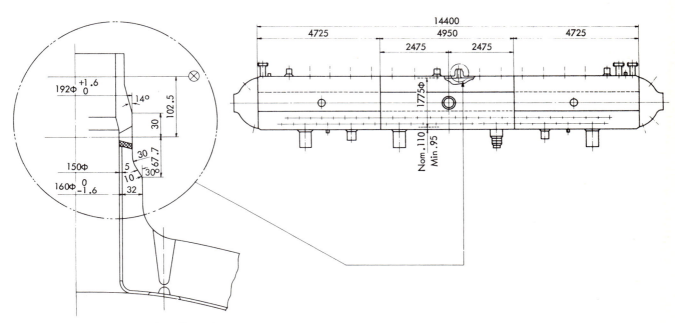

Fig. 1 Steam drum and weld configuration

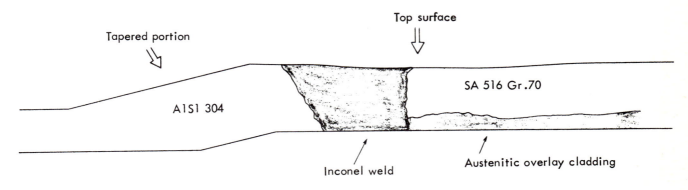

Fig. 2 Surface finishing of test specimen

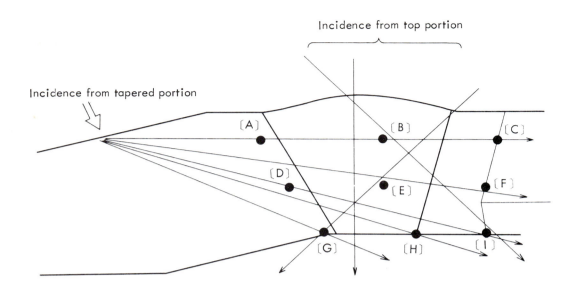

Fig. 3 Multiple probes method aimed at specific parts

Fig. 4 Remote controlled manipulator

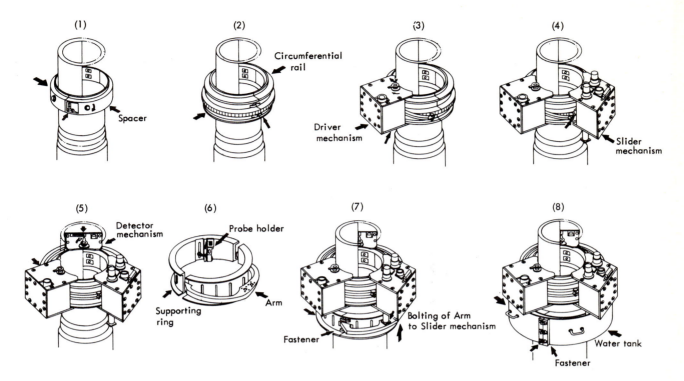

Fig. 5 The procedure for assembly of the manipulator

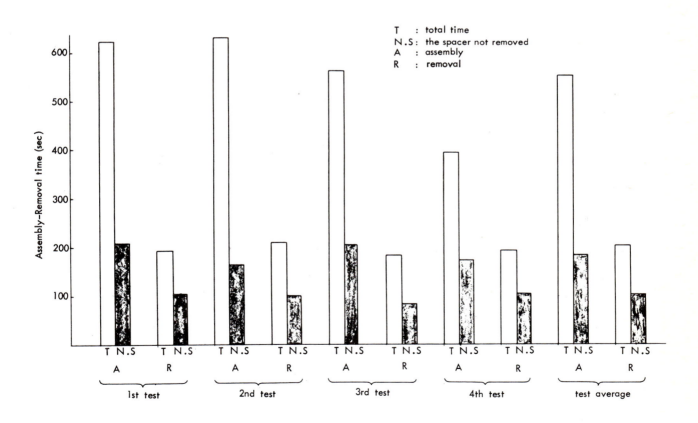

Fig. 6 Assembly-removal time tests

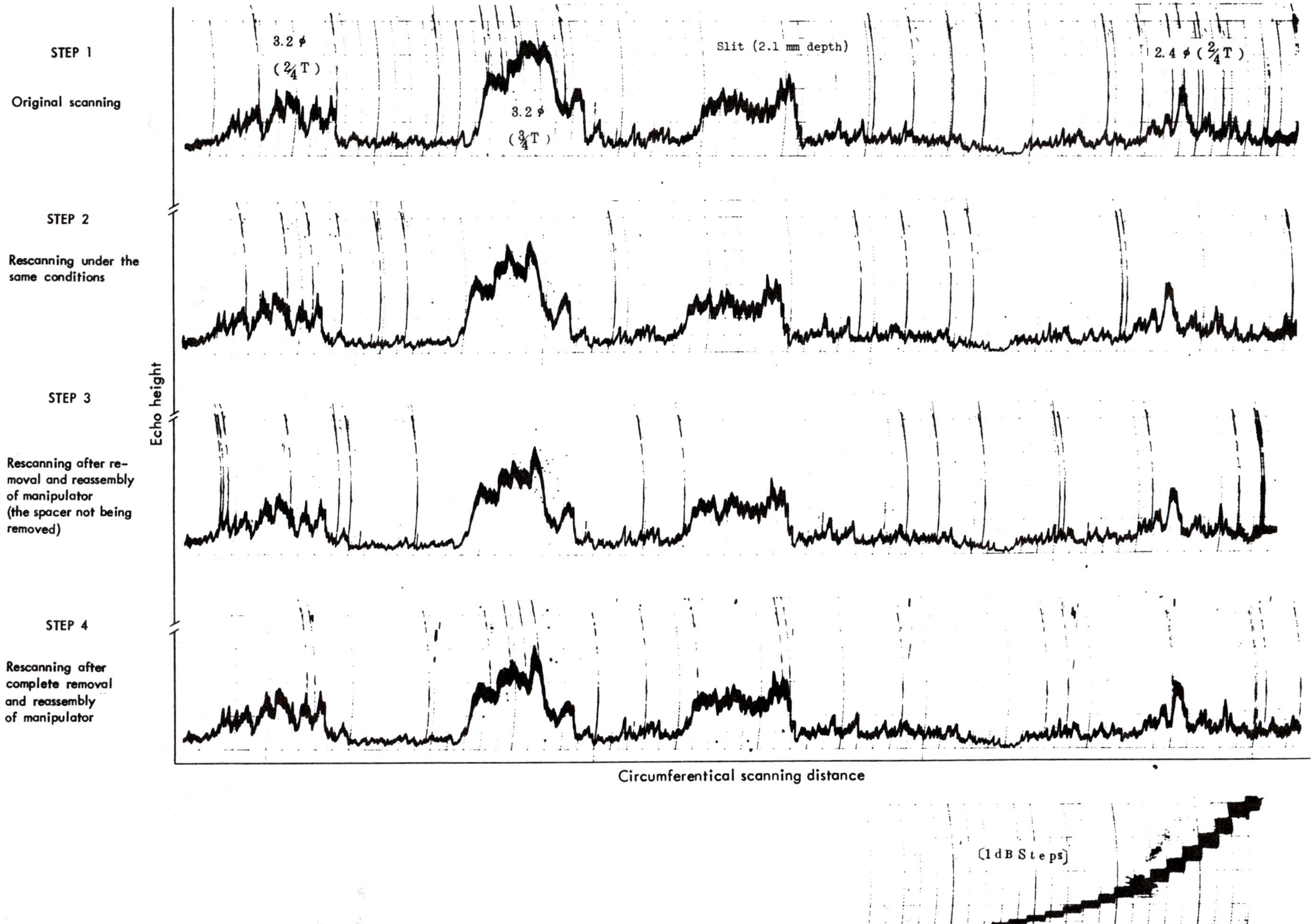
STEP 1
Original scanning
STEP 2
Rescanning under the same conditions
STEP 3
Rescanning after removal and reassembly of manipulator (the spacer not being removed)
STEP 4
Rescanning after complete removal and reassembly of manipulator
Echo height
3.2 φ
(2/4 T)
3.2 φ
(3/4 T)
Slit (2.1 mm depth)
2.4 φ (2/4 T)
Circumferentical scanning distance
[1 dB Steps]

Fig. 7 Results for reproducibility test of equipment (TR5)

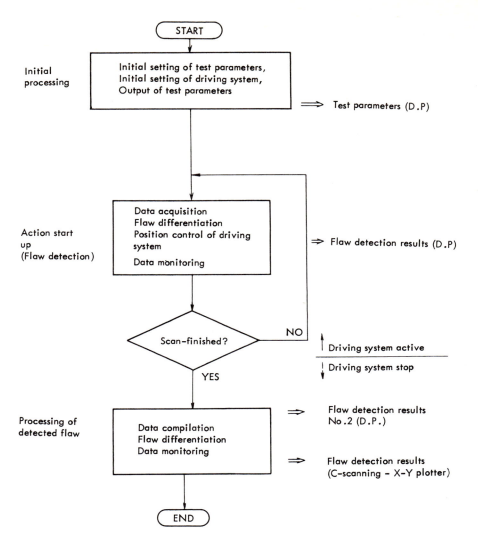

Fig. 8   Computer program flow chart

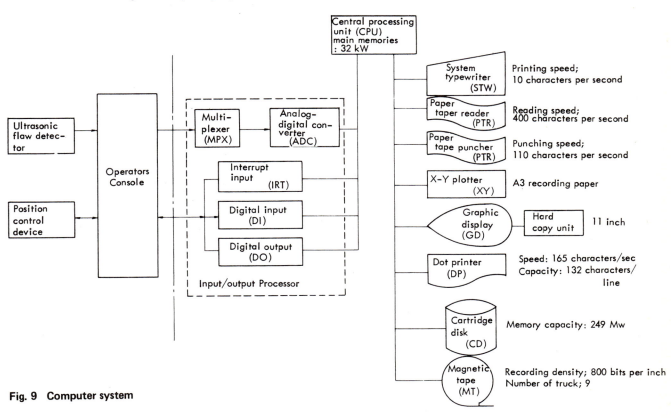

Fig. 9   Computer system

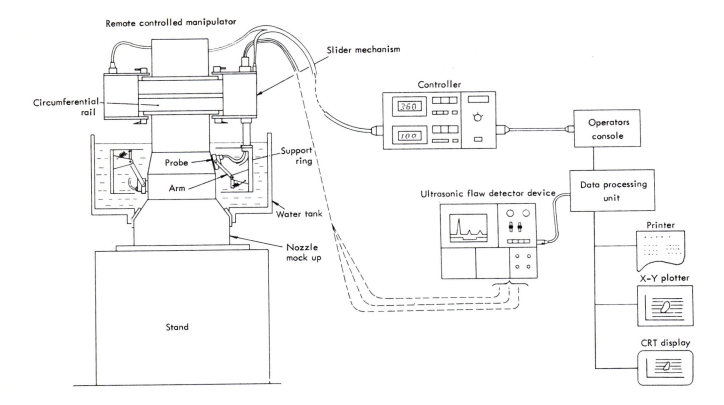

Fig. 10 Schematic diagram of general function test

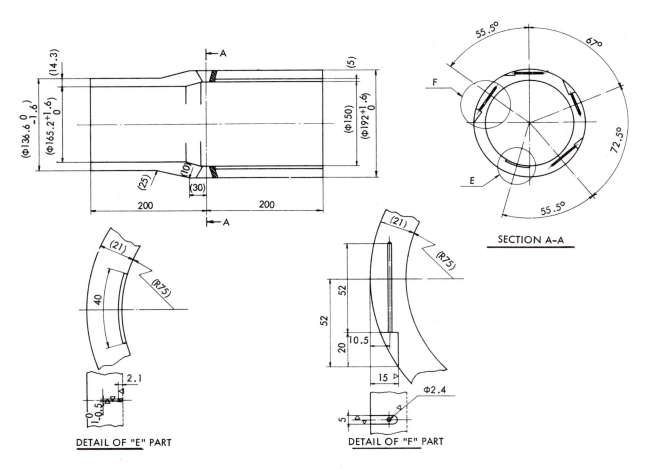

Fig. 11 Artificial flaws in mock-up

244

INSPECTION LIST -- FLAW INFORMATION

X	LEVEL 1 CH.NO.	START	END	LENGTH	LEVEL 2 CH.NO.	START	END	LENGTH	LEVEL 3 CH.NO.	START	END	LENGTH	1	2	3	4	5	6	7	8	9	10	11	12
	2	281	281	1												+								
					2	352	354	3								+								
	2	348	355	8												+								
	2	359	359	1												+								
	5	18	18	1														+						
	5	44	47	4														+						
	5	49	51	3														+						
	5	53	53	1														+						
	5	105	105	1														+						
					5	105	105	1										+						
									5	105	105	1						+						
	5	196	196	1														+						
	5	210	211	2														+						
					5	216	218	3										+						
									5	224	225	2						+						
									5	227	231	5						+						
					5	221	236	16										+						
					5	238	241	4										+						
	5	214	244	31														+						
	5	284	286	3														+						
	5	288	301	14														+						
	5	304	308	5														+						
	5	310	311	2														+						
	5	354	354	1														+						
					5	358	358	1										+						
	5	358	359	2														+						
	4	223	223	1																+				
					4	223	223	1												+				
	4	310	310	1																+				
30	2	256	256	1																		+		
					2	256	256	1														+		

Fig. 12 Output of flaw detection from dot printer (+ 5 mm position)

INSPECTION RESULT

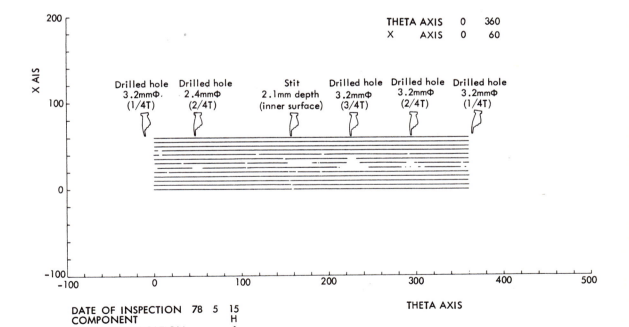

THETA AXIS   0   360
X   AXIS   0   60

Drilled hole 3.2mmΦ (1/4T)   Drilled hole 2.4mmΦ (2/4T)   Stit 2.1mm depth (inner surface)   Drilled hole 3.2mmΦ (3/4T)   Drilled hole 3.2mmΦ (2/4T)   Drilled hole 3.2mmΦ (1/4T)

DATE OF INSPECTION   78 5 15
COMPONENT   H
WELD IDENTIFICATION   1
C-SCAN   25.0

THETA AXIS

Fig. 13 C-scan representation (3 5 mm position)

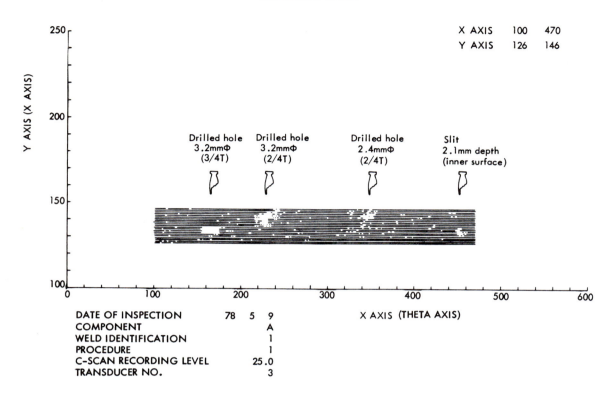

Fig. 14 C-scan representation of step type flaw detection (TR5 . . . . . angle beam method from tapered part)

C34/79

A PRACTICAL COMPREHENSIVE SURVEY ON INSERVICE EXPERIENCE WITH IMPROVED TECHNIQUES

G. ENGL, Dipl Ing and W. MEIER, Dipl Ing
Kraftwerk Union AG, GFR

The MS of this paper was received at the Institution on 20 November 1978 and accepted for publication on 20 February 1979

SYNOPSIS:
The basic philosophy of using searching and - if doubt is left - analyzing techniques demands a high signal significance of the searching techniques. So the improvement of searching techniques has two aims:
A maximum defect detectability and a high ability to classify into relevant and irrelevant or spurious indications. Major contributions for the improvement of the inspection techniques come from various fields. Three of them are shown in this paper:
Theoretical echodynamic models, data acquisition and processing and alternative techniques. As the shown examples and results demonstrate, all three possibilities are of big importance for an optimization of inservice inspections.

INTRODUCTION

With regard to inservice inspection the reactor safety requirements mean a reliable detectability of all relevant - and even hypothetical defects - with a large security factor to the critical size - and the feasibility of a reliable evaluation of the defects found.

Thus for a practicable and realistic ultrasonic inspection concept the following criteria are of most importance:

- high sensitivity and signal to noise ratio

- minimization of sensitivity variations (e.g.V-path-echo variations)

- optimization of detectability for variations of the orientation of planar defects

- minimization of the influence of the cladding and other effects of the structure

- adjustment to the geometrical conditions

- sensitivity variations due to the scanning pattern

- Data processing algorithm for a quick and reliable acquisition and evaluation of the indications

The improvement of inspection techniques complying with these criteria consequently leads to a subdivision of the inspection in two steps (Fig. 1):

The use of searching techniques with a maximum defect detectability is combined with analyzing techniques with a higher potential to determine the actual size of a defect and eventually characterize it into voluminous or planar and if so its orientation (ref.1).

Of most importance within this basic philosophy is the interface of classification based on the informations of the searching techniques.

So the aim of improvement of the inspection system is to gain maximum defect detectability with parallel increasing the potential of classification in relevant and spurious indications (e.g. coherent background).

In this respect the combination of inspection techniques (tandem and single probe) and of angles of incidence are of importance as well as the probe parameters especially frequency, frequency distribution (damping), crystal sizes and transmitter-receiver techniques for special application.

Fig.2 shows a typical probe system for the inspection of the cylindrical part of the reactor pressure vessel. In principle this system consists of a combination of tandem technique with 0° and 45°-single probe and - especially for the surface-near depth zone - the 70° longitudinal transmitter-receiver technique. In addition the system contains sensitivity control functions such as V-path-echo and backwall-echo.

On the basis of extensive investigations and measurements on original components and 1:1-scale large testblocks the parameters for the inspection of components with a wall thickness exceeding 100 mm have been determined als follows:

Frequency: 1 MHz for transversal waves
2 MHz for longitudinal waves

Dimension of the crystals:
in the plane of incidence 20-25 mm, lateral dimension dependent on the depth zone from 8 to 25 mm.

Height of the depth zones:
approximately 30-40 mm dependent on the wall thickness, surface-near depth zone up to 23 mm.

Scanning distance 20 mm.

Theorectical model for systems improvement

For further development and for the adaptation of the system to a special inspection situation (e.g. caused by curvature, wall thickness, most probable defect orientations etc.) theoretical models can be used and help to reduce further experimental investigations on a large scale. Here especially the theoretical model of echo dynamics has to be mentioned (ref. 2, 3, 4). Echo dynamics means the change in the amplitude of a reflector as a function of the probe movement.

This theoretical model introduces sound energy distributions and directivity pattern as factors of transmission. Thus a far field situation for the sound field of the probes and for the reflector is presupposed. The influence of the near field is taken into account (with approximation) by a varied distance dependence and by variation of the divergence of the probe sound field. The model is based upon the integration over the defect area assuming that the reflectors show a partial excitation of their area elements and that the directivity pattern is described by a bessel function of first order and by a variety of angles.

The following three examples show the possibilities of the use of the theoretical deflection model and the comparison with the experimental results:

1) Subdivision of the wall thickness into depth zones for the tandem technique:

The determination of depth zones has to take into account that the sensitivity difference within the depth zone should not exceed 6 dB (as it would if the number of depth zones is to small), so that the sensitivity variation within each depth zone is not too large. On the other hand a relatively large number of depth zones with small sensitivity variation would cause ambiguities and insecurities of location.

In the example in Fig. 3 the task was to determine a tandem depth zone subdivision at a given wall thickness and number of depth zones and given sizes of the crystals, so that the sensitivity difference between the center of each zone and its borders so the next becomes a minimum.
In this figure the curves represent the echo height of a perpendicular 10 mm disc reflector as a function of its depth within the wall thickness.
The calculations based upon the theoretical model have been carried out by

the Bundesanstalt für Materialprüfung (BAM), Berlin and the Institut für zerstörungsfreie Prüfverfahren (IzfP), Saarbrücken. The results of the theoretical models show good correspondence (each with one exception within 2 dB) to the experimental results which have been measured at KWU (ref. 5).

2) Dependence of the echo amplitude on the orientation of the planar reflector

In this respect a very good correspondence (within 3 dB) can be stated within a region of a disorientation of 12° from the normal (=optimum) incidence (Fig.4). This has to be considered with the fact, that the calculation is done with one frequency. On the other hand the probes with high damping have a relatively broad spectrum.

This fact as well as echos from the border of the defect, which are not covered by the theoretical model might be the cause for the differences which are still within 6dB-range.

Similar results were obtained with elliptical planar defects (Fig. 5). Here of course in ideal orientation the echo height is dependent on the defect area (if the defect length 2c does not exceed the beam diameter). The shape of the function of the echo height versus the orientation angle is only dependent on the depth of the defect 2a. This is the cause for the parallel curves for 5×17 mm^2 and 5×34 mm^2 and the nearly identical curves for 10×34 mm^2 and 10×68 mm^2 with a beam diameter of approximately 30 mm.

3) Dynamic behaviour of defect indications in X- and Y-scanning:

The lateral sensitivity distribution (Y-direction) is determining the sensitivity addition in dependence on the scanning distance, which should be as broad as possible with a sensitivity difference as small as possible.

Crystal sizes and a special array of three crystals in parallel have lead to good results.

Taking into account inherent limitations the dynamic behaviour in X-direction (moving the probe towards the reflector) gives hints on orientation and the relevances of a defect, in the way of an indication pattern atlas as a basis of evaluation.
Also here theory and measurement show good correspondence in both scanning directions (Fig.6).

Also for the socalled ILT-technique (included longitudinal twin-probe-technique), which is also of high importance for the inspection of austenitic material, a theoretical model has been worked out at BAM, Berlin.

Theory and experiment shows satisfactory correspondence (Fig.7). A large number of investigations has been carried out comparing the results of the theoretical model with the experimental results. So the theoretical model describes the physical properties very well. Based upon that the theoretical models are of great use especially for the planning of probe systems for special tasks, of limited use only for the evaluation of indications.

Data acquisition and processing

Another most important contribution to the use of improved inspection techniques is the data acquisition and processing system.

This system has been already presented to the public (Ref.6) so here only limits of further improvement will be given.

1) Statistical data presentation:

By this statistical data it is not only possible to determine the sensitivity situation and the necessary sensitivity additions but also to control the reproducibility of the entire inspection equipment: In regular intervals the probe system is repositioned and scanned over the same area. The statistical data of the single measurements are compared. By this comparison the control of all influence factors is feasible: Repositioning (manipulator), probe sensitivity, inspection electronics as well as the computerized data handling itself. Fig.8 shows the statistic data of two measurements on the same reference area. The time between these measurements was approximately 10 hours.

2) Comparison of reduced and original data of an indication

In general about a billion single ultrasonic data occur during e.g. a preservice inspection. This is the cause for the necessity of data reduction. This data reduction is done by a position - correlated averaging parallel to the storing of all data with no respect to a registration level.

This position-correlated averaging means the determination of one average value of the single echo height values over the socalled averaging length, as a part of the scanning path seperatedly for each information channel.
This average length can be chosen and in most cases has been determined with 10 or 20mm.
The average value is compared with a set of thresholds, and each interval in echo height between two thresholds is correlated to a socalled threshold symbol. These symbols are printed out in a map for each inspection technique in each depth zone (depth- and technique-correlated C-scan presentation).

This data reduction gives the possibility to a most clear data presentation, (ref.6) but it also has a potential for background suppression.
Fig.9 shows the comparison of the origi-

nal data from one scanning line from tape (in the above line in analog presentation), where the disturbances can be seen clearly, with the corresponding reduced data (in a threshold symbols' presentation, also called classification atlas). The symbols : and < represent low average values, the symbol * a relatively high one.
All positions with disturbances (some of them reaching the registration threshold) show low average values, a little higher ones are at positions with some dynamic indications below the registration level, and only one indication as such exceeding the registration level shows a high average value.

Alternative techniques

One part of the scope to be inspected inservice are the threads in the stud bore holes in the flange of the lower part of the vessel. The requirements were edited, that an inspection technique has to be developed for this purpose. The application of two techniques seemed possible: Ultrasonic and eddy current.

1) Ultrasonic inspection

Here the ultrasonic waves should propagate parallel to the axis of the basehole. This demands for the application of probes with normal incidence from the upper surface of the flange. This is only possible if there is free access to this surface, which means, that the vessel head has to be removed.

Two types of probes can be taken into consideration: pulse-echo and transmitter-receiver-(twin)probes. As the results show, all thread bottoms up to a depth of ca. 200mm can be inspected with exception of the nearest parts of the thread. This of course does not apply in the case of the twin probe.

The reference reflector, a notch of 2mm depth, 1mm width and a length of 20mm is detectable over the whole thread with sufficient signal to noise ratio. However, an estimate of the depth of a defect seems only possible with big efforts in carrying out the inspection.

2) Eddy current inspection

The use of this techniques allows the movement of the probe in the inside of the borehole along the threads. By that an inspection by manipulator through the boreholes in the head flange is possible. This also leads to a constant defect resolution all over the length of the thread.

For this purpose a special technique for eddy current testing of ferritic material developed at KWU without premagnetisation is used (ref.7).

As the results at a testblock show, the following can be stated:

- A notch of a width of 0,2mm, a length of 20mm and a depth of 0,5mm can be detected (Fig.10)

- By variation of the inspection frequency and the display of the signals in X-Y-presentation an evaluation of the depth is possible up to 10mm (Fig.11)

- Indications caused by changes in the electric or magnetic properties of the material can be distinguished from indications of real defects (Fig.12)

During a revision phase these techniques have been applied for the inspection of a number of boreholes and proved as applicable and reliable tools within inservice inspection.

REFERENCES

1. ENGL G.   Possibilities and limitations of ultrasonic inspection techniques applied at automatically performed inspections; OECD-experts' meeting, Warrington, (1976).

2. KUTZNER J. and WÜSTENBERG H.   Hilfsmittel zur Fehlergrößenabschätzung bei der Wiederholungsprüfung mit Ultraschall; Echodynamik, BAM-report RS2702, Oct.1974, GRS Cologne.

3. WALTE F., WERNEYER R. and HORST B. Zur Bestimmung von Prüfzonen bei der US-Prüfung nach dem Tandemverfahren. Materialprüfung 19, 1977 Nr.5, p.174-177.

4. KUTZNER J., WÜSTENBERG H., MÖHRLE W. and SCHULZ E.   Zonenaufteilung, Empfindlichkeitseinstellung und Prüfkopfhalterung bei der manuellen US-Prüfung mit dem Tandemverfahren. Materialprüfung 17, 1975, Nr. 7, p.246-250.

5. ENGL G., RATHGEB W., WÜSTENBERG H. and WALTE F.   Choice of parameters for ultrasonic probes and probe systems for the automatic ultrasonic inspection by the use of a theoretical deflection model, DGzfP-Conference NDT, Mainz 1978.

6. ENGL G., RUTHROF K., JACOB H. and RATHGEB W.   Verarbeitung, Darstellung und Bewertung von Prüfbefunden bei automatisierten Ultraschallprüfungen dickwandiger Bauteile. DGzfP-Conference NDT, Bremen 1977.

7. MEIER W.   Basic examinations concerning eddy current testing of ferromagnetic materials, DGzfP-Conference NDT, Mainz 1978.

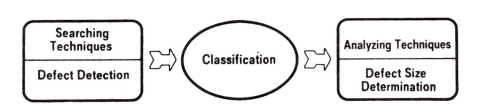

Fig. 1 Basic procedure of mechanical ultrasonic inspection

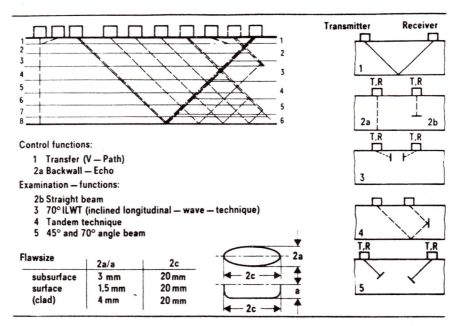

Control functions:

   1 Transfer (V — Path)
   2a Backwall — Echo

Examination — functions:

   2b Straight beam
   3 70° ILWT (inclined longitudinal — wave — technique)
   4 Tandem technique
   5 45° and 70° angle beam

Flawsize	2a/a	2c
subsurface	3 mm	20 mm
surface	1,5 mm	20 mm
(clad)	4 mm	20 mm

Fig. 2 Multiple function UT — probe system for cylindrical part of the RPV and guaranteed detectable flaw size

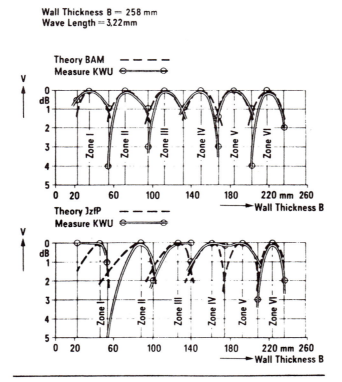

Wall Thickness B = 258 mm
Wave Length = 3,22 mm

Theory BAM   — — —
Measure KWU

Theory JzfP   — — —
Measure KWU

Fig. 3 Tandem depth zone subdivision

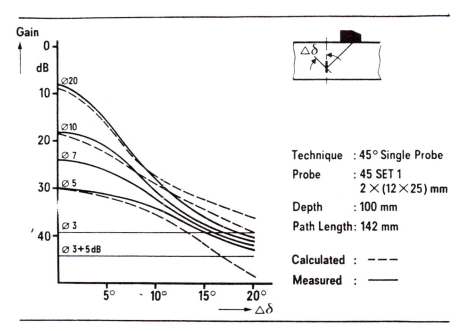

Fig. 4   Indication amplitude as a function of oblique orientation dependence for planar circular flaws

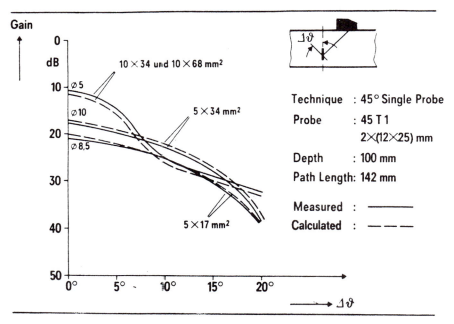

Fig. 5   Orientation dependence for planar elliptic flaws

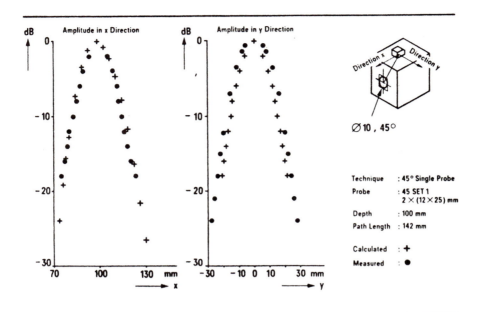

Fig. 6 Ultrasonic echo dynamic

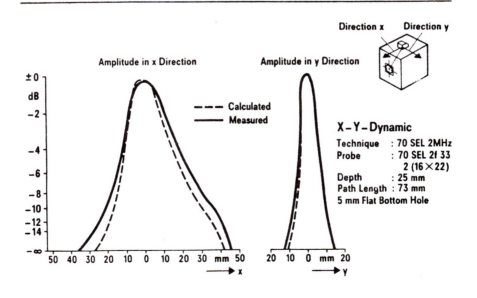

Fig. 7 Ultrasonic echo dynamic

INDEX 16-31

		16	17	18	19	20	21
Number of Measured Values		2307	2307	2307	2307	2307	2307
Arithmetical Mean	DG	155	77	115	87	137	138
	DB	60	30	44	33	53	53
Dispersion	DG	7	8	14	13	13	7
	DB	2	3	5	5	5	2
Variance	DG	43	64	207	188	159	46
Coefficient of Variation	%	4 3	10 5	12 6	15 4	9 3	4 9
Range of Variation	DG	51	58	87	75	109	50
	DB	19	22	33	29	42	19
Maximum of Density	DG	157	76	115	78	142	141
	DB	61	29	44	30	55	54
Median	DG	154	77	114	84	138	138
	DB	60	30	44	32	53	53
Asymmetry	-	0 03	0 16	0 08	0 72	-0 27	-0 00

INDEX 16-31

		16	17	18	19	20	21
Number of Measured Values	-	2205	2205	2205	2205	2205	2205
Arithmetical Mean	DG	155	78	115	90	139	139
	DB	60	30	44	35	54	54
Dispersion	DG	7	9	14	15	13	7
	DB	2	3	5	5	5	2
Variance	DG	49	82	189	228	167	53
Coefficient of Variation	%	4 5	11 6	11 9	16 4	9 3	5 3
Range of Variation	DG	52	60	95	85	106	53
	DB	20	23	37	33	41	20
Maximum of Density	DG	157	75	110	79	141	141
	DB	61	29	42	30	54	54
Median	DG	155	78	115	87	140	139
	DB	60	30	44	33	54	54
Asymmetry	-	0 00	0 03	0 21	0 63	-0 32	0 06

Outprint 1st Measurement

Outprint 2nd Measurement

Fig. 8 Control of reproducibility by statistical data

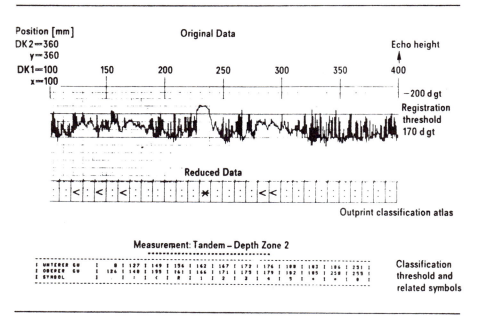

Fig. 9 Effect of data reduction: comparison of original and reduced data by averaging

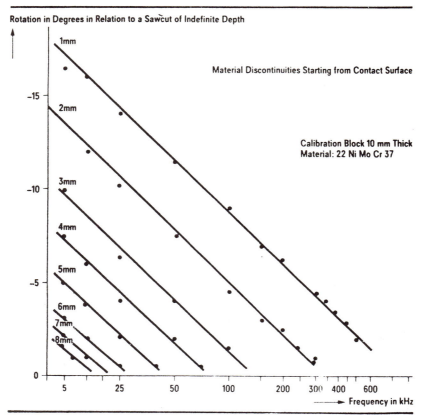

Fig. 10 Amplitude in co-ordinate system as a function of discontinuity depth and frequency

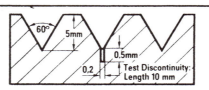

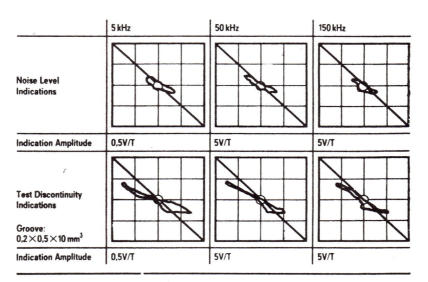

Fig. 11 Test discontinuity indications in the thread root

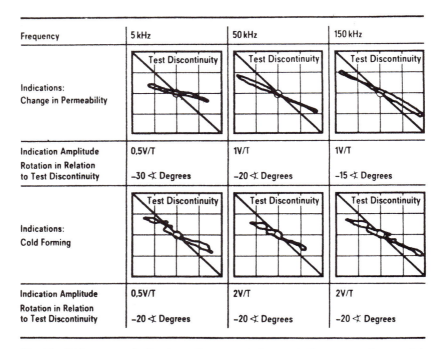

Frequency	5 kHz	50 kHz	150 kHz
Indications: Change in Permeability	Test Discontinuity	Test Discontinuity	Test Discontinuity
Indication Amplitude	0,5V/T	1V/T	1V/T
Rotation in Relation to Test Discontinuity	–30 ∢ Degrees	–20 ∢ Degrees	–15 ∢ Degrees
Indications: Cold Forming	Test Discontinuity	Test Discontinuity	Test Discontinuity
Indication Amplitude	0,5V/T	2V/T	2V/T
Rotation in Relation to Test Discontinuity	–20 ∢ Degrees	–20 ∢ Degrees	–20 ∢ Degrees

Fig. 12 Eddy current indications of structural changes in ferromagnetic materials
(22 NiMoCr 37)

C37/79

ADVANCES IN MULTIFREQUENCY EDDY CURRENT INSTRUMENTATION

T. J. DAVIS, MSEE,
Battelle, Pacific Northwest Laboratories, Richland, Washington

The MS of this paper was received at the Institution on 18 December 1978 and accepted for publication on 22 January 1979

SYNOPSIS   Recent developments in multifrequency eddy current technology have resulted in substantially improved nondestructive inspection capability. A four-frequency test for steam generator tubing in nuclear power plants has demonstrated the potential for highly improved flaw characterization and reduced operator data interpretation. These results are obtained through simultaneous differential and absolute inspection combined with elimination of unwanted test parameters through multifrequency mixing.

INTRODUCTION

1.   Multifrequency eddy current inspection can offer substantial benefits over a conventional single frequency test. The multifrequency method permits cancellation of unwanted test variables which can normally mask or alter the desired information. The cancellation process is effected by linear combination or mixing of signals from different frequencies. The primary thrust of this paper is the application of these techniques to in-service inspection of steam generator tubing in nuclear power plants, wherein multifrequency mixing results in cancellation or reduction of unwanted signals from support plates, probe wobble effects, pilgering, and dents. The work reported in this paper resulted from a research program conducted for the Electric Power Research Institute, Palo Alto, California.

SUMMARY OF RESULTS

2.   A four-frequency test system incorporating both differential and absolute inspection modes has been developed for inspecting steam generator tubing. The test has demonstrated greatly improved capabilities over conventional single frequency testing and over our previous work with multifrequency testing (References 1 and 2). Advantages of the test include:

- A three to fivefold reduction of error in assessing depth of flaws

- A 90 to 95% reduction of indications from tube supports

- An 85 to 90% reduction of probe wobble effects

- A flaw depth sensitivity equivalent to that of the conventional single frequency test

- Continuous profiling of flaws and tubing wall thickness

- Characterization of support corrosion and cracking, even in dented tubes

- Detection and sizing of medium and large wastage flaws in dented regions

3.   The system is currently being prepared for evaluation in a steam generator mockup. Additionally, a plan has been implemented for transfer of the developed technology to direct use in in-service inspection systems.

SYSTEM PERFORMANCE

4.   Advanced mixing techniques, judicious frequency selection and differential-absolute testing provide the system with its present level of performance. The test uses a simultaneous mixing of up to three frequencies to obtain cancellation or reduction of unwanted test variables. A single pair of flaw data outputs is derived from a three-frequency mixer and these outputs are virtually free of probe wobble, pilgering and tube support indications. These outputs possess a flaw depth sensitivity (phase angle versus flaw depth) equivalent to that of the conventional single frequency 400 kHz test. The result is that flaws can now be more accurately detected, sized and profiled at all locations in the tubing. Equally important is that the data interpretation process is simplified because the data is defined by only two output channels. Moreover, the degree of rejection of unwanted test variables is much higher than that achieved in our previous multifrequency tests.

5.   Both differential and absolute multifrequency tests are performed simultaneously by the system. When used for flaw testing, the absolute (single coil) test has demonstrated a three to five times reduction of error in depth assessment of flaws. The system uses absolute testing to profile medium and large wastage flaws, to assess support corrosion and cracking and to inspect dented regions. Flaw profile information, which is not available from a differential test, can be extremely useful in predicting burst pressure of the tubes. Differential (two coil) tests are used for characterization of small wastage flaws in all regions of the tube.

6.   Figure 1 shows inspection of a steam generator tube with details of the types of flaws and anomalies mentioned above.

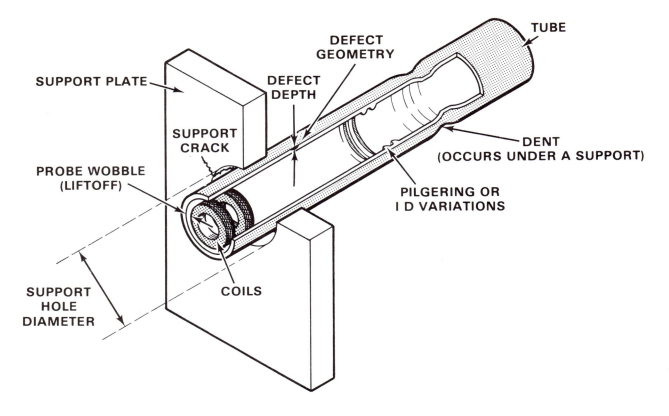

Fig. 1 Steam generator tubing inspection

Elimination of Unwanted Parameters

7.   The capability of the new test for eliminating unwanted parameters is summarized in Figure 2.

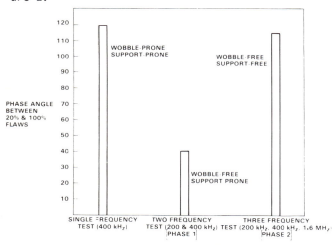

Fig. 2 Flaw depth sensitivity of single and multifrequency tests

This figure compares depth sensitivity of the new test against single frequency and two-frequency tests, and also categorizes the tests for pollution of data by wobble or supports. The flaw depth sensitivity is measured as the phase angle difference between indications from 20% and 100% ASME calibration flaws. These data were taken using tubing representative of a series 44 steam generator. Outer diameter of the inconel tubing was 22.22 mm (0.875 in.) and the wall thickness was 1.22 mm (0.048 in.). The flaws are

located on the outer surface and their depths are quantified in percent of wall thickness.

8.   The single frequency test results shown in Figure 2 are representative of current in-service inspection practice. They enjoy a high level of depth sensitivity but are prone to both wobble and support indications. The middle bar in the graph shows results of a two frequency test which eliminated wobble but which also suffered a 2/3 reduction in flaw depth sensitivity. Results of the present three-frequency test are shown in the right hand bar, in which both wobble and support data are virtually eliminated and the flaw depth sensitivity has been restored to nearly that of the single frequency data.

9.   The support elimination capability of the new test is compared to two-frequency and single frequency testing in Figure 3. The support residue in both output channels of each test is shown normalized to that of the single frequency test, and the probe wobble performance for each test is listed in the chart. Two types of multifrequency mixing may be performed with a two-frequency test. In one case, two wobble-free outputs are generated in which one is 95% support free. In the other, support rejection levels of 70% and 95% are obtained but the data is wobble prone. In the three-frequency case, which utilizes new mixing techniques, support indications are reduced by 90% and 95% in the two output channels and the data is virtually wobble-free .

Differential Flaw Testing

10. The differential test is used primarily for detection and sizing of small flaws such as

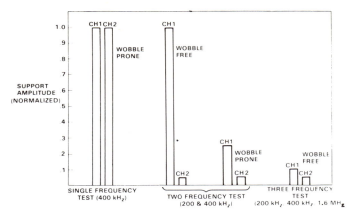

Fig. 3 Support rejection capabilities of single and multi-frequency tests

stress corrosion cracks which have small amounts of wastage (removed metal volume). The differential test can be compensated for wobble using multifrequency techniques to a higher degree than can the absolute test, and thus, is better suited for detecting small flaws.

11. Implementation of a differential test is performed by placing the two coils shown in Figure 1 into opposing legs of an impedance bridge. The bridge is balanced when both coil impedances are equal. An output signal is generated when one but not both of the coils is in the proximity of a defect or other test variation. A bipolar response results when the probe is drawn past a flaw. One polarity corresponds to the leading coil passing the flaw and an indication of the opposite polarity will be generated as the trailing coil passes the flaw.

12. Performance of the three-frequency differential test is shown in Figure 4, compared to that of the single frequency 400 kHz test. The data shown was generated by translating a probe through an ASME calibration tube with a simulated support on one end. The calibration flaws are flat bottom drill holes of the depths indicated in percent of wall thickness. The strip chart recordings are the two instrument outputs for each case and the impedance plane responses were generated by plotting these outputs on an X-Y oscilloscope. Flaw depths are interpreted under current in-service inspection practice by measuring the phase angle of impedance plane responses.

13. Three major improvements are apparent in the multifrequency data of Figure 4. First, the probe wobble visible in Channel 2 of the strip chart data has been virtually eliminated, pro-

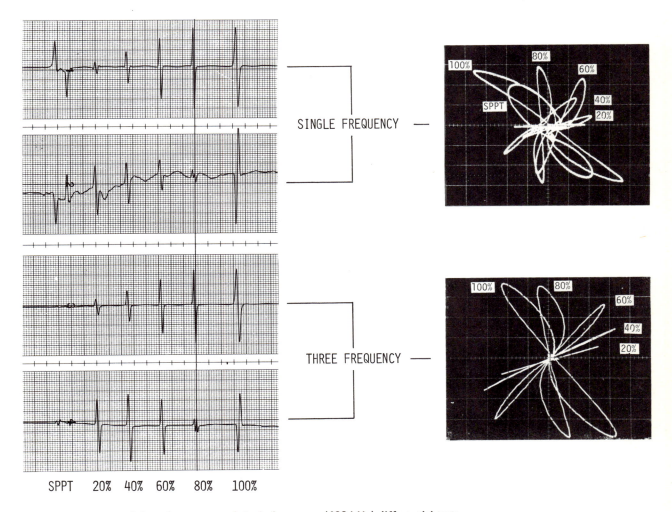

Fig. 4 Comparison of three-frequency and single frequency (400 kHz) differential tests

viding a clean baseline on both traces. Second, the support indications have been reduced to a point that they are barely discernable when compared to the calibration flaw responses. Third, the multifrequency data enjoys a flaw depth sensitivity (phase angle versus flaw depth) on the impedance plane which is comparable to that of the single frequency data. These results mean that flaws can be detected and accurately sized with minimal pollution from wobble or support indications, and that the data interpretation is considerably simplified over previous multifrequency techniques since all the data is contained in two output channels.

14. It should be pointed out that the remaining baseline variation in channel four of the strip chart data was due to the high test frequency (1.6 MHz) response to the end of the tube and not to probe wobble. This phenomena was remedied by shielding of exposed conductors between coils and cables in the probe.

Absolute Flaw Testing

15. Absolute (single coil) testing is used by the system for characterizing medium and large wastage flaws in which the wastage is comparable to or larger than that of the flat bottom ASME calibration holes. The absolute test possesses considerably more accuracy for sizing these flaws than does the differential test. Additionally, a great deal more information on flaw geometry can be obtained since the absolute test generates a continuous profile related to wall thickness. Other uses the system makes of absolute testing (dented region inspection and support characterization) are discussed in following sections of the paper.

16. The absolute test is implemented by extracting the voltage from one of the coils in the differential probe and summing it with a qui-

escent multifrequency signal whose components are equal but out of phase. The resultant nulled signal is amplified and the Fourier amplitude coefficients are detected for each frequency as the probe is translated through the tubing.

17. Performance of the multifrequency absolute test in terms of flaw depth accuracy is compared to the conventional single frequency differential test in Figure 5. The data are in the form of scatter plots which compare actual flaw depth (horizontal axis) to measured flaw depth (vertical axis) for a series of two types of artificial flaws, uniform thinning and elliptical wastage.[1] Averaged errors for both the absolute and differential tests are shown in Table 1. The data shows a 4.65:1 reduction of error for uniform thinning and a 3.3:1 reduction of error for elliptical wastage when using the absolute test.

	Averaged Absolute Value Error in Percentage Points of Wall Thickness	
	Elliptical Wastage	Uniform Thinning
Differential Test	7.9	10
Absolute Test	2.4	2.15

Table 1. Comparison of Sizing Errors for Two Types of Artificial Flaws

[1] The uniform thinning specimens used were a uniform reduction of wall thickness completely around the circumference of the tube with axial lengths ranging from 12.7 mm (1/2 in.) to 5.8 mm (2 in.). The elliptical wastages were manufac-

MULTIFREQUENCY ABSOLUTE TEST

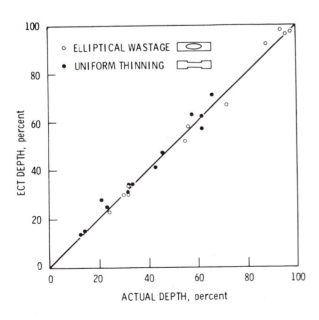

SINGLE FREQUENCY DIFFERENTIAL TEST

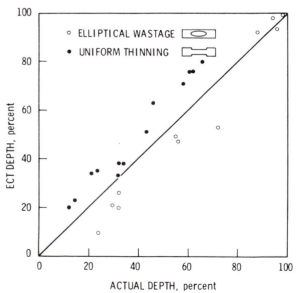

Fig. 5 Comparison of flaw sizing capabilities for single frequency differential testing and multifrequency absolute testing

tured by a technique equivalent to grinding a flaw in the OD with a large diameter circular grinding wheel. When viewed from right angles to the ground surface, the flaw appears as an ellipse.

18. This reduction of sizing error is quite significant to nuclear power plant operations for two reasons. First, tube plugging decisions may be affected with a much higher degree of precision, and second, growth of flaws between inspection outages may be meaningfully quantified.

19. The main reason for the improved performance of the absolute test is that it generates a continuous measurement of the minimum wall thickness encountered in the tube. In contrast, the differential test is primarily sensitive to the edges of flawed regions where the wall thickness is undergoing an abrupt change. A gradual or tapered change in wall thickness can be missed or erroneously sized by the differential test.

20. The ability of the multifrequency absolute test to generate continuous profiles of flaw depth is shown in Figures 6 and 7. Absolute data are compared with 400 kHz single frequency differential data for two uniform thinning specimens in the figures. Advantages of the absolute test as shown by this data are: 1) flaw depth at any axial position along the flaw may be detected by the absolute test whereas the differential test is primarily edge sensitive, and 2) the absolute data is much easier to interpret in both the strip chart and impedance plane data. This is particularly true of the impedance plane data of Figure 7, in which the differential pattern tends to have the appearance of a multiple or compound flaw.

21. An additional point of significance is emphasized by the data of Figure 7 in which a convex flaw was used, e.g., the flaw was shallower in the middle than it was at either end. This information is easily detected by the absolute test but is omitted by the differential. The same relationships would exist for a concave flaw which was deeper in the middle than at the ends. Thus this type of flaw will be erroneously sized at a shallower-than-actual depth by the differential test.

Dent Inspection

22. Research is currently in progress to provide the prototype system with means for rapid inspection of dented regions and present results are encouraging. A dent is shown in Figure 1

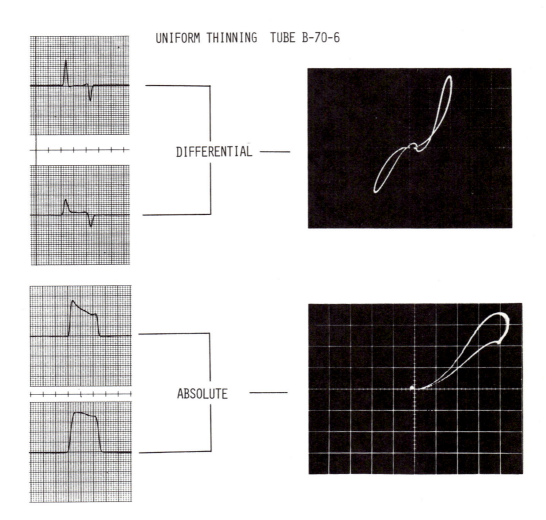

UNIFORM THINNING TUBE B-70-6

DIFFERENTIAL

ABSOLUTE

Fig. 6 Comparison of differential and absolute test results for a uniform thinning specimen with 50% flaw depth

and is a diameter reduction located under a support. The reduction is caused by volumetric expansion of corrosion products from the support. A typical dent can produce an indication 50 times larger in amplitude than those of the ASME drill holes when using a single frequency differential test with conventional bobbin coils as shown in Figure 1.

23. One successful single frequency method currently used for in-service inspection of dented regions employs a point probe which is held in direct contact with the tubing ID via spring loading. This reduces the liftoff effect produced by the dent. The small dimensions of the probe also improve the flaw-to-dent signal ratio since the flaw now occupies a much larger percentage of the total search field. However, the time required to perform a helical scan of a dented region with this probe is prohibitive due to limitations of rotational speed for the direct contacting probe.

24. Multifrequency testing opens the possibility of performing a high speed helical scan test of dented regions with a noncontacing point probe. Work is being conducted on development of a helical scan probe in which a small inspection field could be rotated rapidly enough to permit translation of the probe at the conventional throughput speed of 0.3 m (1 foot) per second. Multifrequency mixing can be used to compensate against liftoff variations caused by dents, pilgering or ordinary probe wobble.

25. Results of inspecting simulated dents with a noncontacting point probe are shown in Figure 8. The test specimen contained four dents approximately 25.4 mm (1 in.) long in the axial direction with an ASME calibration flaw machined in the middle of each dent. Data for both single frequency and multifrequency absolute tests are shown in the figure. The data was taken without rotation of the probe. It is immediately obvious from the data that the multifrequency test can readily detect and size defects and that the defect indications have substantial amplitude in comparison to those from the dent. In contrast. the flaw information is completely obliterated in one channel of the single frequency information.

26. An added advantage of performing all inspection with a helical scan test would lie in the area of flaw profiling. This test could generate much more information on flaw geometry than does the present absolute test with bobbin coil probes. Moreover, a properly designed helical scan test would be insensitive to flaw orientation.

Support Characterization

27. The ability of an absolute multifrequency test to detect both cracks and corrosion of the support plate has been demonstrated. A support crack as shown in Figure 1 generally results from planar stress in the support due to build-up of corrosion products between the supports

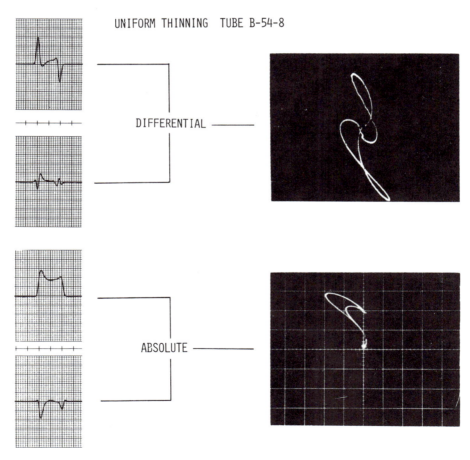

UNIFORM THINNING TUBE B-54-8

DIFFERENTIAL

ABSOLUTE

Fig. 7 Comparison of differential and absolute test results for a uniform thinning specimen with 80% flaw depth

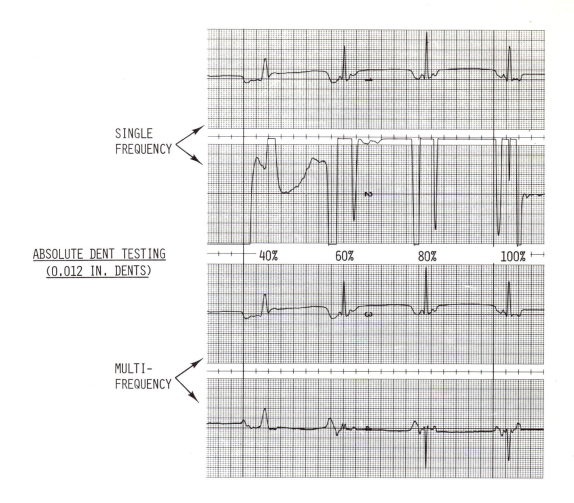

SINGLE
FREQUENCY

ABSOLUTE DENT TESTING
(0.012 IN. DENTS)

MULTI-
FREQUENCY

Fig. 8 Comparison of single and multifrequency results for inspecting dented regions with a point probe

and the tubes. The corrosion itself results in increased diameter of the support hole.

28. Two combinations of two frequency testing have been evaluated for support characterization. One used a 400 kHz test which was compensated for wobble and dents by 1.6 MHz information, and the other used a 25 kHz test which was compensated for dents with 200 kHz data. The 200-25 kHz combination provided better dent compensation.

29. Performance of a two-frequency support crack test is shown in Figure 9. The curve of phase angle versus actual crack depth was taken from the data of Figure 10, which is a run through a clean tube having supports with simulated cracking. The crack depths are measured in percentage of the spacing between support holes, which for these specimens was 3.05 mm (0.120 in.). The cracks were simulated by EDM notches 0.15 mm (0.006 in.) wide, which extended along the complete width of the 19 mm (0.75 in.) thick support.

30. A pronounced two-pronged pattern results on one channel of the strip chart data for cracked supports. This pattern is easily identifiable when compared to the pattern for a normal support, and the depth of the crack may be obtained by taking the arctangent of the two channels at a point located in the center or

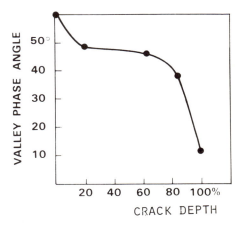

Fig. 9 Calibration curve for a two-frequency absolute support crack test. Crack depth is measured as a percentage of spacing between support holes (3.05 mm for this test)

valley of the support response. Likewise, the two channels could be plotted on an X-Y oscilloscope to obtain the phase angle.

31. The ability of a two-frequency absolute test to measure support corrosion is shown in Figure

263

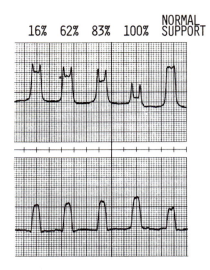

Fig. 10 Two-frequency absolute test data for supports with simulated cracking (25 kHz, 200 kHz)

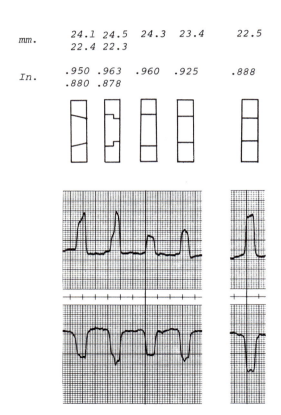

Fig. 11 Two-frequency absolute test data for supports with simulated corrosion (400 kHz, 1·6 MHz)

11. Above each strip chart response is sketched a detail of the support hole with diameters listed.   Actual correlation of the upper channel with the hole diameter variation is observable for the two supports at the left of the illustration.   These pieces feature a gradual taper and a step change and these profiles are visible in the strip chart data.   Straight diameter reductions in the other two test pieces are easily discernable in the chart data when compared to the indication for a normal support shown at the right of the illustration.

264

MULTIFREQUENCY TEST SYSTEM

32. A block diagram of the four frequency inspection system appears in Figure 12.   Primary components include the multifrequency instrument, a 16 channel magnetic tape recorder, a calibration memory unit, a mixing unit, and a strip chart recorder and X-Y oscilloscope for data display.

33. Fourier amplitude coefficients for differential and absolute tests at individual frequencies are output by the multifrequency instrument and may be analyzed in real time or recorded for post-test analysis on the magnetic tape recorder.   The calibration memory unit incorporates a microprocessor and a solid state memory to store up to 10 seconds worth of data from eight input channels.   This unit can be used either to store calibration data and repetitively play it back into the mixing unit, or to pass the input data directly through to the mixer.   The calibration memory unit is extremely helpful in set-up of the mixer since it can capture a calibration tube run from either the test instrument or magnetic tape playback and then can repetitively play all or selected portions of the data into the mixer.

34. The system mixing arrangement appears in more detail in Figure 13, which shows the mixing system acquiring data directly from the multifrequency instrument.   Individual mixing is performed on three-frequency differential data and on four-frequency absolute data.   Frequencies used for 22.22 mm (0.875 in.) diameter tubing are:

F1	25 kHz
F2	200 kHz
F3	400 kHz
F4	1.6 MHz

A three-frequency mix is performed in both differential and absolute mixers on the three highest frequencies (F2, F3 and F4) to obtain flaw data which is wobble-free and support-free. The differential data is used to detect and size small wastage flaws and the absolute data is used for characterizing the medium and large wastage flaws.   An absolute mix is also performed with frequencies F1 and F2 for the purpose of obtaining a dent-free assessment of support cracking and corrosion.

35. More details of the system will be contained in Reference 3.

CONCLUSIONS

36. A four-frequency differential-absolute eddy current test for steam generator tubing has demonstrated a dramatic improvement over earlier work described in References 1 and 2.   Tubing flaws may be accurately detected and sized with minimal pollution of data from unwanted parameters such as supports, probe wobble and pilgering.   Multifrequency data interpretation is simplified since a given flaw may be sized using a single pair of test outputs such as is done with the conventional single frequency test.   The differential test provides sensitive detection of small wastage flaws while the absolute test results in accurate sizing and profiling of medium and large wastage flaws.

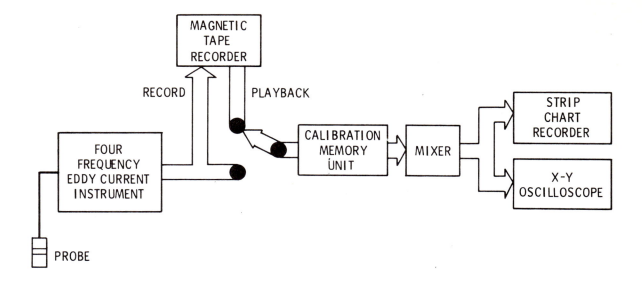

Fig. 12   Block diagram of four-frequency inspection system

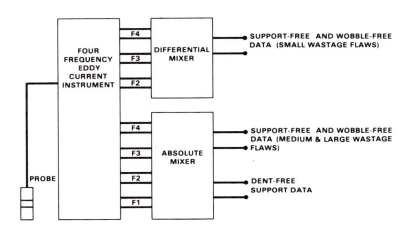

Fig. 13   Arrangement of multi-frequency mixing employed by new system (other system components omitted for simplicity)

Support cracking and corrosion in the presence of dents can also be measured with the absolute test. Finally, defects in dented regions can be detected and sized using an absolute multifrequency test with a noncontacting point probe.

37. Further work under the research program will place heavy emphasis on implementation of a high speed helical scan test with multiple frequencies. This test could be used for flaw detection at all points in the tubing, including dented regions. Additionally, the test would be less sensitive to flaw orientation than current tests which employ bobbin coil probes. Work is also planned in the area of developing computerized pattern recognition for absolute test data.

REFERENCES[a]

1. DAVIS, T.J., Multifrequency Eddy Current System for Steam Generator Tubing Inspection, Vol. I, Progress Summary; EPRI NP-758. Nov. 1977

[a]Copies of all references may be requested from: Research Report Center, Electric Power Research Institute, P. O. Box 10090, Palo Alto, California 94303

2. BROWN, S.D., An Evaluation of Eddy Current Inspection Methods for PWR Steam Generator Tubing, EPRI NP-636, Oct. 1978

3. DAVIS, T.J., Multifrequency Eddy Current System for Inspection of Steam Generator Tubing, Interim Report, EPRI Project RP403-2 (In Preparation)

C39/79

ESTIMATION OF FLAW SIGNIFICANCE FROM ACOUSTIC EMISSION DATA - A PROGRESS REPORT

P. H. HUTTON, BSME, E. B. SCHWENK, MSMetE and R. J. KURTZ, BSPhysMet
Pacific Northwest Laboratory, Richland, Wa

The MS of this paper was received at the Institution on 18 December 1978 and accepted for publication on 27 February 1979

SYNOPSIS The purpose of this work is to evaluate the feasibility of detecting and analyzing flaw growth on an operating reactor using acoustic emission (AE) data. Work to date has produced two preliminary empirical AE-flaw growth models. One relates AE rate to ΔK and the other relates cumulative AE to K_{max}. Evidence of temperature and flaw geometry effects on AE has also been observed. High temperature AE sensor testing has identifed at least one available sensor which appears suitable for long term monitoring of a reactor.

INTRODUCTION

1. In spite of an outstanding safety record to date in nuclear facilities, efforts are continuing to provide even greater safety assurance. The work described here, sponsored by the U.S. Nuclear Regulatory Commission, is a part of that effort.

2. The purpose of the program is to develop an experimental/analytical evaluation of the feasibility of detecting and analyzing flaw growth in reactor pressure boundaries by means of continuously monitoring acoustic emission (AE). Major objectives are:

- Develop criteria to distinguish flaw growth AE from innocuous acoustic signals

- Develop an AE-flaw growth model for relating inservice AE to flaw significance

- Demonstrate application of program results through both off-reactor and on-reactor testing

3. The program was initiated July 1, 1976 and is scheduled to continue through FY-83 contingent on intermediate results achieved. The desired end result is a demonstrated method and a prototypic hardware system for detecting and evaluating flaw growth by continuous AE monitoring.

4. This paper discusses results achieved to date toward program objectives.

APPROACH

5. The investigation is devoted exclusively to ASTM Type A533, Grade B, Class 1 material. The basic approach to interpretive model development is through laboratory testing of 1 to 1½ inch (25.4 to 38 mm) thick specimens in both tension-tension fatigue crack growth (FCG) and fracture at both room temperature and 550°F (288°C). Figure 1 shows a typical FCG specimen used for laboratory testing. Six parameters are measured for each AE signal and related to fracture mechanics variables. AE data from fracture testing of 6 inch (152 mm) thick pressure vessels is also incorporated in the analysis. The type of vessel monitored is illustrated in Figure 2.

6. We are taking precautions to standardize the AE measurement method from test to test. The same sensors (200 kHz to 1 MHz bandwidth) are used in each test and they are calibrated on the specimen both before and after each test, using an alumina grit blast broadband excitation technique. The sensors are cooled during high temperature testing to avoid sensitivity changes. An advanced AE monitor system (see Fig. 3) is used on all tests. It provides for isolation of the signals originating from the flaw location and measures six AE parameters, plus three mechanical parameters-total load cycles, number of load cycles producing AE and crack opening displacement. On a fracture test, load values are measured in place of load cycles. All test measurements have been made with 90dB gain and a voltage threshold of 0.9 to 1.1 volts peak. All of the data is recorded on digital cassette tape from which the information can be transferred to a computer for analysis and/or printed out to provide a hard copy for reference purposes (see Fig. 4). Referring to the data columns in Figure 4, Delta T (time) identifies a valid AE event plus showing the difference in time of signal arrival at the two sensors in microseconds relative to the midpoint between the sensors (zero Delta T). Energy is a volts²-time measure of area under the signal envelope. Peak Time shows two parameters-polarity of the first half cycle of the signal (+ or -) and the time in microseconds from first threshold crossing to reach peak signal amplitude. Pulse Ht. is the amplitude in volts of the signal peak. Cycle No. is the load cycle number that produced the AE signal in the context of a running count of total load cycles applied. COD-Max. and COD-Min. are voltage analogs of the maximum and minimum crack opening displacements associated with maximum and minimum load. Load Pos. gives a definition of the location within a load cycle

where the AE signal occurred. The load cycle is divided into 100 equal increments with zero starting where the positive going load crosses mean load in a sinusoidal load waveform. Thus, maximum load corresponds to a number of 25 and minimum load is 75.

7. With the above system, all measured parameters from an individual AE signal are available to a computer to enable correlations with fracture mechanics variables such as the crack growth rate ($\frac{da}{dn}$), the stress intensity factor (K) and the crack opening displacement (COD). A cassette tape interface to the computer was selected instead of direct computer input to keep the data acquisition system portable. The same measurements can be made regardless of the proximity of the test to a computer facility.

8. A separate but closely related part of the program is being conducted in parallel with specimen testing. This involves testing and evaluation of high temperature AE sensors suitable for the 550°F (288°C) environment that would be experienced on an operating reactor system. The approach being used is to screen available sensors by testing in a furnace in the laboratory for a period of time. Sensors that do not degrade under temperature alone will then be tested on a reactor where they will be exposed to the full environment of temperature, radiation and humidity. In both phases of this work, the sensors are mounted on a plate by pressure coupling, which would be usable on a reactor system.

RESULTS

9. Primary emphasis to date has been in the area of developing AE-flaw growth empirical relationships. Development of a reliable AE-flaw growth model for interpretation of flaw significance by AE is the program objective with the highest uncertainty and it is essential to the successful completion of the program.

10. AE-mechanical test results to date show that increasing amounts of AE are consistently associated with increasing flaw severity during fatigue crack growth and fracture of A533B C1.1 pressure vessel steel. This in itself implies the technical feasibility of using AE in a continuous monitoring mode as a measure of flaw severity. The results also begin to show possible effects of factors such as temperature and flaw geometry.

Fatigue Crack Growth

11. AE-fatigue crack growth data derived from ten conditions (2 thicknesses, 2 flaw geometries, 2 R-ratios, 2 temperatures and 2 metal pre-test conditions) are given in Figure 5. The information is presented in terms of AE rate versus crack growth rate. Two of the specimens involved plastic deformation prior to testing. Specimen 1-2A-1B was plastically bent to remove specimen warp and Specimen 1-1A-2A was prestrained 3% as a part of the test design. The data shows a consistent increase in AE with increasing crack growth rate at constant temperature. Increased temperature appears to decrease the amount of AE but the trend of the data is similar. R-ratios of 0.1 and 0.5 produced similar data.

12. One of the AE-flaw growth empirical models considered is illustrated by the data in Figure 5. This involves using AE rate at a known temperature to estimate crack growth rate or indirectly, the stress intensity factor range (ΔK), which provides a measure of flaw severity. We are very conscious of the fact that the data scatter presently shown is too great to represent an application relationship. Work over the next year includes several specific areas of investigation which we hope will reduce the scatter. This is discussed briefly later.

13. Another method of handling AE-fatigue crack growth data is also being considered. Figure 6 shows how the summation of AE data relates to K. Here the relationship is linear, whereas the same parameters based on rate data are log-linear. This approach is attractive because of reduced data scatter and the linear behavior. It may, however, require an initial definition of flaw shape and stress field once the AE indicates a growing flaw. For both approaches, the effect of additional variables such as load cycle rate (da/dt vs. da/dn), R-ratio, simulated service loading, material volume and temperature must be further defined.

14. We have examined our data in light of several models previously proposed by others (ref. 1-7). Data from certain specimen types fit some of these models but none of the models describe the total data based on theoretical plastic zone volume, plastic zone size and crack area.

15. Development of AE-flaw growth relationships focuses on fatigue crack growth because surveys of pressure boundary defects in general (ref. 8, 9, 10) indicated that fabrication defects coupled with operation-induced fatigue crack growth were the dominant cause of failure. Fracture testing, however, has also produced useful results.

Fracture

16. Two tests have been performed on intermediate scale pressure vessels through the cooperation of the Heavy Section Steel Technology Program, sponsored by the U.S. Nuclear Regulatory Commission at Oak Ridge National Laboratory, Oak Ridge, Tennessee. A typical vessel tested was shown in Figure 2. The two tests represent substantially different conditions in temperature and flaw dimensions. The tests are described in greater detail in other reports (ref. 11, 12). Briefly, the V-7B test was performed at 200°F (93.3°C) with a longitudinal O.D. flaw 18 inches long by approximately 5-3/8 inches deep (457 x 136.5 mm) in the 6 inch (152 mm) vessel wall. The flaw was located in the heat-affected zone of a repair weld. Test V8 was performed at -5°F (-20.6°C) with a longitudinal O.D. flaw 8 inches long by 2 inches deep (203 x 51 mm) located in an area of high residual stress in a repair weld.

17. Test results are summarized in Figure 7 which shows that increasing amounts of AE parallel increases in COD. Accumulated AE during both tests appeared to be sensitive to flaw plasticity and flaw growth associated with pop-ins (V8), and crack growth leading to leakage (V-7B and V8). Pop-in precursor activity (V8) was also sensed by the AE system. During the

most dynamic part of the pop-ins (including leakage), the AE system locked out. This is due to a design feature to avoid multiple counts from one signal. Another significant feature is the apparent similarity of the AE/COD curves for the two tests in spite of the differences in test conditions.

18. Comparing results from fracture testing of laboratory specimens with the vessel test results suggests a possible flaw geometry effect, i.e., part through wall versus through wall. Data in Figure 8 from fracture testing a surface notch laboratory specimen show distinct similarity to data from the vessel tests with a generally linear or bi-linear relation between AE and COD. Results from fracture testing a single edge notch specimen shown in Figure 9, however, are quite different. This is typical of results from three single edge notch fracture tests where much of the AE data occurs after the specimen is well into gross yielding and ductile crack extension is occurring. Flaw-rolling plane orientation was constant for lab specimens.

High Temperature Sensors

19. In the context of this program, availability of a high temperature AE sensor for reactor monitoring is critical. Three surface mount sensors and a metal wave guide sensor have been subjected to 550°F (288°C) temperature testing for 2600 hours this year. In Figure 10, the response of these test sensors to the same broadband input pulse from a nearby transducer is shown. The data shown was taken after 960 hours of 550°F (288°C) testing. A subsequent check at 2600 hours of high temperature testing showed no significant change. The earlier data is shown because it includes a comparison with room temperature sensors being used in laboratory specimen tests. An important aspect of this evaluation is sensor response versus increasing lower frequency limit. It appears that continuous monitoring of a vessel during operation would likely need to be done with a lower frequency limit of about 400 kHz to avoid hydraulic noise interference. In this sense, Sensor B is the most promising of the surface mount high temperature sensors. It retains a good degree of its broadband sensitivity above 400 kHz. The signal-to-noise ratio is comparable to that of the laboratory sensor which is known to be effective for detecting AE.

SUMMARY

20. The successful application of AE to pressure vessel monitoring depends on two major factors. First, a consistent empirical relationship between damage accumulation, such as fatigue crack growth and the AE produced is necessary. Second, a data acquisition method suitable for long term exposure to a reactor environment is required. Results to date indicate that credible relationships do exist between AE and fatigue crack growth parameters. The key element in a data acquisition method is the AE sensor. Results from this year's work have identified at least one surface mount sensor in addition to a metal waveguide which shows good evidence of being serviceable for an extended period at 550°F (288°C).

FUTURE DEVELOPMENTS PLANNED

21. Recognizing the need for refinement of the AE flaw growth relationships, we plan in the coming year to:

• Further define the effects of variables such as cycle rate, R-ratio and temperature

• Investigate AE signal characterization through pattern recognition analysis and incorporation of new measured AE signal parameters

• Initiate development of an AE-flaw growth model application concept

• Initiate development of a data acquisition concept for reactor application

We also plan to install high temperature AE sensors on an operating reactor for testing under the total reactor environment.

REFERENCES

1. MORTON, T.M., HARRINGTON, R.M. and BJELETICH, J.G., Eng. Frac. Mech., 5 (1973) 691.

2. HARRIS, D.O. and DUNEGAN, H.L., Exp. Mech. 14 (1974) 71.

3. MORTON, T.M., SMITH, S. and HARRINGTON, R.M., Exp. Mech. 14 (1974) 208.

4. LINDLEY, T.C., PALMER, I.G. RICHLARD, C.E., Mater, Sci. Eng., 32 (1978) 1.

5. PALMER, I.G. and HEALD, P.T., Mater. Sci. Eng., 11 (1973) 181.

6. PALMER, I.G., Mater. Sci. Eng. 11 (1973) 227.

7. SINCLAIR, A.C.E., CONNERS, D.C., FORMBY, C.L., Mater. Sci. Eng. 28 (1977) 263.

8. GIBBONS, W.S. and HACKENY, B.D. "Survey of Piping Failures for the Reactor Primary Coolant Pipe Rupture Study", GEAP-4574, General Electric Company, San Jose, CA, May 1964.

9. SCHWENK, E.B. and JAMES, L.A., "Estimate of the Fatigue Cracking Behavior of the Fast Test Reactor Primary Piping", Battelle Northwest, Richland, WA, October, 1969, pp. 75-84.

10. PHILLIPS. C.A.G. and WARWICK, R.G., "A Survey of Defects in Pressure Vessels Built to High Standards of Construction and Its Relevance to Nuclear Primary Circuit Envelopes", United Kingdom Atomic Engergy Authority, AASB(S)R 162, 1968.

11. WHITMAN, G.D. and BRYAN, R.H., "Heavy-Section Steel Technology Program, Quarterly Progress Report for July-September, 1977" NUREG/CR-0206, ORNL-TM166, April, 1978.

12. BRYAN, R.H., MERKLE, J.G. and WHITMAN, G.D., "Quick-Look Report on Test of Intermediate Test Vessel V8-Test of Flaw in Residual Stress Field", ORNL/SST-3, August, 1978.

HSST V7-B TEST VESSEL
~8 FT. HIGH
39 INCHES O.D.
6 INCH WALL
NUCLEAR
PRESSURE
VESSEL STEEL

Fig. 1   Single edge notch fatigue crack growth specimen

Fig. 2   Typical HSST vessel monitored for AE

Fig. 3   Multiparameter AE monitor with micro-processor interface to digital tape

BATTELLE NORTHWEST ACOUSTIC EMISSION TEST PROFILE

AE TEST: 1-2A-6A
DATE: 11-30 78
TEMP: HT 260 C
LOCATION: ZONE AE
INSTR: MICROPRO
UPDATE: 100
LOAD: 6/60 KIP
LOAD RATE: 2HZ
DELTA CYCLES: 7500

PROGRAM: A

BLOCK # 0001 (000000-000024)

DELTA T	ENERGY	PEAK TIME	PULSE HT	CYCLE NO	COD=MAX	COD=MIN	LOAD POS
+ 225	034	+ 001	03.98	000000	01.79	01.75	015
◆ TIME=00000 SECS							
+ 190	028	+ 002	03.98	000000	01.79	01.75	029
+ 199	020	+ 001	03.98	000000	01.79	01.75	241
+ 001	080	− 001	05.23	000000	01.79	01.75	198
+ 074	014	+ 000	03.98	000041	02.03	01.99	042
− 140	074	− 001	04.29	000047	02.03	01.99	043
+ 063	050	− 001	04.14	000052	02.03	01.99	043
+ 001	022	+ 001	02.73	000054	02.03	01.99	077
+ 001	082	− 001	04.45	000060	02.03	01.99	027
+ 080	022	− 001	03.98	000068	02.03	01.99	043
+ 077	016	+ 001	03.98	000075	02.03	01.99	043
+ 060	056	− 003	04.29	000078	02.03	01.99	043
+ 061	068	− 003	04.21	000080	02.03	01.99	043
+ 063	062	− 000	04.14	000089	02.03	01.99	043
+ 079	032	− 001	03.98	000097	02.03	01.99	043
◆ TIME=00077 SECS							
+ 061	054	− 006	04.06	000100	02.03	01.99	044
+ 068	070	− 001	04.14	000110	02.03	01.99	027
+ 070	074	− 000	04.29	000111	02.03	01.99	044
− 003	080	− 003	04.45	000118	02.03	01.99	027
+ 054	080	− 015	04.76	000125	02.03	01.99	027
+ 061	060	− 003	04.14	000133	02.03	01.99	044
+ 058	082	+ 006	04.53	000141	02.03	01.99	044
+ 076	102	+ 003	04.45	000145	02.03	01.99	027

BLOCK AVG: TOTAL AVG:
    +(DELTA T)=0070.52   +(DELTA T)=0070.52
    −(DELTA T)=0006.21   −(DELTA T)=0006.21
    ENERGY=0054.86   ENERGY=0054.86
    PEAK TIME=0002.39   PEAK TIME=0002.39
    PULSE HT=04.18   PULSE HT=04.18
    LOAD POS=0054.65   LOAD POS=0054.65

BLOCK SUM: TOTAL SUM:
    +POLARITY=08   +POLARITY=08
    −POLARITY=15   −POLARITY=15

TOTAL VALIDS=000023

Fig. 4 Sample data printout of cassette tape record from AE monitoring

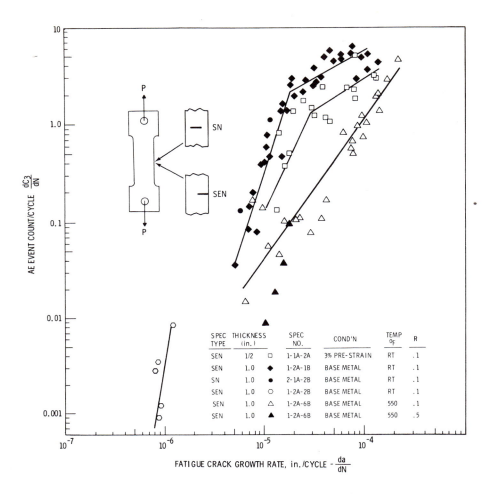

Fig. 5 Composite AE rate versus fatigue crack growth rate

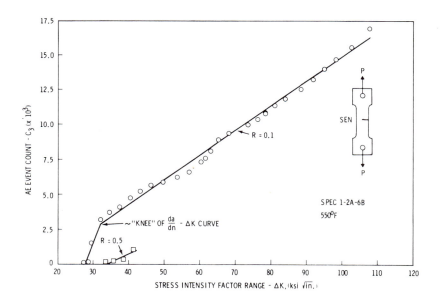

Fig. 6 Summation AE event count versus ΔK, specimen 1-2A-6B, 550° F

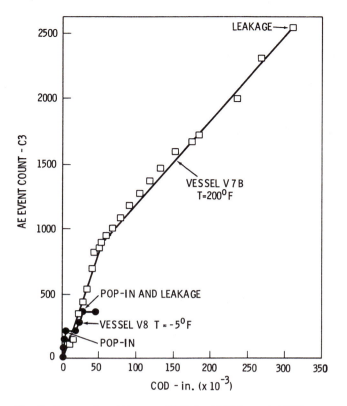

Fig. 7 Summation AE versus COD for HSST vessels V-7B and V8

Fig. 8 Load and AE event count versus COD, SN specimen 2-1A-2B, RT

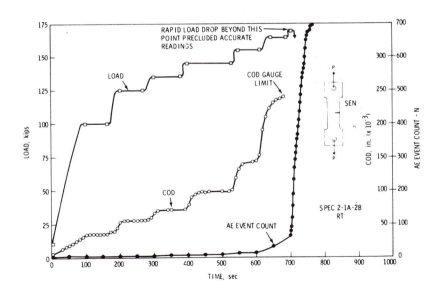

Fig. 9 Load and AE event count versus COD, SEN specimen 1-2A-6B, 550° F

HIGH PASS FILTER - KHz

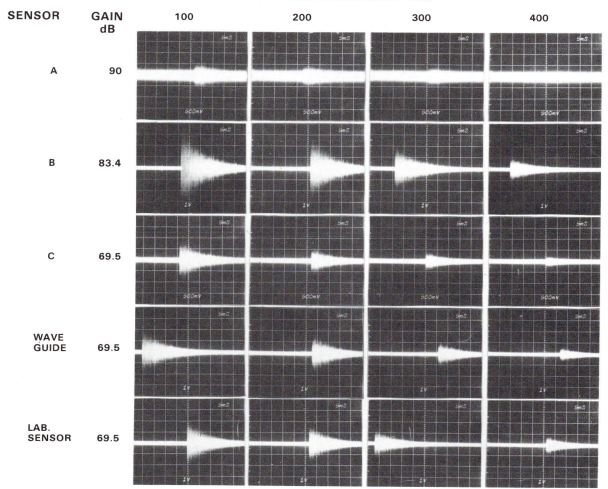

Fig. 10   High temperature sensor response after 550° F (288°C) exposure

C40/79

RELATIONSHIPS BETWEEN ACOUSTIC EMISSIONS AND MICROSTRUCTURES

G. V. RAO, BEMech, MTech, PhD and R. GOPAL, BEE, MS, PhD
Westinghouse Nuclear Technology Division

The MS of this paper was received at the Institution on 11 December 1978 and accepted for publication on 27 February 1979

SYNOPSIS   Results of a systematic study of 'microstructure-deformation-acoustic emission' relationships on two widely used pressure retaining component materials, namely A533-B nuclear pressure
vessel steel and a 7075 aluminum alloy, are presented. The study consists of conducting acoustic
monitored tensile tests on a variety of quenched and aged microstructures in the two alloy systems
and extensive microstructural characterization of test specimens by light optic and electron microscopy techniques. The results suggest a consistent relationship between acoustic emissions and
microdeformation mechanisms. The role of specific microstructural constituents in generating acoustic
emissions in the two alloys is discussed.

INTRODUCTION

1. Acoustic monitoring is becoming a valuable
inservice inspection technique for the surveillance of the integrity of nuclear pressure
boundaries and other pressure retaining components. Significant advances (Ref. 1, 2, 3)
have been made at Westinghouse in the past
several years in developing signal detection and
processing capabilities and acquiring experiences
in monitoring operating nuclear power plants.

2. A critical step in bringing the acoustic
monitoring technology to its fullest practical
value is the establishment of the physical
significance of acoustic signals and the
development of signal interpretation methodology,
in order that defect growth characterization
capabilities can be achieved.

3. Acoustic emission originates from flaw tip
plasticity and bears a direct relationship
(Ref. 4) with the elastic-plastic fracture
mechanics parameters that characterize significance of the active flaw at a pressure boundary.
The exact nature of the relationship, however,
is controlled by the metallurgical condition of
the microdeformation mechanisms operating at
various strain levels within the crack tip
plastic zone. Although acoustic monitoring is
being widely utilized for non-destructive
inspection, the microscopic sources of emission
and the mechanisms of acoustic generation in
specific pressure retaining component materials
are not generally well understood. A clear
knowledge of the contributions to acoustic
emission from specific microdeformation process
occurring at various strain levels, is very
valuable and is fundamental to the development
of signal interpretation methodology. The
current study is undertaken to meet this
objective.

4. A systematic study of 'acoustic emission-
microstructure-deformation mechanism' relationships is presented here on two widely used
pressure retaining component materials; namely

A533-B nuclear pressure vessel steel and a
commercial grade 7075 aluminum alloy. The
purpose of the current study is to investigate
the physical significance of acoustic signals
generated from the deformation of a variety of
controlled microstructures in the two alloy
systems and establish mechanisms of acoustic
generation. It is hoped that this will provide
a sound basis for the development of flaw
growth characterization methodology for
pressure boundary integrity monitoring.

TEST MATERIALS AND HEAT TREATMENTS

5. Both the materials utilized for the current
investigation are obtained from commercial
sources. For pressure vessel steel, a reactor
vessel nozzle cutout, consisting of A-533
Grade B Class I plate material, is employed
(Figure 1). Sheet tensile specimens are
fabricated from 1/4 T location of the nozzle
cutout, with the specimen loading axis parallel
to the rolling direction and the specimen width
parallel to the surface of the nozzle cutout.
In the case of 7075 aluminum, a commercial heat
of 7075 aluminum alloy sheet material (.06 inch
thick) is obtained from Reynolds Metals Company,
from which sheet tensile specimens are fabricated
with their loading axis parallel to the rolling
direction of the sheet. A mill chemical
analysis of the reactor vessel steel and a wet
chemical analysis of the 7075 aluminum alloy
specimen are listed under Tables I and II,
respectively. The geometry of the sheet tensile
specimens employed in the acoustic monitored
tests are shown in Figure 2. The fact that a
slightly smaller gage section is employed in the
case of 7075 aluminum alloy specimen has no
special significance other than material con-
servation reasons.

6. Heat treatment of 7075 aluminum specimens
are carried out in an open boat placed in a tube
furnace under argon flow. For pressure vessel
steel specimens, the heat treatment of test
specimens is conducted in vycor capsules back-
filled with argon. Listings of the series of

quenching and annealing treatments employed to obtain controlled microstructures in the two alloy systems are included in Tables I and II.

MICROSTRUCTURAL CHARACTERIZATION TECHNIQUES

7. Changes in the microstructure accompanying various quenching and aging treatments in the two alloys are closely monitored by extensive microstructural characterization studies employing light optic, transmission electron, and scanning electron microscopy techniques. A 200 KV Phillips 300 electron microscope and a Cambridge Mark IIA stereoscan are employed, respectively, in conducting transmission and scanning electron microscopy. Preparation of pressure vessel steel specimens for thin foil microscopy included mechanical grinding of specimens to about 0.003 inch followed by electrolytic thinning by window method until break through occurs and jet polishing disc punchouts from thin regions of the window sample. A modified Fishione twin jet polisher with a chromic acid-acetic acid electrolyte is employed for the final stage of thinning. The procedure is designed to cut down the mass of thin foil sample so that magnetic effects during imaging are minimized. Final sample preparation for the aluminum alloy consisted of jet polishing of .004 inch thick disc punchouts with an electrolyte consisting of 250 ml nitric acid, 750 ml methanol, and two drops of HF acid at -20°C.

ACOUSTIC MONITORING TESTS

8. The transducer utilized for the acoustic emission monitoring tests is a Westinghouse laboratory size transducer specially designed to produce a 300 to 700k Hz frequency response that is similar to the response characteristic of a Westinghouse field transducer. A water soluble ultrasonic couplant is used to couple the transducer to the test specimen throughout the tests. The amplified and filtered signals from the transducer are passed through a train of instrumentation in order that a number of acoustic emission parameters can be on-line monitored during the tests. Among the parameters monitored are: ringdown total counts, emission rate, total acoustic energy, amplitude distribution, and energy distribution. An amplifier gain of 80 dB and a discriminator threshold of 100 mV are employed throughout the tests. A special module, designed to measure the area under the squared envelope of an input acoustic puluse, is utilized for monitoring the energy content of acoustic signals while a computer-based, multichannel pulse height analyzer is employed for generating amplitude and energy distribution plots of acoustic signals. However, for the purpose of current discussion, only data on total counts and emission rate are considered due to lack of space.

9. The mechanical tests are conducted on a hard-beam Instron testing machine employing strain rates of 8.33×10^{-4} sec^{-1} in the case of pressure vessel steel specimens and 1.11×10^{-3} sec^{-1} in the case of 7075 aluminum specimens. Prior to testing, all the sheet tensile specimens are lightly polished with a 600 grit silicon carbide polishing paper to ensure that no oxide films are present on either the specimen surfaces or around the loading pinholes. Tests are conducted on three different specimens corresponding to each heat treatment and typical data from one of the specimens is considered for discussion under the current investigation.

RESULTS AND DISCUSSION

A533-B pressure vessel steel

10. Data from acoustic monitored tensile tests on as-quenched and quenched and tempered pressure vessel steel specimens are illustrated in Figures 3 through 9. The data clearly shows that:

i) Acoustic emission activity varies significantly with the degree of tempering of the as-quenched martensite upon deformation.

ii) The emission is predominant only during the yield region of the stress-strain curve with the activity rapidly decreasing to negligible levels beyond plastic strains of up to about 8%; little or no acoustic emission is present during the entire plastic range beyond 8% strain.

iii) Significantly higher rates of emission are observed from the yielding of slightly tempered martensites (i.e., martensites tempered at 750°F, 850°F, and 950°F; Figures 4, 5, and 6) as compared to the emission rates resulting from the yielding of martensites tempered at temperatures greater than 950°F (Figures 7, 8, and 9).

iv) Tempering of martensite in the range between 950°F to 1250°F resulted in a distinct drop in the acoustic emission rate during yielding, with the emission progressively decreasing with increasing tempering temperatures (Figures 7, 8, and 9).

11. Carbon steels employed for reactor vessel construction are either ferrite-pearlite aggregates or have largely bainitic microstructure depending on the thickness and heat treatment (Ref. 5). The hardenability of A-533B steel is controlled so that a primarily bainitic microstructure is produced following quenching and tempering during fabrication. Bainite is a preferred constituent since it has a lower ductile-brittle transition temperature. Figure 10 is an electron micrograph of pressure vessel steel material taken from 1/4 T location of reactor vessel nozzle cutout, following a stress relief heat treatment. The microstructure is characterized by a dispersion of carbide spheroids in ferrite matrix containing dislocation networks and low angle grain boundaries. Figure 11 illustrates the acoustic emission data resulting from the stress annealed microstructure.

12. Water quenching following austenitization treatment results in primarily a lath type martensitic structure in A533-B steel (Figure 12). The as-quenched substructure is generally characterized by a heavy density of dislocation tangles in a super-saturated ferrite matrix where the body centered tetragonal ferrite is in a highly strained state with carbon atoms occupying interstitial sites. Tempering at relatively low temperatures in steels usually promotes diffusion in interstitial carbon atoms whereby tetragonal martensite becomes cubic relieving the internal strains of b.c.t. lattice. Diffusion of carbon at these low tempering temperatures also promotes the forma-

tion of a plate-like transition precipitate known as epsilon carbide (Ref. 6) (ε-carbide). The actual tempering temperatures vary with alloy composition and Ms temperature (Ref. 7). Tempering at intermediate range of temperatures results in the formation of cementite type of carbide accompanied by the dissolution of ε-carbide, while at higher tempering temperatures (usually between 1000°F to 1300°F) alloy carbides are formed (Ref. 6). The matrix substructure changes from heavy dislocation tangles to dislocation networks and low angle boundaries following tempering at temperatures in the range 1000°F to 1300°F. The significance of these microstructural changes in controlling the yield process will be examined in conducting the interpretation studies of acoustic emissions observed from the tensile deformation of various quenched and tempered specimens.

13. Figures 13a and 13b are the scanning electron micrographs of the fracture surfaces in the as-quenched and 850°F tempered test specimens, respectively. The fracture of as-quenched microstructure indicates a brittle, cleavage mode of failure with a small amount of plastic flow. The mobility of a dislocation in the as-quenched structure is severely restricted by the interstitial strengthening effect of the carbon atoms, which is the primary source of strength, and by the fine, transformation-induced, substructure. Figure 14 is a transmission electron micrograph of the as-quenched microstructure. An interesting observation is that regions of thin platelets of oriented carbides are occasionally seen within the martensite grains. Figure 15 illustrates the microstructure of 850°F tempered martensite. Oriented platelets of matrix precipitates are extensively present although some of them appear to have begun to dissolve as the grain boundary carbides (cementite) appear to form. In fact, transmission electron microscopy of tempered specimens confirmed that oriented carbide platelets are persistent for tempering temperatures of up to 950°F beyond which only elongated or spheroidal blockey carbides (cementite and alloy carbides) are seen. Microstructures resulting from tempering at temperatures beyond 950°F (i.e., 1050°F, 1150°F, and 1250°F tempers) did not show any evidence of oriented carbide platelets. The 1050°F and 1250°F temper microstructures are shown in Figures 16a and 16b, respectively. The carbide platelets observed here in the as-quenched martensites and in the martensites tempered at temperatures up to 950°F are believed to be the Widman-Statten type ε-carbides since these are the first precipitates that are likely to come out of martensite crystals at low annealing temperatures, and since they bear an orientation relationship with the matrix. The formation of ε-carbides during quenching is not totally surprising considering the fact that the Ms temperature is relatively high (∼700°F) for A533-B steel (Ref. 5, 6, and 7). It is well known (Ref. 14) in the tempering of martensite that ε-carbide precipitates extremely rapidly from super-saturated ferrite and that it is subsequently replaced by cementite as the tempering temperature is increased.

14. As can be seen from Figures 3 through 9, significantly higher levels of acoustic emission are generated from the yielding of as-quenched martensites as well as martensites tempered at 750°F, 850°F, and 950°F. A distinct drop in

emission rate is clearly seen as the tempering temperature is increased beyond 950°F. This is consistent with the observation of ε-carbide platelets, which are first seen to form during quenching, are also persistent in martensites tempered at temperatures of up to 950°F. The acoustic emission activity is somewhat less pronounced in the as-quenched microstructure where significant population of ε-carbide platelets are replaced by spheroidal cementite carbides. Highest emission rates are observed from the yielding of 750°F and 850°F tempered microstructures where maximum number of ε-carbide platelets are seen. These observations emphasize the direct role of ε-carbide platelets in generating acoustic emissions during yielding in these microstructures. It is known that the geometric constraint of elongated platelets can result in high enough local stresses at the particle-matrix interface that can induce fracture of ε-carbide platelets during yielding. The above observations clearly support that fracture of ε-carbide platelets is the likely mechanism by which acoustic emissions are generated in the as-quenched and inadequately tempered microstructures.

15. In martensites tempered at temperatures higher than 950°F, ε-carbide platelets are replaced by more stable spheroidal carbides (cementite and alloy carbides). The fracture of these carbide spheroids is highly unlikely during yielding. Experimental evidence from earlier works (Ref. 8 and 9) shows that very high plastic strains are required to initiate fracture in spheroidal carbides so that any contribution to acoustic emission from the fracture of these carbides should appear at very high plastic strains. A distinct drop in acoustic emission rate is seen as the tempering temperature is raised from 950°F to 1050°F with the emission rate progressively decreasing with increasing tempering temperatures beyond 1050°F (Figures 7, 8, and 9). Yield serrations (resulting from the break-away of dislocations from solute clusters) in the stress-strain curve are often marked by the appearance of peaks in the emission rates. The above observations clearly show that the primary source of acoustic emission from martensites tempered at temperatures above 950°F is the dislocation-solute atom interaction. The observation that the emission activity decreases progressively with increasing tempering temperatures is very much consistent with the fact that progressive depletion of solute atoms (from the matrix) occurs with increasing tempering temperatures, where nucleation and growth of spheroidal carbides is expected to occur.

16. It follows that there are basically two mechanisms by which acoustic emissions are generated during the yielding of as-quenched, and quenched and tempered pressure vessel steel microstructures; namely, fracture of ε-carbide platelets and dislocation break-away from solute atmospheres. In martensites, tempered at temperatures below 950°F as well as in as-quenched martensites, fracture of ε-carbide platelets appears to be the dominant mechanism by which acoustic emission is generated; although contribution from dislocation-solute atom interaction is also expected to play some minor role. The progressive increase of emission rate from the onset of yielding until the fracture point in the as-quenched martensite suggests that emission

comes mainly from fracture of ε-carbide platelets. The relatively lower emission rate observed here (Figure 3) appears to be due to the fracture of low population of ε-carbide platelets present in the as-quenched microstructure. Interstitial carbon atoms in b.c.t. lattice of the as-quenched martensite are severely immobile so that contribution from solute atmospheres is not a major factor here. In martensites tempered at higher temperatures (>950°F), dislocation-solute atom interaction appears to be the primary source of acoustic generation. Peaks in the emission rate accompanying yield serrations and decreasing levels of emission activity with increasing tempering temperatures clearly support this argument.

7075 aluminum alloy

17. Acoustic emission data from the deformation of five distinctly different microstructures of 7075 aluminum alloy (listed in Table II) are illustrated in Figures 17 through 21. It can be clearly seen from the data that the emission activity significantly varies with heat treatment, with the emission rate peaking at different regions of stress-strain curves depending on the initial microstructure and deformation mechanism.

18. The aging sequence and the kinetics of precipitation in 7075 aluminum alloy system have been extensively studied and well documented in recent literature (Ref. 10, 11, and 12). The general aging sequence follows:

Super-Saturated Solid Solution	→	Guinier Preston (G.P.) Zones	→	Intermediate η' Precipitate ($Mg\ Zn_2$)	→	Equilibrium η Precipitate ($Mg\ Zn_2$)

Due to the difficulties in imaging small precipitates of less than 50A, transmission electron microscopy studies conducted earlier were mostly on latter stages of precipitation. Recent studies of early stages of precipitation (Ref. 13 and 14) by high resolution electron microscopy suggested evidence of ordered G.P. zones or some form of spherical η' precursors (precipitate clusters) preceding the precipitation of η' plates. Most commercial 7075 aluminum comes in what is known as T651 temper (usually referred to as T6 temper) where it is given a slight plastic stretch (up to 3%) prior to aging at 248°F (120°C) for 24 hours following solution treatment. The microstructure is primarily coherent G.P. zones with 5% η' precipitates. The stretching operation enhances precipitation of G.P. zones (at dislocation sites) and results in superior mechanical properties.

19. Electron micrographs of the aged 7075 aluminum alloy specimens utilized for the current investigation are shown in Figures 22 and 23. Figure 24 is the microstructure of solutionized and as-quenched specimen where some precipitation of platelets of η' is already seen, formed during quenching. Figures 22a, 22b, and 22c illustrate fine structure of G.P. zones formed during aging for 48 hours at 248°F (120°C) following quenching. Extensive dark field electron microscopy analysis has also been performed to resolve the structure of these precipitates and the analysis will be reported elsewhere. The microstructure resulting from 9-hour aging at 347°F (175°C), of the as-received commercial alloy specimens (known as T7351 overage temper), is shown in Figure 23. The aging produces primarily a mixture of η' plates and η rods (laths). No evidence of G.P. zones is seen in this heat.

20. Acoustic emission data from the deformation of as-quenched (solutionized at 520°C followed by water quench) specimens shows an interesting pronounced peak in the emission rate occurring within a narrow region of yield strain followed by a relatively lower level of emission in the work hardening region at higher strains (Figure 19). Electron metallography conducted on this specimen (Figure 26) shows the formation of few η' plates during quenching. It is unlikely that fracture of these plates can occur during the yield region. In fact, it is shown in a later discussion on deformation of overaged specimens (Figure 23) that fracturing the η' plates requires plastic strains in the work hardening range well beyond yield strain. A number of recent investigations (Ref. 13 and 14) on the earlier stages of precipitation in the alloy system confirm the formation of some form of solute clusters known as preprecipitates (or precursors) prior to the precipitation of η' platelets. These observations suggest that the peak emission rate observed in the yield region of the as-quenched specimens is the result of either dislocation-solute atom interaction or the result of shearing of the precursor precipitates (not resolved in the Figure 24) which are known to form preceeding the precipitation of η' particles during quenching. The low emission activity observed at larger plastic strains (in the work hardening range) in the as-quenched specimens is due to the result of the fracture η' platelets formed during the quenching. This is supported by the data (Figure 21) from overaged specimens (discussed in Section 22) where deformation of η' and η containing matrix resulted in acoustic emissions exclusively in the work hardening range.

21. Figures 18, 19, and 20 illustrate the acoustic emission generated from the deformation of specimens corresponding to the three 248°F (120°C) aging treatments. The microstructures for the three treatments (listed in Table II) are dominated by massive number of fine coherent G.P. zones. Figure 22 illustrates contrast from the fine G.P. zones in a 48-hour aged specimen. The acoustic emission data shows that emission is predominant through most part of the work hardening range, starting from the yield strain. The emission rate rapidly increases with plastic strain beyond yield point, peaks at earlier stages of work hardening, and gradually decreases as the ultimate tensile strength is reached prior to fracture. Two significant observations become apparent from the test data. Increase in emission activity is observed in specimens aged for longer times (48 hours as compared to 16 hours) and in specimens stretched prior to aging. This is a positive confirmation since this implies that specimens containing increasing amounts coherent precipitates consistently contributed to increasing levels of acoustic emission upon deformation. The acoustic emission data from the deformation of the specimen in as received is also illustrated in Figure 25. These observations demonstrate that the major contribution to the high acoustic emission activity generated from the deformation of 248°F (120°C) aged specimens comes from the dislocations shearing the massive number of fine

coherent G.P. zones present in these microstructures.

22. The acoustic emission data from the tensile deformation of an overaged specimen (as-received plus 9 hours aging at 347°F) is illustrated in Figure 21. Figure 23 is an electron micrograph of the overaged specimen. The microstructure consists of premarily η' and η precipitates distributed within the matrix. In contrast with the emissions observed from the 248°C (120°C) aged specimens, the emission from the overaged specimens is significantly lower and is persistent through most poart of work hardening range. Matrix yielding, which is primarily controlled by the dispersion strengthening effect of hard η' and η particles, did not generate any appreciable emissions. Instead, acoustic emissios are generated at higher plastic strains where strain induced particle fracture is likely to occur. A comparison of acoustic emission behavior and microstructure suggests that fracture of η' and η particles is the major source of acoustic emission in the overaged specimens. The marked decrease in the acoustic activity from the overaged specimens as compared to the activity from the 248°F (120°C) aged specimens, appears to be directly related to the decrease in the population of precipitates generating acoustic emission (relatively fewer number of η' and η precipitates as compared to the massive number of fine G.P. zones).

23. In summary, correlation studies of the observed acoustic emissions and the as-quenched and aged microstructures in 7075 aluminum alloy specimens clearly show that there are primarily three basic mechanisms by which acoustic emissions are generated in this alloy system; namely:

i) Dislocations cutting through some form of fine solute clusters, which are precursor η' precipitates formed during quenching,

ii) Dislocation cutting through G.P. zones, and

iii) Fracture of η' and η precipitates.

CONCLUSION

24. Test results from acoustic monitored tensile tests on as-quenched and quenched and tempered A533-B pressure vessel steel microstructures demonstrate that:

i) Acoustic emission varies significantly with the degree of tempering of the as-quenched martensite, upon deformation.

ii) The acoustic emission is predominant only during yielding; practically no emission is observed beyond plastic strains of about 8%.

iii) Significantly higher levels of acoustic emission are observed from as-quenched and quenched and inadequately tempered (tempering at temperatures <950°F) microstructures.

iv) The primary source of acoustic generation from the deformation of as-quenched martensites and martensites tempered at temperatures <950°F is the fracture of epsilon carbide platelets that are shown to be formed during quenching and during earlier stages of tempering.

v) In martensites tempered at temperatures exceeding 950°F, the only contribution to the acoustic emission appears to be the dislocation-solute atmosphere interactions accompanying yield serrations.

25. Results from acoustic monitoring tests on as-quenched and aged specimens of 7075 aluminum alloy demonstrate:

i) Consistent agreement between acoustic emission and microdeformation mechanisms exists with the acoustic emission rate peaking at distinctly different regions of the stress-strain curve, depending upon the initial microstructure and deformation mechanism.

ii) Correlation studies of acoustic emissions and the as-quenched and aged microstructures in 7075 aluminum alloy suggests there are primarily three basic mechanisms of acoustic generation in the alloy system, namely:

(a) Dislocations cutting through precursor η' precipitate clusters formed during quenching,

(b) Dislocations cutting through G.P. zones, and

(c) Fracture of η' and η precipitates.

iii) Significantly higher level of acoustic emissions are generated from the shearing of spherical G.P. zones in aged specimens, as compared to the emission from the fracture of η' and η precipitates in overaged specimens.

ACKNOWLEDGEMENTS

26. The authors wish to acknowledge the assistance of R. L. Burkhart of Westinghouse Electric Corporation in conducting acoustic monitored tests and R. J. Davis, J. J. Haugh, and C. W. Hughes, also from Westinghouse Electric Corporation, for their assistance in conducting the optical, transmission, and scanning electron microscopy. The work reported here is funded under a joint program sponsored by Westinghouse Electric Corporation and Empire State Electric Energy Research Corporation.

REFERENCES

1. R. Gopal, J. R. Smith, and G. V. Rao, "Experience in Acoustic Monitoring of Pressurized Water Reactors", Third Conference on Periodic Inspection of Pressurized Components, Institute of Mechanical Engineers, London, September 1976.

2. G. V. Rao and R. Gopal, "Characterization and Quantification Studies of Pressure Boundary Leaks for Pressurized Water Reactor Applications", Paper Submitted for the 1979 ASTM Symposium on Applications for Acoustic Monitoring.

3. R. Gopal, J. R. Smith, and G. V. Rao, "Acoustic Monitoring Instrumentation for Pressurized Water Reactors", Twenty-Third International Instrumentation Symposium, Instrument Society of America, May 1977.

4. H. L. Dunegan, D. O. Harris, and C. A. Tatro, "Fracture Analysis by Use of Acoustic Emission", Engineering Fracture Mechanics, 1968, Vol. 1, pp. 105-122.

5. R. H. Sterne, Jr. and L. E. Steele, "Stress for Commercial Nuclear Power Reactor Pressure Vessels", Nuclear Engineering and Design, Vol. 10, 1969, pp. 259-307.

6. M. A. Grossman and E. C. Bain, "Principles of Heat Treatment", ASM, Metals Park, Ohio, 1964, pp. 129-176.

7. A. R. Cox, "The Effects of Copper, Silicon, and Manganese on the Strengthening Process in Medium Carbon Steels", Trans. J. Inst. Metals, 1968, Vol. 9 Supplement, p. 118.

8. W. E. Swindlehurst, "On Carbide Cracking as a Source of Acoustic Emission in Steel", J. Materials Science, Vol. 13, 1968, p. 209.

9. A. R. Rosenfield, G. T. Hahn, and J. E. Embury, Met. Trans., Vol. 3, 1972, p. 2799.

10. P. N. Adler, R. DeIasi, and G. Geschwind, Trans. AIME, Vol. 3, December 1972, p. 3191.

11. G. Thomas and J. Nutting, "The Aging Characteristics of Aluminum Alloys", J. Inst. Metals, Vol. 88, 1959-60, p. 81.

12. C. E. Layman and J. B. Vandersande, Trans. AIME, Vol. 7A, August 1976, p. 1211.

13. Peter J. Brofman and Gary Judd, Trans. AIME, Vol. 9A, March 1978, p. 457.

14. F. E. Werner, B. L. Averback, and M. Cohen, Trans. ASM, Vol. 49, 1957, p. 823.

TABLE I

REACTOR PRESSURE VESSEL STEEL

LISTING OF CHEMISTRY AND HEAT TREATMENTS

A533 Grade B Class I Plate Material

Mill Chemical Analysis - Wt%

Heat No.	C	Mn	P	S	Si	Ni	Mo	Cu	Fe
A9406-1	.25	1.37	.010	.020	.27	.61	.55	.13	Bal.

As Received Condition

i)   Austenitized at 1650°F/1 hour per inch thickness and water quenched

ii) Tempered at 1225°F/-3/4 hour per inch thickness

iii) Stress relieved at 1150°F/20 hours; heat and cool at ∿100°F/hour

Specimen Heat Treatments

Specimen ID Numbers	Heat Treatments
AR15	As received
Q9 & Q14	Austenitized at 1650°F/0.5 hours, water quenched
QT3 & QT4	Quenched and tempered at 750°F/1 hour
QT5 & QT8	Quenched and tempered at 850°F/1 hour
QT11 & QT12	Quenched and tempered at 950°F/1 hour
QT7 & QT1	Quenched and tempered at 1050°F/1 hour
QT6 & QT10	Quenched and tempered at 1150°F/1 hour
QT13 & QT2	Quenched and tempered at 1250°F/1 hour

TABLE II

7075 ALUMINUM ALLOY (WLM 7075)

LISTING OF CHEMISTRY AND HEAT TREATMENTS

Wet Chemical Analysis (Wt%)

Element	Zn	Mg	Cu	Fe	Cr	Si	Ti	Zr	Mn	Al
Wt%	5.68	2.77	1.60	.24	.20	.049	.028	<.01	.069	Bal.

Specimen Heat Treatment and Microstructure

Heat Treatment*	Microstructure
520°C/2 hr/water quench	Precursor precipitates & fewer number of η' platelets
A +/120°C/16 hr	Primarily G.P. zones
A +/120°C/48 hr	GPZs + η' precipitates
A +/1% stretch/120°C/48 hr	More of GPZ + η'
As received/175°C/9 hr	η' + η (no GPZs)

---

*(520°C ≃ 968°F, 120°C ≃ 248°F, 175°C ≃ 347°F)

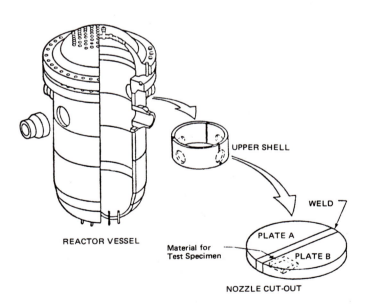

Fig. 1   Test material location for A533-B pressure vessel steel
specimens

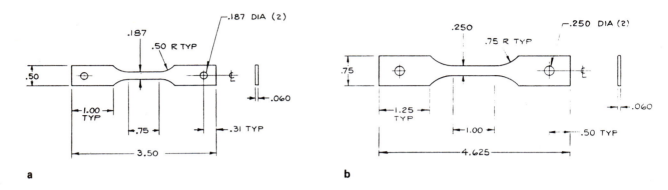

a

b

Fig. 2 Geometry of test specimens utilized for acoustic monitored tests on:
    a 7075 aluminium alloy and
    b A533-B pressure vessel steel

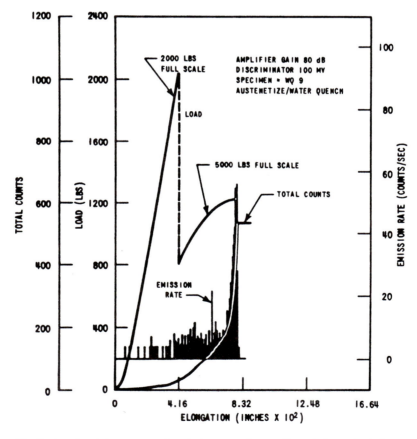

Fig. 3 Load, total counts, and emission rate of acoustic emissions plotted as function of elongation for pressure vessel steel sheet tensile specimen, austenitized and water quenched

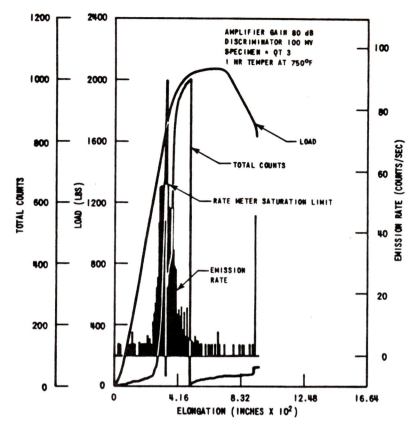

Fig. 4 Load, total counts, and emission rate of acoustic emissions plotted as function of elongation for pressure vessel steel sheet tensile, quenched and tempered for 1 h at 750° F

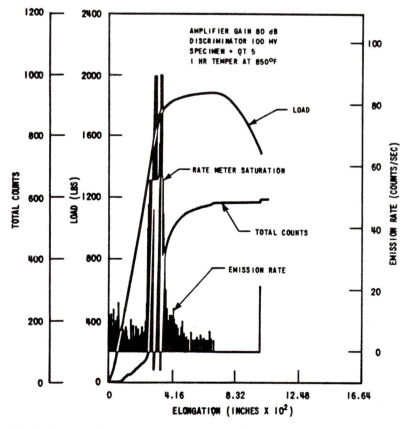

Fig. 5 Load, total counts, and emission rate of acoustic emissions plotted as function of elongation for pressure vessel steel sheet tensile, quenched and tempered for 1 h at 850° F

283

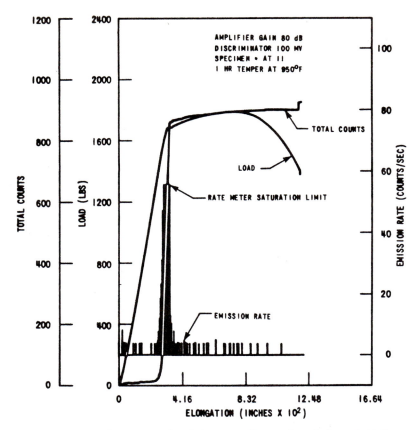

Fig. 6 Load, total counts, and emission rate of acoustic emissions plotted as function of elongation for pressure vessel steel sheet tensile, quenched and tempered for 1 h at 950° F

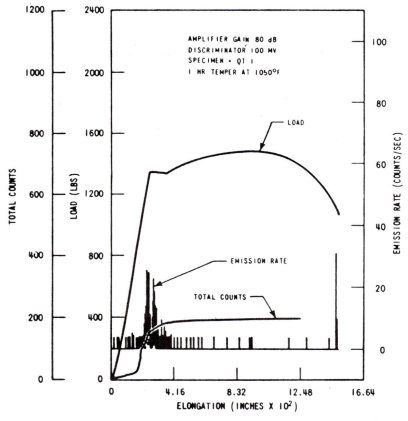

Fig. 7 Load, total counts, and emission rate of acoustic emissions plotted as function of elongation for pressure vessel steel sheet tensile, quenched and tempered for 1 h at 1050° F

284

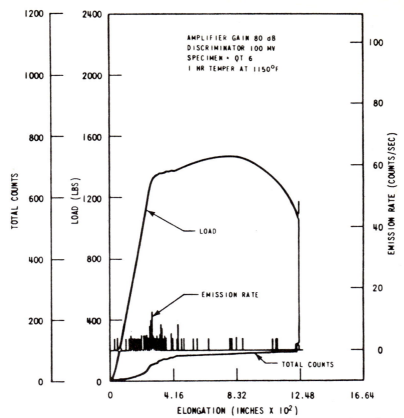

Fig. 8　Load, total counts, and emission rate of acoustic emissions plotted as function of elongation for pressure vessel steel sheet tensile, quenched and tempered for 1 h at 1150° F

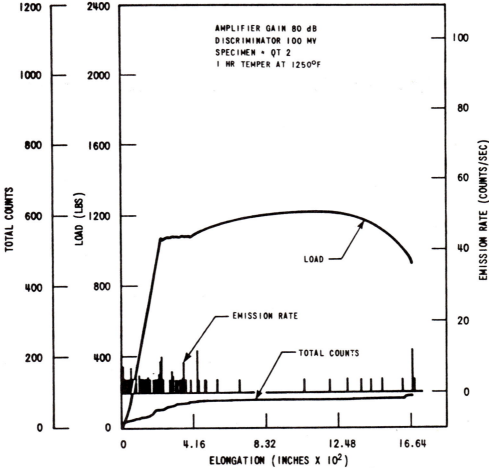

Fig. 9　Load, total counts, and emission rate of acoustic emissions plotted as function of elongation for pressure vessel steel sheet tensile, quenched and tempered for 1 h at 1250° F

285

Fig. 10 Transmission electron micrograph of stress annealed
A533-B pressure vessel steel

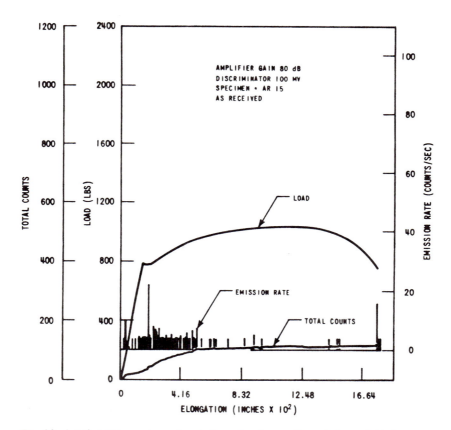

Fig. 11 Load, total counts, and emission rate of acoustic emissions plotted as
function of elongation for stress-annealed pressure vessel steel sheet
tensile taken from 1/4 T location

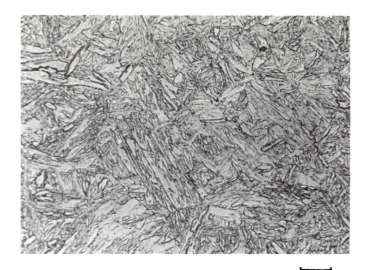

Fig. 12   As-quenched martensite in A533-B steel (optical-nital etch)

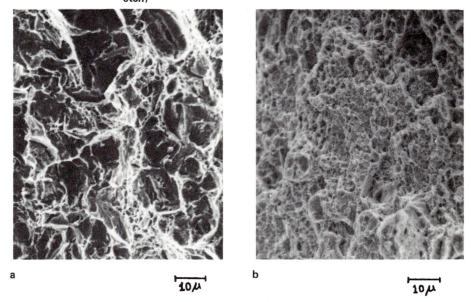

a

b

Fig. 13   Fracture in pressure vessel steel specimens (scanning electron micrographs)
a As-quenched
b 850° F tempered

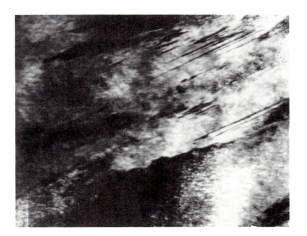

Fig. 14   Transmission electron micrograph of as-quenched microstructure in A533-B steel

1 μ

Fig. 15   Transmission electron micrographs of 850° F tempered martensite in A533-B steel

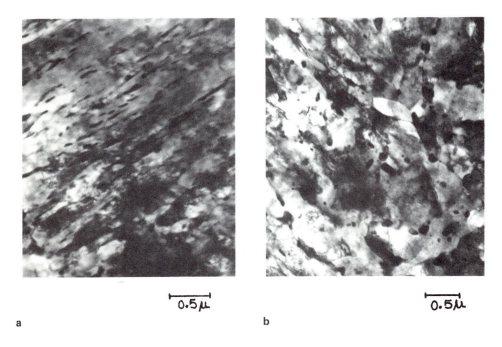

a   0.5 μ   b   0.5 μ

Fig. 16   Transmission electron micrographs of martensites in A533-B steel
         a 1050° F tempered
         b 1250° F tempered

288

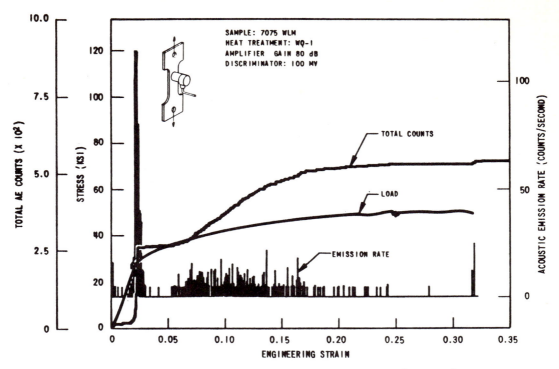

Fig. 17 Acoustic emission from tensile deformation of 7075 aluminum alloy (post-test heat treatment: 550°C/2 h/water quench)

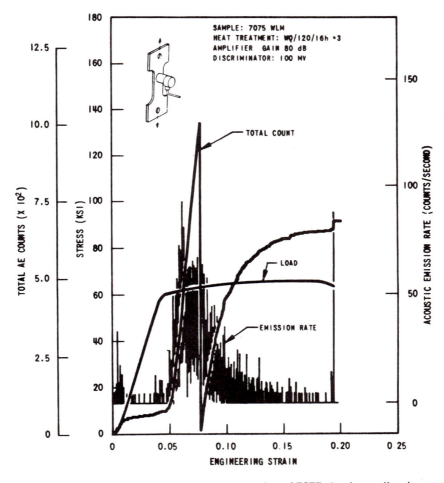

Fig. 18 Acoustic emission from tensile deformation of 7075 aluminum alloy (post-test heat treatment: 550°C/2 h/water quench/120°C/16 h)

289

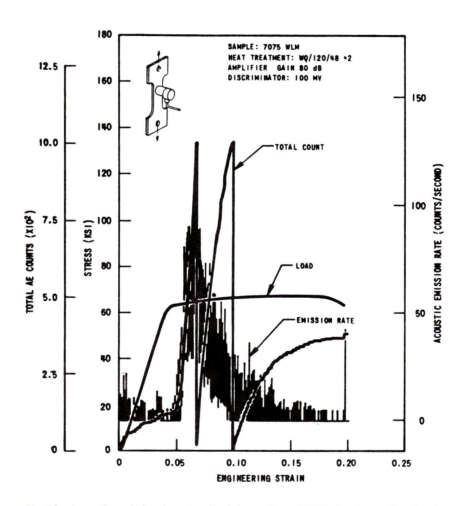

Fig. 19 Acoustic emission from tensile deformation of 7075 aluminum alloy (post-test heat treatment: 550°C/2 h/water quench/120°C/48 h)

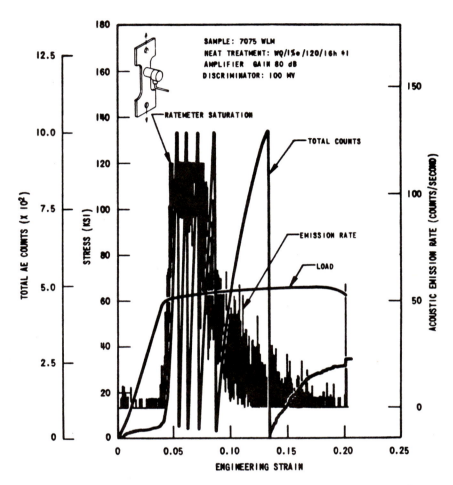

Fig. 20 Acoustic emission from tensile deformation of 7075 aluminum alloy (post-test heat treatment: 550°C/2 h/water quench/1% stretch at RT/120°C/16 h)

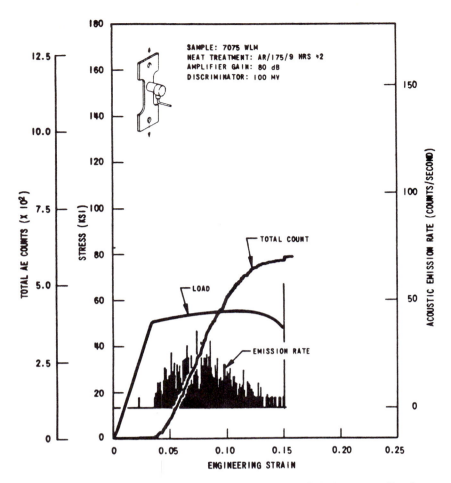

Fig. 21 Acoustic emission from tensile deformation of 7075 aluminum alloy (post-test heat treatment: as-received T651/175°C/9 h)

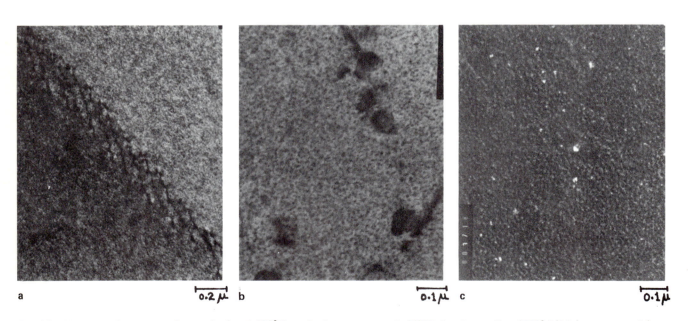

a   0·2 μ   b   0·1 μ   c   0·1 μ

Fig. 22 Electron microscopy of quenched and 120°C aged microstructures in 7075 aluminum alloy (520°C/2 h/water quench/ 120°C/16 h), showing contrast from spherical GP zones

         a and b Bright field micrographs   c Dark field micrograph of a reflection from GP zones

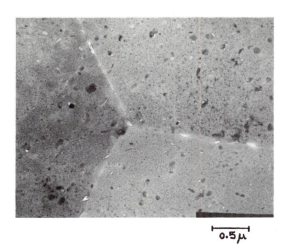

Fig. 23 Transmission electron micrograph of 7075 aluminum alloy in overage temper (T7351) showing contrast from η' and η precipitates

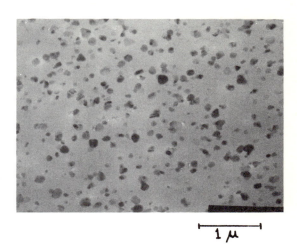

Fig. 24 Transmission electron micrograph of solutionized and as-quenched microstructure in 7075 aluminum alloy showing the formation of η' plates during quenching

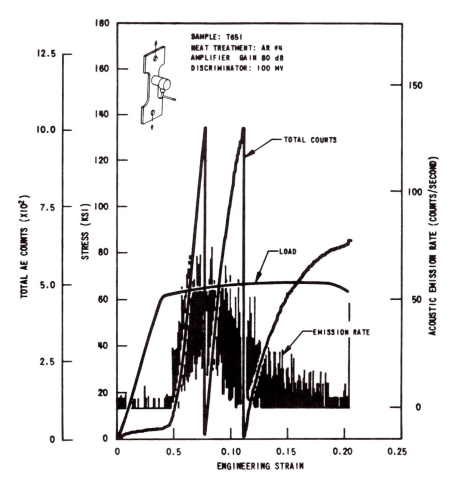

Fig. 25 Acoustic emission from tensile deformation of 7075 — T651 aluminum alloy, tested in as-received condition

293

C103/79

AN ULTRASONIC LINEAR ARRAY SYSTEM FOR PULSE ECHO AND HOLOGRAPHIC IMAGING OF FLAWS IN SOLIDS

G. J. POSAKONY, BSEE, B. P. HILDEBRAND, PhD, T. J. DAVIS, MSEE, S. R. DOCTOR, PhD,
V. L. CROW, BSEE, and F. L. BECKER, BSc
Battelle, Pacific Northwest Laboratories, Richland, Washington, USA

The MS of this paper was received at the Institution on 27 February 1979 and accepted for publication on 2 March 1979

SYNOPSIS   Intensive research into the use and application of linear ultrasonic arrays has been carried out over the past several years. Most of the research has been concentrated in the medical field and many companies have commercialized sequentially pulsed [1], phase steered [2], and/or phase focused [3] linear array systems for medical diagnostic ultrasound. Application of current medical ultrasonic imaging for nondestructive evaluation of "flaws" in solids appears impractical as boundary conditions, material velocities and intrinsic differences in the nature of information displayed make direct conversion of medical technology difficult. [4]   However, with the tremendous strides that have been achieved in the medical field, we clearly must evaluate linear array technology and establish the usefulness of arrays for inspection of structural materials and weldments.

This article describes an ultrasonic linear array system developed for the inspection of weldments in nuclear reactor pressure vessels. The imaging system utilizes a sequentially pulsed, phase steered linear array to develop pulse echo images and a line focused illuminating transducer in conjunction with a linear receiver array to develop holographically reconstructed images. The results recorded from the computer based experimental system demonstrate the capability of array technology. Excellent results from both the pulse echo and holographic modes of operation have been achieved. Pulse echo images of flaws in weldments are displayed in B-scan, C-scan or isometric presentations. Reconstruction of the phase or holgraphic images are compared with pulse echo results and demonstrate the enhancement potential for the holographic procedure. We have demonstrated that: (a) linear array technology can duplicate conventional pulse echo procedures, (b) linear arrays are an effective means for developing holographic images, (c) one major advantage of the array system is speed of inspection, and (d) potential for specific applications is extremely encouraging.

INTRODUCTION

For the past three years, the Electric Power Research Institute (EPRI) has sponsored a research program at Battelle Northwest in Richland, Washington to "develop an ultrasonic imaging system to rapidly and accurately detect and characterize flaws in heavy section steel structures of nuclear reactor pressure vessels." The program objective was to develop improved means for ultrasonic nondestructive evaluation (NDE) of nuclear power plant components. Under the EPRI contract, RP 606-1, Battelle proposed use of ultrasonic linear array technology as an advanced means for obtaining pulse echo and holographic images of subsurface flaws. ASME Section XI [5] (III-2110 Instrumentation) specifies that a pulse echo ultrasonic flaw detector will be used for those examinations which employ the ultrasonic test method. However, the Section XI document (IWA-2240) makes provision for substitution of alternative examination methods or newly developed techniques provided that the results are demonstrated to be equivalent or superior to those currently specified. Battelle's research, therefore, was aimed at design, fabrication and demonstration of an ultrasonic system that:

- described the capability of phase steered ultrasonic linear arrays for satisfying the Code requirements for pulse echo examinations, and,

- established the "comparative performance" of ultrasonic holographic techniques with pulse echo results to provide an initial basis for validation of acoustic holography as a potential Code accepted procedure.

Linear arrays have a technical advantage over single transducer elements in that high speed electronic switching can be used to control and/or direct the ultrasonic sound beam thus providing means for substantially reducing the time required to perform an inspection. At the same time, array technology can provide improved inspection procedures for better display and characterization of subsurface flaws. Battelle's research to date has been aimed at inspection of the weld and weld volume of nuclear reactor pressure vessels. The technology, however, can be used for the inspection of a variety of welds or components. This report highlights the ultrasonic imaging system and the test results achieved under the EPRI research program.

ULTRASONIC IMAGING SYSTEM

As the ultrasonic inspection system is presently configured, the pulse echo mode of operation is designed as a high speed search, detect, and locate examination to identify ultrasonic reflectors which exceed an inspection level established by ASME Codes. The holographic mode of operation is designed to size and characterize the nature of the ultrasonic reflection detected by the pulse echo examination. The design strategy followed for the ultrasonic imaging system was to demonstrate that array technology can duplicate the pulse echo inspection capability of existing flaw detector systems. We further planned to demonstrate that arrays can produce an order of magnitude decrease in the time required to perform an inspection. Since acoustic holography, even with array technology, is relatively slow, we elected to use this inspection mode for flaw characterization and did not intend that holography be employed in a "scanning" mode.

The system developed for EPRI is shown in Figure 1. The system consists of:

- a 120 element piezoelectric array for pulse echo examinations,

- a 120 element receiver array and line focus insonifying transducer for holographic imaging,

- a PDP 11/34 host computer and terminal,

- the pulse echo flaw detector, electronics, scan converter and TV display,

- a mechanical scanning bridge (microprocessor controlled) for manipulating the arrays, and,

- peripheral power supplies, hardware and fixturing to complete the system.

The experimental model was designed for the evaluation of the weld volume of both vertical and horizontal welds of nuclear reactor pressure vessels from an OD surface. As the principal aim of the research program was to establish and demonstrate feasibility for array technology, the system was designed for experimental versatility under test bed or laboratory environment rather than more hostile environments attendant with inservice inspection conditions.

PULSE ECHO OPERATION

A simplified interpretation of ASME Section XI, 1974 Code requirements is shown in Figure 2.

The ultrasonic technique employed must examine the weld volume with several (typically 0, $\pm 45°$, $\pm 60°$) sound beams. Under the present practice, discrete transducers (e.g., 0, 45° and 60°) are mechanically moved back and forth across the weld and indexed along the weld by elaborate mechanical remote inspection systems. The signal amplitude of echos returning from subsurface reflectors is recorded and the extent or length of the reflection is mapped to provide an interpretation of the size of the reflector.

As part of the present research, we have replaced the individual discrete transducers with a linear array which is capable of electronically steering the sound beam to specific selected angles. The array was placed on a plastic wedge which was designed to ride the surface of the structure. The array used for pulse echo operation was relatively long (e.g., 90mm - 3.6 in.); however, only a portion of the array is used at any one time. The sound beam projected from the array duplicates that which would be generated by a discrete transducer of the type used by conventional flaw detector systems. The sound beam is electrically switched from one grouping or sector of the array to the next. By switching sequentially the sound beam is "stepped" long the length of the array. Thus a single pass of the array provides a wide inspection swath. This reduces the length of time needed to mechanically manipulate the transducer(s) over the surface area required to achieve the appropriate inspection.

Ultrasonic Array

The characteristic performance of the ultrasonic linear array establishes the design criteria for the acoustic, electronic, computer, display and mechanical components of the system. During the initial engineering and modeling phases of our research, several array designs were considered including:

1. hybrid design [6] would utilize a single array to accomplish both pulse echo and holographic imaging,

2. individual array designs [7] which divided the pulse echo and holographic functions into separate inspections modes.

In the final design for the ultrasonic imaging system, two arrays were used. One was designed for the pulse echo mode and a second for the holographic mode of operation.

The ultrasonic linear array used for pulse echo imaging is shown in Figure 3. It consists of 120 individual piezoelectric elements set side by side with the narrow dimension of each of the elements perpendicular to the long axis of the array. This array is 90mm x 25mm (3.6 in. x 1 in.). Each of the elements of the array is 0.56 mm (0.022 in.) wide by 25.4mm (1.0 in.) long. The center-to-center spacing of the elements is 0.76mm (0.030 in.) and the spacing between the elements is 0.2mm (0.008 in.). The piezoelectric material is PZT 5A with a cut frequency of 2.7 MHz and a peak operating frequency of 2.1 MHz. Each of the elements of the array is separately addressable; that is, each can act as a separate transmitter or receiver. The sound field generated is a function of the radiation field projected from each element of the array. The sound beam pattern from the array is dependent on the number of individual elements (groupings) used at any one time and the phase relationship between each of the elements in a given grouping. For the pulse echo operation, a group of 16 elements is used at any one time to generate the sound beam. The sound beam pattern is equivalent to that projected from a solid piezoelectric plate 12mm x 25.4mm (0.48 in. x 1.0 in.). Beam steering is achieved by adjusting the phase or time relationship for launching (or receiving) the energy from the individual elements of a selected group.

The sketches of Figures 4a and 4b describe schematically the development of an in-phase and phase steered sound beam from the array.

Figure 4a shows an in-phase excitation which produces an 0 beam. The sound beam from "n" elements of the array are electronically sequenced along the length of the array in a fraction of a second. Figure 4b describes schematically the effect of changing the phase relationship of the electronic drive voltage to the "n" elements. The sound beam is steered to an angle Θ_0 by the amount of time (phase) delay introduced between elements. The delays are chosen to produce specific angles of propagation (e.g., 35°, 40°, 45°, 50°, etc.) as determined by the computer and the electronic networks.

The equation for the total radiation pattern of "n" identical elements in an array is given as follows:

$$|F,(\Theta)|^2 = \left(\frac{\sin^2 \frac{\pi a}{\lambda} \sin \Theta}{\frac{\pi a^2}{\lambda} \sin^2 \Theta}\right)\left(\frac{\sin^2 \left\{\frac{n\pi d}{\lambda}(\sin\theta-\sin\Theta_0)\right\}}{n^2 \sin^2\left\{\frac{\pi d}{\lambda}(\sin\theta-\sin\Theta_0)\right\}}\right)$$
(eq. 1)

Where:

F = pressure

n = number of elements excited,

a = actual element width

λ = wave length at operating frequency,

Θ = measurement angle, = angle to which array is steered.

The array is designed so that each of the individual elements of the array has a wide radiation pattern. The vectorial summation of pressures at a chosen angle determine the amplitude of the sound beam at that angle. Electric or acoustic crosstalk or coupling between elements will adversely influence the radiation pattern and reduce the energy available for steering.

The steered energy from a given grouping of "n" elements is defined as follows:

$$\sin \Theta = \frac{fN\lambda t}{d} \pm \frac{k\lambda}{d}$$
(eq. 2)

Where:

f = frequency

N = time interval number (e.g., 1, 2, 3, --)

t = time interval (phase relationship)

d = center-to-center spacing

k = grating lobe number (e.g., 1, 2, --)

The first term in equation (2) describes the primary steered beam while the second term describes the rating lobe. The beams from an array can be steered from 0 through critical longitudinal and beyond through critical shear beam inspection angles. The grating lobe, which is an undesirable sound beam, is a phase reinforcement of the sound field that is associated with the center-to-center spacing of the array. The linear array shown in Figure 3 has been designed so that the center-to-center spacing and operating frequency result in elimination of the grating lobe at angles below a 60° shear inspection beam.

Pulse Echo System

The functional block diagram of the pulse echo system is shown in Figure 5. The system is controlled by the host computer which establishes the electronic steering and receiving protocol, the spatial position of the array on the

mechanical bridge, and computes the spatial coordinates of each of the signals returned from ultrasonic reflectors so that the displayed image describes the "defects" at their proper location. There are 120 elements in the array, and each element may act as either a transmitter or receiver. Since each of the elements must be individually addressed in both the transmit and receive operations, there are 120 transmitters and 120 receivers in the system. In its search, detect and locate mode the information from three inspection angles are superpositioned and recorded on the isometric display. In a typical sequence, the 0°, 45°, and 60° inspection information is recorded. To produce the 0° beam, the transmitter electronics excite a 16 element group simultaneously (zero phase shift) producing the 0° beam. A pulse echo pulse repetition cycle is completed for each array grouping before the sound beam is sequenced to the next grouping. The energy reflected from internal defects and from the back surface are recorded as received, stored in the analog scan converter, and recorded on the TV display. A small receiver group (3 to 5 elements) in the center of the transmit group are used as receivers. When each pulse echo cycle is complete, the computer selects an adjacent grouping of 16 elements and the cycle is repeated. As an example, cycle 1 would use elements 1 through 16; cycle 2 would use elements 4 through 20; cycle 3 would use elements 8 through 24, etc.

As presently designed, the first step is to sequence the 0° inspection beam the length of the array. In the second step, the 45° inspection beam is sequenced the length of the array, and this is followed by sequencing the 60° inspection beam. The computer then instructs the system to return to the 0° inspection beam and the sequence is repeated. Completion of the full sequence provides a single-plane examination or B-scan of the zone directly in front of the array. As the array is 90mm (3.6 in.) long an inspection swath that approximates the length of the array is produced. Several hundred miliseconds are required to electronically sequence the sound beam the length of the array. Echos returning to the receive array elements are phased according to the beam angle and summed to increase the system's sensitivity and to discriminate against unwanted artifact signals. The receiver elements are chosen in the center of the transmit grouping so that the spatial coordinates of the transmitted beam and the spatial position of the returning echos can be computed and positioned at their proper location on the TV display. The pulse echo system can produce B-scan, C-scan, or isometric (B-C scan) displays of information reflected from subsurface targets.

The mechanical scanning bridge is designed to move the array in a rectangular or orthogonal scan path. The speed of scan of the array is dictated by the sound beam. A 50% beam overlap is required for the pulse echo inspection. In our present experiments, the array is moved at speeds as fast as 75mm (3 in.) per second. This speed is dictated by the pulse echo time required to complete each cycle sequence, the electronic sequencing along the length of the array, and the requirement for a minimum of 50% beam overlap to assure complete coverage. The mechanical scanning bridge was designed to scan an area of approximately 50mm x 1500mm (20 in. x 60 in.). Scanning this surface area provides the potential for full volumetric scan and display of a 500mm (20 in.) length of weld in a 250 mm (10 in.) thick steel section.

Pulse echo data are acquired from the 0°, 45°, and 60° inspection angle and these data are stored in the scan converter. Since the computer controls the scanning angle and establishes whether the recorded information is received from the longitudinal or the shear angle beam inspections, the computer computes the corrected coordinates so that the spatial location of any reflected energy appears at a correct position when the image is displayed. The operator views the television monitor and can observe the image as it is developed. The isometric display shows the top surface and the bottom surface as a secondary "gray-level" while flaw echos are shown in full black. A back surface gate records "loss of back" which provides a shadowgraph image of any targets within the inspection volume. The A-scan display is used for setup and for point-by-point evaluation of specific targets. The high speed search, detect, and locate mode of operation is capable of providing an ASME Section XI inspection of the weld volume at lineal rates approaching 300mm (1 ft.) per minute.

Figure 6 shows experimental test results of heavy section steel weldment used as part of a "round robin" experiment. The test block was provided by Babcock & Wilcox Corporation and contained both large and small ultrasonic reflectors. This view is but one of many taken during the evaluation; however, the results typify the performance capability of the pulse echo tests. The test plate was 1500mm x 1650mm (5 ft. x 5-1/2 ft.) and was approximately 145mm (5.75 in.) thick. The weldment contained several intentionally introduced welding defects. A section of the block contained cladding approximately 9mm (.375 in.) thick. The plate was flat and both surfaces were ground smooth. Figure 6 shows a section of the weld volume 450mm x 150mm wide x 145mm thick (18 in. x 6 in. x 5.75 in.). The inspection was performed through the weld cladding thus the influence of the clad is recorded. The top left hand picture is the isometric view as recorded on the television monitor. This view shows the top and bottom surfaces, the dark black area from the direct reflection of the weld defect and the white area in the back surface shadowgraph. The shadowgraph information provides additional detail concerning the size and location of the defect. This particular welding defect appears to have three indications from the direct reflection but only two from the back surface shadowgraph. Greater detail concerning the flaw is shown in Figure 9 of the holographic results. The top right hand recording is the 0° beam, C-scan, taken from the same data. The bottom left hand recording shows the side view perpendicular to the weld line while the bottom right hand shows the location of the welding defect as seen from an end view. It should be noted that no angle beam information appeared at the 50% DAC recording level used for this examination, hence, no data were recorded. Even though the pulse echo mode of operation is intended to provide high speed search, detect and locate information, substantial detail can be obtained from the recorded data.

THE HOLOGRAPHIC SYSTEM

In the ultrasonic imaging system developed for EPRI, the pulse echo mode of operation was designed to match the performance of existing inspection systems and to demonstrate that array technology could be used to decrease the time required to perform an ultrasonic examination. Once a pulse echo ultrasonic indication has been identified, either through direct or shadowgraph information on the isometric display, the image is switched to the holographic mode of operation and a detailed examination is performed of that specific zone. Technically, the phase information reflected from an ultrasonic target should contain more intrinsic data for the sizing and characterization of material flaws. The array used for the holographic mode of operation is separate from the pulse echo array and the network design is substantially different in that two transducer assemblies are used. The first is a line focused source transducer designed to insonify the zone under evaluation. The second array is a multi-element linear array containing 120 small receiver elements which record the phase information reflected from ultrasonic targets. The holographic array network is shown in the sketch of Figure 7. The line focused transducer contains a 150mm x 25mm (6 in. x 1 in.) piezoelectric, 2.25MHz element which is lens-focused to produce an effective wedge of sound energy for insonifying the inspection zone. The receiver array contains 120 elements and is 150mm (6 in.) long; however, the effective dimension of each of the individual receiver elements is only 1.0mm x 1.5mm (0.04 in. x 0.06 in.). The center-to-center spacing of the elements in the receiver array is 1.25mm (0.05 in.) and the spacing between elements is 0.25mm (0.01 in.). The small size of the receiver elements assures that the phase coherence of the energy obtained from the subsurface reflectors is maintained. The arrays can be used for both 0° beam and angle beam (longitudinal and shear) holographic examinations. Best results to date have been obtained by angling the illuminating and receiving array to achieve the desired inspection view.

Figure 8 is a block diagram of the holographic reconstruction system. The major components (such as the host computer and the mechanical scanning bridge) are the same as those used for the pulse echo operation. The system has been designed to obtain information from a 150mm x 150mm (6 in. x 6 in.) aperture. The length of the array provides one axis of the aperture. The mechanical scanning bridge, indexing the array across the inspection zone, provides the second aperture dimension. In operation the transmitter power amplifier provides a 10 to 30 cycle tone burst to the insonifying array. Energy reflected from internal targets is received at each of the 120 elements in the linear array and is processed to extract the phase information. The phase information at each of the receivers is recorded, amplified, and stored digitally in the host computer. The receiver switching electronics samples each element of the array for each of the mechanical index locations in the 150mm x 150mm (6 in. x 6 in.) aperture. The phase data in a given aperture contains a minimum of 120 x 120 data points.

Reconstruction of the phase information to provide an image of the holographic data is to be performed by a computer procedure using either backward wave reconstruction algorithms or Fourier holographic reconstruction algorithms. Preliminary work performed by Battelle on backward wave reconstruction procedures is most encouraging; however, specific data were not available to include in this report. The final system will utilize computer reconstruction to provide curvature corrections and compensation so the images will be accurate representations of the flaw. The array processor provides fast Fourier transforms and enhances the speed for developing the reconstruction. The disc storage of the PDP 11/34 is used to store the corrected data for output to a two dimensional refresh memory. In contrast to conventional holographic systems, the computer reconstruction approach does not involve the generation of an optical hologram nor does it require the laser-optical reconstruction system generally associated with acoustic holography. The holographic display shows only the image of the ultrasonic reflector. The computer reconstruction system and procedure are still being developed; however, it appears likely that the reconstructed image from a 150mm x 150mm (6 in. x 6 in.) aperture could be obtained in approximately 4 minutes. This allows 2 minutes for mechanical scanning and 2 minutes for the computer reconstruction time.

While no test results are yet available from the computer reconstruction system, Battelle has demonstrated the feasibility of using array technology for generating digital holograms and for reconstructing the holographic data by conventional optical procedure. (8) Figure 9 is an optically reconstructed image of the phase data obtained from the same flaws shown in Figure 6 of the pulse echo results. Figure 9 was obtained using 0° illumination and sampling a 120 x 240 aperture matrix. The image was obtained through the cladding and clearly defines the weld defect as two separate defects with a possible third zone (below). Comparing the results of Figure 9 with those of Figure 6 indicate that holographic procedures can provide flaw characterization that substantially exceeds present Code accepted pulse echo techniques.

CONCLUSIONS

Ultrasonic linear array technology can be used for both pulse echo and holographic image reconstruction. The principle advantage demonstrated to date is one of speed. It has been demonstrated that the linear array operating in the pulse echo mode can reduce the time required to perform an inspection in accordance with ASME Section XI codes by a factor of 5 to 10. The ability to electronically steer and control the sound beam and the techniques developed for displaying the image have enhanced the detection probability and the inspection reliability. While the pulse echo display techniques (B-scan, C-scan, and isometric display) could be used with equal advantage with conventional inspection procedures, the array provides the opportunity for increasing the speed and versatility of the inspection. The major disadvantage of the pulse echo array technique is related to the complexity of the electronic system required to perform an inspection. A fundamental requirement for phase steered arrays is that each of the elements of the array must be separately addressable. This necessitates a large number of pulsers and receivers and involves precise control of the delay networks, amplifiers, pulsers, array elements and each subcomponent of the system.

The greatest advantage for the technology developed under the EPRI program appears to be related to high speed acoustic computer reconstruction of the phase or holographic data. The relative simplicity of the line focused insonifying transducer, operated in conjunction with the multi-element linear array and combined with computer reconstruction, offers the promise for substantial increase in the ability to rapidly and accurately characterize ultrasonic reflectors in solid materials. The combination of the insonifying transducer plus linear receiver array reduces the requirements for precise mechanical manipulators used in current acoustic holographic instrumentation. The computer reconstruction eliminates the need for the film hologram and optic reconstruction system and converts the phase information directly to a reconstructed image of the flaw.

While further research and development are necessary to refine array technology and adapt inspection systems to specific examination requirements, the program sponsored by the Electric Power Research Institute has clearly demonstrated that linear arrays can be successfully and beneficially applied to the inspection of nuclear power plant components as well as to other ultrasonic examinations.

REFERENCES

1. Wilcox, M.H., "Ultrasonic Cross-Sectional Imaging System," U. S. Patent 3,881,466, U. S. Patent Office, Washington, D. C., May 6, 1975.

2. Beaver, W. L., Signal Processor for Ultrasonic Imaging, U. S. Patent 4,005,382, U. S. Patent Office, Washington, D. C., January 25, 1977.

3. Kossoff, G., Linear Array Ultrasonic Transducer, U. S. Patent 3,936,791, U. S. Patent Office, Washington, D. C., February 3, 1976.

4. Posakony, G. J., Acoustic Imaging - A Review of Current Techniques for Utilizing Ultrasonic Linear Arrays for Producing Images of Flaws in Materials, AMD-Volume 29, Elastic Waves and Non-Destructive Testing of Materials procedings, ASME Annual Meeting, December 1978.

5. ASME Boiler and Pressure Vessel Code, Section XI, Division 1, "Rules for Inservice Inspection of Nuclear Power Plant Components," 1977 edition, ASME United Engineering Center - New York, N. Y. 10017.

6. Becker, F. L., Crowe, J. C., Crow, V. L., Davis, T. J., Hildebrand, B. P., Posakony, G. J., "Development of an Ultrasonic Imaging System for the Inspection of Nuclear Reactor Pressure Vessels - Third Progress Report," September 1977, prepared for Electric Power Research Institute, Palo Alto, CA.

7. Becker, F. L., Crow, V. L., Davis, T. J., Doctor, S. R., Hildebrand, B. P., Lemon, D. K., Posakony, G. J., "Development of an Ultrasonic Imaging System for the Inspection of Nuclear Reactor Pressure Vessels" - Fourth Progress Report, September 1978 - prepared for the Electric Power Research Institute, Palo Alto, CA.

8. Note: Holosonics Inc., under EPRI Contract 606-2, is developing the computer reconstruction system to interface with the system developed by Battelle under Contract 606-1. No results were available at the time of preparation of this article.

Fig. 1 Ultrasonic imaging system

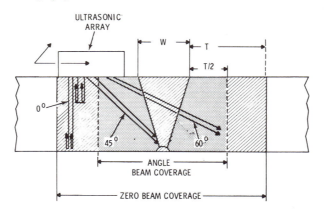

Fig. 2 Weld volume to be inspected

Fig. 3 Pulse echo linear array

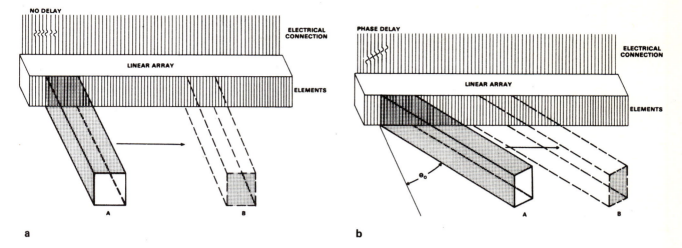

Fig. 4 Schematic of phase controlled array patterns
    a In-phase excitation
    b Delayed phase excitation

PULSE ECHO MODE

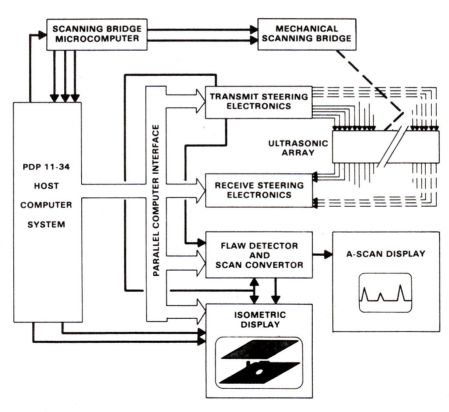

Fig. 5 Functional block diagram of pulse echo system

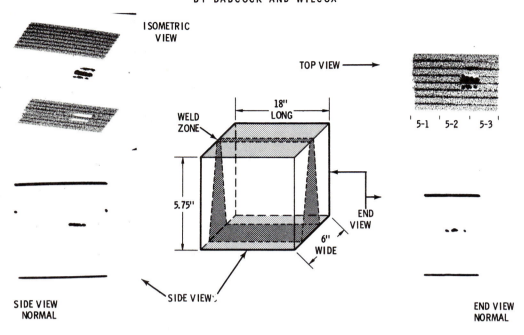

Fig. 6 Pulse echo test results from flaw in weldment

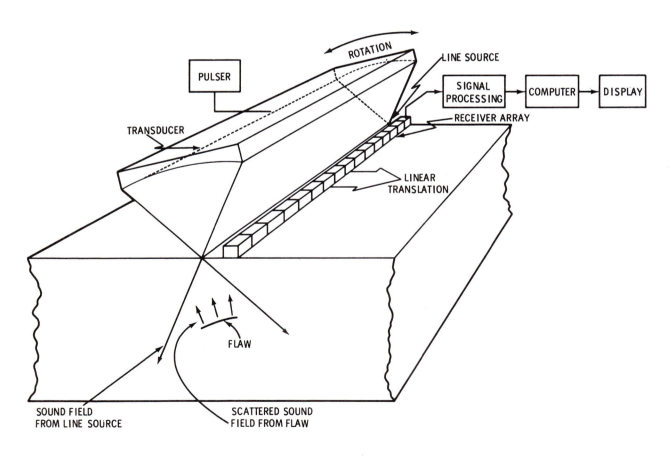

Fig. 7 Holographic line focused transducer and linear receiver array

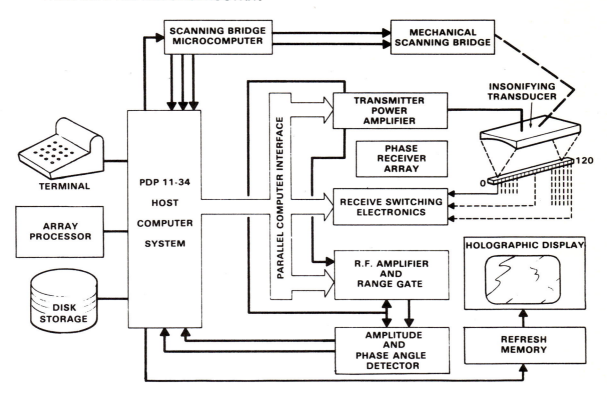

Fig. 8   Functional block diagram of holographic imaging system

Fig. 9   Reconstructed hologram of flaw in weldment